SUPRAMOLECULAR ENZYME ORGANIZATION

SUPRAMOLECULAR ENZYME ORGANIZATION

QUATERNARY STRUCTURE AND BEYOND

BY

PETER FRIEDRICH, M.D., D.Sc. (Biol.)

Institute of Enzymology, Biological Research Centre,
Hungarian Academy of Sciences, Budapest, Hungary

PERGAMON PRESS, OXFORD   AKADÉMIAI KIADÓ, BUDAPEST

Pergamon Press is the sole distributor for all countries, with the exception of the socialist countries.

HUNGARY — Akadémiai Kiadó, Budapest, Alkotmány u. 21. 1054 Hungary

U.K. — Pergamon Press Ltd., Headington Hill Hall, Oxford OX3 0BW, England

U.S.A. — Pergamon Press Inc., Maxwell House, Fairview Park, Elmsford, New York 10523, U.S.A.

CANADA — Pergamon Press Canada Ltd., Suite 104, 150 Consumers Road, Willowdale, Ontario M2J 1P9, Canada

AUSTRALIA — Pergamon Press (Aust.) Pty. Ltd., P.O.Box 544, Potts Point, N.S.W. 2011, Australia

FRANCE — Pergamon Press SARL, 24 rue des Ecoles, 75240 Paris, Cedex 05, France

FEDERAL REPUBLIC OF GERMANY — Pergamon Press GmbH, Hammerweg 6, D-6242 Kronberg-Taunus, Federal Republic of Germany

Copyright © 1984 Akadémiai Kiadó, Budapest

Library of Congress Cataloging in Publication Data

Friedrich, Peter.
  Supramolecular enzyme organization.

  Bibliography: p.
  1. Enzymes. I. Title. II. Title: Quaternary
structure and beyond.   [DNLM: 1. Enzymes—metabolism
QU 135 F911s]
QP601.F75 1984   574.19'25   84–11080
ISBN 0–08–026376–3

British Library Cataloguing in Publication Data

Friedrich, Peter
  Supramolecular enzyme organization.
  1. Enzymes   2. Molecular structure
  I. Title
  547.7'58 QP601

  ISBN 0–08–026376–3

Pergamon Press   ISBN 0 08 026376 3
Akadémiai Kiadó ISBN 963 05 3369 3

Printed in Hungary

PREFACE

Ever since I embarked on the writing of this book I have longed to come to the preface. This desire, of course, has been motivated by the circumstance that prefaces are written at the end, with the main bulk of work done, rather than by the inherent pleasures of preface-writing. In fact, these misplaced epilogues are in most cases utterly useless. They are superfluous to a good book, but do not help a poor one. Nevertheless, there are intermediate cases where some preliminary explanations may be handy. I entertain the hope that the present volume lives up to that standard.

This book reflects my scientific interest over the past ten years. Hence, it is a highly subjective treatise, particularly as regards the delimitation of subject matter. It may appear unusual to include the enzymes' quaternary structure in a monograph essentially aimed at their supramolecular organization or, the other way round, once preoccupied with subunit structure to trespass so far into the murky field of the "beyond". Perusal of the Contents may suggest to the reader that the author's meandering interest gave the book a backbone suffering from scoliosis.

Let me admit that, however arbitrary it may appear, the scope of the text was made such deliberately. Indeed, I believe it makes sense to start scrutinizing enzyme interactions at the quaternary level and then to proceed to higher orders of organization. In this way conceptually related phenomena can be placed in a common perspective. Numerous monographs, reviews and the like have already been written on the subunit structure of enzymes, but none of them really ventured to roam outside the safe walls of Fort Enzymology. However, structural and functional complementarities are not confined to individual enzyme molecules. To my mind, enzyme organization is a continuum and it is in this spirit that the present book was conceived.

Even if the above reasoning is acceptable, the selection of material was arbitrary of necessity. As stressed at too many points in this volume, I am afraid, my aim was to illustrate and not to give comprehensive lists of all observations for the various organizational modes. Obviously, more emphasis is given to areas and objects related to my own research work, perhaps at the expense of more meritorious sources. I apologize to all those who feel unfairly left out by my choices. There was certainly much more material available than I could cope with. Let the shortcomings and omissions of this book be a challenge to others to write their version of the story!

This book is intended, first of all, *not* for the specialist, who knows much more about the individual questions than the text offers, but rather for the general biochemical reader interested in enzyme organization. There is practically no mathematics involved. Instead, quite elementary things are described, so that students and biologists whose bread is not buttered with enzymes should also be able to join in. To facilitate further reading references have been amply given.

Let me finally say a word about the scientific credit of the book's content. I think, in this respect it is more heterogeneous than the average. It discusses in brief solid text-book facts while, on the other hand, it reports about the frontiers of research in a highly uncertain and deceptive field. The author is tantalized to know how many and which of the present trends will stand the test of time. Already now, I suffer from "journalophobia", a mental state characterized by avoidance of current journal issues in the fear of finding yet more new facts. For this reason readers are warned not to take a fundamentalist attitude towards what is written here, especially when it comes to provocative new ideas, but rather to follow up developments since the completion of the manuscript. Even in this scientometric world of ours, the impact of this book should not be quantitated by the number of workers it has led astray.

With these premonitions advanced, I wish the reader a useful journey through the land of quaternary enzyme structure — and beyond.

Peter Friedrich

ACKNOWLEDGEMENTS

I wish to thank all colleagues who gave permission to reproduce their figures in this book; several of them also called my attention to recent developments. Their names are indicated in the legends to the figures. I am particularly obliged to Drs Pál Elődi, Tamás Keleti and John Londesborough for their thorough criticism of the manuscript. John Londesborough gave me invaluable help also by improving the English of the text. Whatever faults, scientific or linguistic, remain in the book, they lie entirely with me. I thank my coworkers Drs Magda Solti, János Hajdú, Ferenc Bartha, and all members of the Institute of Enzymology (directed by Brunó F. Straub) of the Hungarian Academy of Sciences who contributed to the scientific atmosphere in which this book was conceived. I am sincerely indebted to Mrs Katalin Radnai, Mrs Ágnes Külföldi, Mrs Ágnes Csurgó and Mrs Szilvia Kövécs for their devoted technical assistance. I remember with gratitude the helpful librarians and the peaceful guestroom of the Biological Research Centre at Szeged. Last but not least, I thank my family for tolerating my whims, while writing, with patience and love.

CONTENTS

LIST OF ABBREVIATIONS

ADH	alcohol dehydrogenase
ADP	adenosine diphosphate
ALD	aldolase
AMP	adenosine-5'-phosphate
ATCase	aspartate carbamoyl transferase
ATP	adenosine triphosphate
B-3-P	band-3-protein of the erythrocyte membrane
BPG	bis-phosphoglycerate
cAMP	cyclic adenosine-3',5'-monophosphate
cGMP	cyclic guanosine-3',5'-monophosphate
CoA-SH	coenzyme A
CoQ and $CoQH_2$	oxidized and reduced forms of coenzyme Q (ubiquinone), respectively
DAHP	3-deoxy-D-arabinoheptulosonate-7-phosphate
DCCD	dicyclohexyl carbodiimide
DEAE	diethylaminoethyl-
DHQ	5-dehydroquinate
DHS	5-dehydroshikimate
EGTA	ethylene glycol bis(β-aminoethyl ether)-N,N,N',N'-tetraacetic acid
FAD	flavin adenine dinucleotide
FBP	fructose-1,6-bisphosphate
GAP	D-glyceraldehyde-3-phosphate
GAPD	D-glyceraldehyde-3-phosphate dehydrogenase
GDP	guanosine diphosphate
GTP	guanosine triphosphate
Hb	haemoglobin
IMP	inosine monophosphate
LDH	lactate dehydrogenase
MAO	monoamine oxidase
NAD and NADH	nicotinamide adenine dinucleotide, oxidized and reduced forms, respectively
NADP and NADPH	nicotinamide adenine dinucleotide phosphate, oxidized and reduced forms, respectively
NMR	nuclear magnetic resonance
OMP	orotidine-5'-phosphate
OSCP	oligomycin sensitivity conferring protein
–P	phosphoryl group
P_i	inorganic phosphate

PDC   pyruvate dehydrogenase complex
PFK   phosphofructokinase
PGK   3-phosphoglycerate kinase
SDS   sodium dodecyl sulphate
Tris   tris(hydroxymethyl)amino methane
UDP   uridyl diphosphate

THE HIERARCHY OF ENZYME STRUCTURES

"Synopsis of previous chapters:
there are no previous chapters."

Stephen Leacock: Gertrude the
Governess or Simple Seventeen
(1911)

Any treatise on enzyme organization should be based on a thorough consideration of the properties of protein structure. The pertinent knowledge, the content of unwritten previous chapters to this book, calls for a monograph in its own right. Fortunately, there are excellent such monographs at our disposal (e.g. [515, 1072]) where the reader can get a sure foothold on protein chemistry. On the following pages only a bird's-eye-view will be given of this vast field, just to serve as an introduction to our subject matter.

STRUCTURE OF MONOMERIC ENZYMES

Enzymes share the property with other globular proteins that they are linear polymers of amino acids folded in an irregular but highly specific manner to form fairly compact three-dimensional structures. One polypeptide chain usually gives rise to one globular unit. The simplest enzymes, such as lysozyme, trypsin or ribonuclease, consist of a single polypeptide chain or at least are derived from one chain. The latter case is exemplified by chymotrypsin, which is synthesized as single-chain chymotrypsinogen and then converted to α-chymotrypsin through several proteolytic fissions. The α-chymotrypsin molecule finally consists of three unlike polypeptides in one globular entity.

The structural organization within the above monomeric enzymes and other globular proteins has distinct levels. According to the classical categorization, the *primary structure* is the sequence of amino acid residues in the polypeptide chain. The *secondary structure* comprises various ordered arrangements of the polypeptide backbone, such as α-helices, reverse turns and β-pleated sheets. The term *tertiary structure* denotes the overall three-dimensional conformation of the polypeptide chain. More recently, as a result of the accumulation of data obtained by X-ray crystallography, two further levels of organization emerged between the secondary and tertiary ones. *Supersecondary structures* are aggregates of secondary structures apparently preferred in several proteins for reasons of thermodynamic or kinetic stability. Examples of them found in globular proteins are the $\beta\times\beta$-unit (two parallel strands of a β-sheet connected by segment x), the Rossmann-fold (a $\beta\alpha\beta\alpha\beta$-unit, i.e. two α-helical segments intercalated between three parallel β-strands), and the β-meander (a sheet of three antiparallel β-strands). *Structural domains* are somewhat vaguely defined entities separated by clefts in the overall structure and usually built up of a continuous part of the polypeptide chain. A globular protein consists of one or more structural domains. Different domains as a rule have different functions, while the same type of domain may be found in various enzymes (e.g. the NAD-binding domain of dehydrogenases). Sometimes the same domain occurs more than once even in a monomeric protein, conferring an element of quasi-symmetry on the structure. Such domain linkage might have been produced by gene duplication, whereas the splicing of different domains is assumed to have occurred through gene fusion. It appears that enzymes, and other globular proteins too, have been constructed on a *modular basis:* the various elements needed for biological function were joined to give a single polypeptide chain.

Detailed treatments of the above structural principles can be found in advanced texts on proteins [515, 1072].

ENZYME STRUCTURES BEYOND THE MONOMERIC STAGE

Most enzymes, however, are neither single-chain proteins nor multichain ones in the sense that chymotrypsin is. Rather, they are composed of two or more poly-peptides each folded into a separate spheroidal particle (Fig. 1.1). Since the whole enzyme molecule as it is prepared and characterized physicochemically is regarded as *the* (structural) *unit,* the component folded polypeptide chains have been named *subunits.* If the subunits are identical, the protein is an *oligomer,* whose subunits are called *protomers.* An enzyme may be composed of unlike subunits as well, the differ-ent subunits (not protomers!) having different functions. If only one type of sub-unit is involved in catalysis whereas the other(s) fulfil non-catalytic functions (e.g. regulatory), we speak of a *complex enzyme.* On the other hand, if the unlike subunits carry out different, in most cases linked, catalytic reactions, we deal with a *multi-enzyme complex.* A special case of the juxtaposition of two or more enzymes is when the component enzymes share a common polypeptide chain. This arrangement, termed *multienzyme conjugate* [1307], is akin to the organization of a monomeric enzyme into different structural domains. The distinctive feature is that in multi-enzyme conjugates there is a separate active site on each "domain" catalysing different chemical reactions; in fact, such conjugates are sometimes resolvable into the com-ponent enzymes by mild proteolysis.

The enzyme structures discussed so far have the property in common that they are free to move about without breaking down the organizational level. This is not so with *scaffolded enzyme arrays* where the topology of enzymes is maintained by

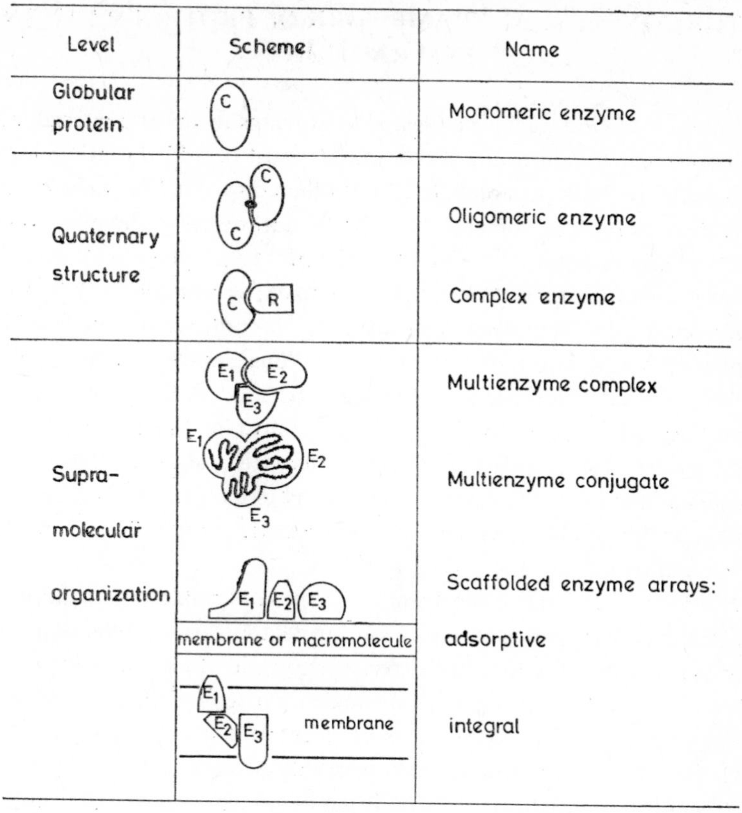

Level	Scheme	Name
Globular protein		Monomeric enzyme
Quaternary structure		Oligomeric enzyme
		Complex enzyme
Supra-molecular organization		Multienzyme complex
		Multienzyme conjugate
		Scaffolded enzyme arrays: adsorptive
		integral

Fig. 1.1. Structural organizational levels of enzymes. Individual contours correspond to globular entities, usually composed of not more than one polypeptide chain. C and R denote catalytic and regulatory subunits, respectively. E_1, E_2 and E_3 are enzymes catalysing three consecutive steps in a pathway. In the multienzyme conjugate the continuity of polypeptide chain over $E_1 \rightarrow E_3$ is indicated.

some kind of support or matrix. One may distinguish *adsorptive arrays* in which enzymes are bound to the surface of a membrane or another macromolecule, and *integral arrays* incorporated into the lipid bilayer of a membrane. Adsorptive arrays are considered to be more dynamic structures than integral ones: the former may readily decompose or exist in a state of equilibrium between free (randomized) and bound (ordered) enzymes, whereas the latter are fairly stable, subject mainly to lateral sliding in the plane of membrane.

THE CONCEPTUAL FRAMEWORK OF PROTEIN–PROTEIN INTERACTIONS

In the previous section we have classified the various forms of quaternary structure and supramolecular organization of enzymes. It must be added that the terminology adopted is not universally accepted, let alone adhered to. There is some inconsistency in the literature as regards nomenclature that is not merely a semantic issue but has both practical and conceptual roots. For example, tryptophan synthetase, a simple and popular archetype of a multienzyme complex, is usually described as having α and β *subunits* and a *quaternary structure* $\alpha_2\beta_2$. In contrast, no one would describe the complex of, say, two glycolytic enzymes as "quaternary structure". The reason for this usage is that tryptophan synthetase is isolated as a complex, whereas glycolytic enzymes are typical individual proteins, most of them well-established oligomers. However, the association of two different catalytic proteins to promote their combined reaction is the same, distinct, level of enzyme organization, whatever the stability of the complex, which may vary greatly even among tryptophan synthetases from different sources.

Quaternary structure and most forms of supramolecular organization have the common feature that they are based on protein–protein interactions, in which definite areas on the protein surface recognize each other when forming specific aggregates. Instructions, "messages", may be transmitted through the *contact surfaces* by molecular movements recognized as conformational changes. If we look from this viewpoint, there is no difference in principle between enzyme–enzyme interactions in a complex, subunit interactions in an oligomer, or even the interaction of structural domains within a monomeric enzyme. All are subject to the stipulations discussed in the next chapter.

CHAPTER 2

CHEMISTRY OF PROTEIN ASSOCIATIONS

THE CONCEPT OF COMPLEMENTARY SURFACES

In order for two proteins to associate, they must have surface areas complementary to each other. Here complementarity means that the association partners share the same van der Waals' envelope over some area [840]. This envelope is the surface tangent to the van der Waals' spheres of all atoms in the given area (Fig. 2.1) [724]. Complementary surfaces need not be continuous; they readily tolerate holes. In contrast, a protuberance on either of the surfaces destroys precise fit and hence complex formation (Fig. 2.2).

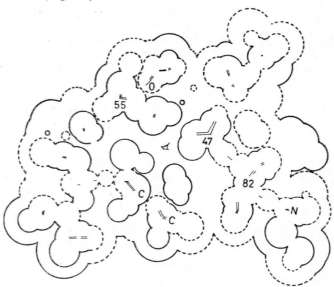

Fig. 2.1. The van der Waals' envelope and accessibility contour of a section of ribonuclease-S. The polar oxygen and nitrogen atoms are drawn by dotted lines, the apolar carbon and sulphur atoms by solid lines. The accessibility contour is the outermost covering line divided into dotted and solid parts according to the type of atoms beneath, and was "drawn" by the centre of a solvent molecule (of radius 0.14 nm) as it rolled along the protein making the maximum permitted contact. The length of arc above a given atom is the measure of accessibility of the atom in that plane. (From Lee and Richards [724].)

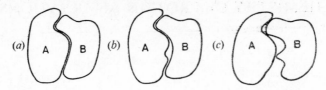

Fig. 2.2. Geometric complementarity as a condition of association of two proteins. (a) Proteins A and B share a common van der Waals' envelope over a major area of their surfaces. Extensive complementarity secures a stable complex. (b) A hole in the contact surface of A, though lessening binding strength, still allows association. (c) A protuberance on the contact surface of protein B (or A) interferes with extensive fit; a stable AB complex will not form.

Identical globular proteins, such as the protomers of oligomers, may have *self-complementary* surfaces. Association through such surfaces gives rise to dimers with a two-fold rotational symmetry (Fig. 2.3). It has been predicted *a priori* [834]

Fig. 2.3. The Chinese symbol, *T'ai ki*, illustrating self-complementary surfaces with the dyad axis indicated. In a real dimer the same dyad relates contours of various shapes, each possessing twofold symmetry, in successive sections of the protein. (From Morgan *et al.* [840].)

and also shown in proteins of known structure [840] that a self-complementary surface itself possesses two-fold symmetry in sections perpendicular to the dyad axis (cf. Fig. 2.3). If this symmetry is violated by a group of atoms protruding beyond the symmetry surface, as done by His-103 at the α-α interface of haemoglobin [840], association is prevented. A special case of symmetry violation, referred to as "dynamic asymmetry" [970, 1243] was found in the chymotrypsin dimer*, where protruding Phe-39 takes different conformations in the two subunits to allow the surfaces to fit together. This is, in fact, a particular type of "quaternary constraint", a term introduced by Monod *et al.* [834] to denote a certain amount of rearrangement of tertiary structure in the subunits upon association. Here, however, constraint acts to destroy rather than to maintain local symmetry.

* The dimerization of α-chymotrypsin at acidic pH has been extensively studied both in crystal and solution [37, 85, 333]; "subunit-exchange" chromatography [22], a procedure based on specific protein–protein interactions, was elaborated with this dimer. However, since chymotrypsin works in the alkaline region, the acid dimer is physiologically irrelevant. Near its pH optimum ($\sim$pH 8) α-chymotrypsin tends to polymerize after dimerization, but this dimer is distinct from the acidic one and probably corresponds to the enzyme-substrate intermediate preceding autolysis [908, 1099, 1214]. For these reasons chymotrypsin is not to be regarded as an oligomeric enzyme.

COMPLEMENTARY PROTEIN SURFACES ARE TIGHTLY PACKED

Richards [1001, 1002] has demonstrated that the protein interior is as tightly packed as crystals of amino acids. This conclusion was reached by calculating for small atomic groups in proteins the ratio of the combined atomic volume to the volume of the corresponding Voronoi polyhedron, i.e. the polyhedron whose faces separate the particular group of atoms from all its neighbours. Packing densities thus obtained were, on the average, 0.75 for proteins and 0.74 for crystals of small molecules. (Also 0.74 for equal-sized hard spheres.) By using the same approach Chothia and Janin [215] have determined the volume of residues buried in protein interfaces and compared them to those in the protein interior and in amino acid crystals (Table 2.1). It is seen that there is no systematic trend toward higher volumes at the interface; the residues, on the average, occupy the same volume at all three types of location.

TABLE 2.1.

Comparison of residue volumes at protein interfaces, interiors and crystals of amino acids

Residue	V_S	V_P	V_C
Gly	68.5	66.4	66.5
Phe	196.3	203.4	–
Ala	89.4	91.5	96.6
Val	141.1	141.7	143.4
Pro	128.9	129.3	124.4
Leu	161.3	167.9	–
Ile	175.2	168.8	169.7
Ser	99.7	99.1	102.2
Tyr	205.1	203.6	201.7
Cys	108.9	105.6	108.7
Trp	241.6	237.6	–
His	167.6	167.3	166.3
Gln	165.5	161.1	148.0
Asn	147.4	135.2	–
Lys	172.9	171.3	–
Asp	127.2	124.5	122.0
Average: 16 residues	149.8	148.4	–
11 residues	134.3	132.6	131.8

V_S: average volume of residue in protein interfaces;
V_P: average volume of residue in protein interior;
V_C: volume of residue in crystals of amino acids.
Volumes are given in $nm^3 \times 10^3$; standard deviation about 6%.
(From Chothia and Janin [215].)

ABOUT THE PREDOMINANT ROLE OF HYDROPHOBIC INTERACTIONS

The free energy of dissociation, ΔG_d, and the dissociation constant, K_d, of a complex are related by the equation

$$\Delta G_d = -RT \ln K_d \qquad (2.1)$$

Thus, for a very stable complex of $K_d = 10^{-13}$ M, such as the trypsin–trypsin inhibitor complex [1270], $\Delta G_d = 75$ kJ/mole, whereas for a loose complex of $K_d = 10^{-5}$ M, such as the insulin dimer [923], $\Delta G_d = 30$ kJ/mole at 27°C. Attempts have been made to establish how the various non-covalent forces operative in proteins (Table 2.2) contribute to the binding energy.

TABLE 2.2.

Non-covalent forces operative in protein structure

Type	Example		Binding energy kJ/mole	Free energy of transfer water→ethanol kJ/mole
van der Waals' interaction[a]	aliphatic H... aliphatic H	—C—H···H—C—	−0.1	
	aliphatic C... aliphatic C	C···C	−0.5	
	amide N... amide N	N···N H H	−0.7	
Electrostatic	salt bridge	—COO⁻···H₃N⁺—	−21	−4
interaction[c]	two dipoles	—C=O···O=C—	+1.3	−13[d]
Hydrogen bond[b]	protein backbone	N—H···O=C	−16	
Hydrophobic force	side chain of Phe			−10

[a] van der Waals' interactions actually correspond to the E_m parameter of Lennard–Jones 6–12 potentials, which unite London's 6th power term of inverse distance for attractive dispersion forces [750] and Lennard–Jones' 12th power term of inverse distance for electron shell repulsion [597]. The values are the average of those given by Momany et al. [829], Lifson and Warshel [737] and Warshel et al. [1286].

[b] From Kresheck and Klotz [686].

[c] Calculated by assuming a dielectric constant $\varepsilon = 4$, as found in amide polymers [509].

[d] The values −4 and =13 kJ/mole refer to external [933] and internal [373] salt bridges, respectively.

Pauling and Pressman [919] suggested that the multiplicity of hydrogen bonds and van der Waals' interactions over a large surface area produce strong bonding. Alternatively, Kauzmann [622] proposed that entropic forces in aqueous media favour the formation of a hydrophobic core, resembling an oil droplet, in the protein interior and, likewise, such "hydrophobic forces" would be the major factor in sticking proteins together. Extending this idea Chothia and Janin [215] argued that electrostatic interactions (including hydrogen bonds) can equally be established with water and dissolved ions, which drastically reduces the net contribution of these forces, much below the values shown in Table 2.2. On the other hand, as 1 nm^2 of protein surface area when buried gives rise to 10.5 kJ/mole hydrophobic free energy [213], the 13.9 nm^2 surface area buried in the trypsin–trypsin inhibitor complex [214, 215] amounts to 147 kJ/mole of binding energy.

Ross and Subramanian [1025] have recently questioned the general predominant role of hydrophobic interactions in protein associations. Having examined a number of known cases these authors propose that protein associations proceed in two steps: the mutual penetration of hydration layers which causes disordering of the solvent, followed by further short-range interactions. It seems that in addition to the hydrophobic effect, the strengthening of hydrogen bonds in the low dielectric macromolecular interior and the newly created van der Waals' interactions also play a substantial role in the stability of protein complexes.

ELECTROSTATIC EFFECTS MAINLY CONTRIBUTE TO THE SPECIFICITY OF PROTEIN ASSOCIATIONS

The high degree of specificity found in protein complexes is incurred, in addition to the geometric complementarity of contact surfaces, by the proper positioning of polar groups. In effect, to yield a stable association all charged side chains at protein interfaces should form salt bridges and all hydrogen bond formers are to find their donor or acceptor counterparts [622, 934]. While the formation of a salt bridge has small net contribution of -4 to -13 kJ/mole (cf. Table 2.2) to the free energy of binding [373, 933] the transfer of a single charge to a hydrophobic environment is energetically very unfavourable (about 40 kJ/mole). It is worth mentioning that in haemoglobin even water-accessible salt bridges contribute to the maintenance of the "tense" (deoxy) conformation [42, 934]. Although the reason for this is unclear, the explanation might be similar to that given by Warshel [1285] for the greater stabilization of the polar transition state by lysozyme than by bulk water: proteins can apparently combine the effects of permanent and induced dipoles in a highly favourable manner.

Salt bridges, shielded from water, play a major role in the thermostability of enzymes isolated from thermophilic microorganisms. Comparison of the three-dimensional structures of D-glyceraldehyde-3-phosphate dehydrogenases from lobster [839] and *Bacillus stearothermophilus* [82] revealed that in the latter there are four additional salt bridges between the four subunits. The remarkable stabilizing effect of a few salt bridges is not inconsistent with their small overall contribution to binding energy, if we consider that a ten-fold decrease in denaturation rate at 70°C only requires 6.6 kJ/mole extra energy for stabilization.

In conclusion, protein–protein associations invoke the complementarity of protein interfaces, ensured by the proper fit of van der Waals' envelopes and the precise pairing of charged groups and hydrogen bond donors/acceptors, as well as an appropriate amount of hydrophobic forces, to attain reasonable bond strength.

CHAPTER 3

INTRAMOLECULAR ORGANIZATION: THE QUATERNARY STRUCTURE OF ENZYMES

"Behold, how good and how pleasant it is for brethren to dwell together in unity!"

Psalms, CXXXIII, 1.

OLIGOMERIC ENZYMES

According to the definition given in Chapter 1, an oligomeric protein is composed of identical subunits called protomers. On closer inspection this statement must be qualified by saying that the protomers need not be strictly identical, only quasi-identical. For instance, lactate dehydrogenase, a tetrameric enzyme, has two main types of subunit, H and M (standing for heart and muscle, respectively). Since the H and M subunits can more or less freely recombine under suitable conditions, beside the homotetramers H_4 and M_4 three hetero- or hybrid tetramers—H_3M, H_2M_2 and HM_3—also form both in nature and in the test tube. Obviously, in the hybrid tetramers subunits are not identical, they differ in primary structure and consequently in their functional properties as well, yet they catalyse the same reaction: the reversible oxidation of lactate by the aid of NAD. The hybrid lactate dehydrogenases are therefore oligomers consisting of quasi-identical protomers. Another common, though non-enzymic, example is haemoglobin. It has an $\alpha_2\beta_2$ structure, i.e. two different polypeptides build up the tetramer, yet all four subunits do the same work: the reversible binding of oxygen through the haem group. It is worth recalling that the similar structure designation ($\alpha_2\beta_2$) for tryptophan synthetase mentioned above has a quite different physical meaning: here the unlike proteins catalyse altogether different chemical reactions, hence this assembly is a multienzyme complex.

NUMBER OF SUBUNITS IN OLIGOMERIC ENZYMES

The overwhelming majority of oligomeric enzymes are *dimers* and *tetramers*. Toward the end of 1980 there were about 250 enzymes conclusively or tentatively identified as dimers and some 150 as tetramers. These figures are uncertain for several reasons. First, the available evidence on subunit stoichiometry may not be convincing. In general, more than one method is required to prove conclusively a given quaternary structure. The current methodology of protein chemists, if fully exploited, is already powerful enough to solve the question in cases not fraught with special difficulties, such as latent proteolytic contaminants, labile peptide bonds, etc. Such spectacular metamorphoses as the "promotion" of rabbit muscle aldolase, the single

TABLE 3.1.

*List of enzymes claimed to be trimers**

Enzyme	Source	Protomer mol. wt	Type of evidence	Remark	References
Adenine phosphoribosyl transferase, EC 2.4.2.7	Human erythrocytes	11,100	SDS-PAGE, G-CH, ULCE	May be active as dimer or hexamer	1217
ATP-AMP-phosphotransferase (Adenylate kinase), EC 2.7.4.3	Rat liver	23,000	G-CH, ULCE	Monomer-dimer-tetra-mer rapid equilibrium	268
Amino acyl-tRNA binding enzyme	Rabbit reticulocytes	62,000	SDS-PAGE, ULCE		805
4-Amino butanal dehydrogenase, EC 1.2.1.19	Pseudomonas	75,000	SDS-PAGE, G-CH		159
Arginase, EC 3.5.3.1	Saccharomyces cerevisiae	39,000	CL, SDS-PAGE		926
	Rat liver	35,000	CL, SDS-PAGE		926
Aspartate carbamoyl transferase, EC 2.1.3.2	B. subtilis	33,500	CL, SDS-PAGE, ULCE		125
Carboxylic ester hydrolase, EC 3.1.1.1	Pig liver	55,000	SDS-PAGE, G-CH, ULCE, CL		601
	Ox liver	56,000	SDS-PAGE, G-CH, ULCE, CL		601
Glyceraldehyde-3-P dehydrogenase, EC 1.2.1.12	Spinach leaves	37,000	SDS-PAGE, G-CH	Trimer of dimers	1149
Histidine decarboxylase	Micrococcus		X-S		462
2-Keto-3-deoxy-6-phosphogluconic (KDPG) aldolase, EC 4.1.2.12	Pseudomonas	24,000	X-D, ULCE		795 506
L-Ornithin carbamoyl transferase, EC 2.1.3.3	Saccharomyces cerevisiae	37,500	CL, SDS-PAGE	Enzyme only "highly purified"	926
	E. coli W	35,000	CL, SDS-PAGE		728
Purine nucleoside phosphorylase, EC 2.4.1.2	Calf spleen, human erythrocytes	28,000	ULCE, SDS-PAGE		332
	Human placenta	31,000	CL, SDS-PAGE		441

Quinolinate phosphoribosyltransferase EC, 2.4.2.a	Rat liver	54,000	ULCE		903
Succinic semialdehyde dehydrogenase, EC 1.2.1.16	Pseudomonas	55,400	SDS-PAGE, G-CH, ULCE	Also forms hexamers	1022
Urease, A_1 EC 3.5.1.5	Jack bean	80,000	ELMI	Also forms hexamers (α-urease), and polymers of α	375

Type of evidence: SDS-PAGE: sodium dodecylsulphate polyacrylamide gel electrophoresis
          CL: chemical crosslinking
          ULCE: ultracentrifugation
          ELMI: electron microscopy
          G-CH: gel chromatography
          X-D: X-ray diffraction
          X-S: small angle X-ray scattering

* See also the list of trimeric enzymes in carbamoylphosphate metabolism by Vickers [1266]

apparently well-documented example of a trimeric enzyme in the early sixties [295, 681, 1037, 1180], to a tetramer [625], are no longer likely to occur. The second source of ambiguity in subunit stoichiometry is the inherent property of some oligo- mers to exist in a dynamic equilibrium of different aggregational states. Thus, for example, glycogen phosphorylase, basically a dimeric molecule [382, 1290], tends to form tetramers under the effect of modifiers such as AMP or covalently linked phosphoryl groups (cf. [472]). It is therefore inaccurate, though done in many a text- book, to regard phosphorylase *b* as dimer and phosphorylase *a* (the phosphorylated form) as tetramer. All the more so, since glycogen, the natural support medium for the enzyme in muscle [815], dissociates all kinds of phosphorylase tetramers to dimers [318, 496, 1281]. The third reservation concerning oligomer assignments comes from association phenomena observed under unphysiological conditions. We have already referred to α-chymotrypsin, which at acidic, i.e. unphysiological, pH exists as a dimer [37, 333] but presumably acts as a monomer *in vivo*. It is due to the above ambiguities that comprehensive lists on the subunit stoichiometries of proteins (e.g. [286]) are often in error at several places and are useful for providing us with an overall picture rather than with specific information about one par- ticular protein.

The next most abundant group of oligomers comprises *hexamers*. Some forty enzymes have so far been claimed to belong to this family. Like some dimers, hexa- mers also tend to further associate giving rise to higher aggregates (cf. next section).

The number of *octameric* enzymes borders twenty, among them the chloroplast enzyme ribulose-1,5-bisphosphate carboxylase, probably the most abundant enzyme on earth.

In the order of decreasing occurrence *trimeric* enzymes follow, which are the simplest oligomers of uneven protomer number. In the aftermath of the decline of the trimeric status of fructose-bisphosphate aldolase, and spellbound by the allo- steric enzyme model of Monod *et al.* [834], many investigators in the late sixties questioned the very existence of uneven numbered oligomers. However, by now we know of well-defined cases for trimeric enzymes. Table 3.1 lists enzymes that have been claimed to be trimers, along with the type of evidence presented. *Pentameric* enzymes are rare, whereas no *heptamer* has so far been found. It is therefore justi- fiable to say that even-numbered oligomers are much more common than those having an uneven number of protomers, but the latter also exist. In the next section we will examine in some detail why this is so.

Very few true oligomeric enzymes contain ten, twelve or more subunits. They will be mentioned in connection with oligomer symmetry.

SUBUNIT ARRANGEMENTS IN OLIGOMERIC ENZYMES

After establishing the number of protomers found in an oligomeric protein, the next question is how these protomers are arranged within the oligomer. The spatial disposition of subunits usually gives rise to structures of various degrees of symmetry, as we shall see below.

Open and closed structures

When two protomers (or any two globular proteins) associate, some surface area on each will be buried forming an interface (cf. Chapter 2). Monod et al. [834], referring to oligomeric proteins, suggested the terms "binding set" for the spatially organized collection of all residues on a protomer that bind another protomer, and "domain of bonding" for two linked binding sets through which two protomers are associated.

If we consider how identical subunits can associate, it is useful to define two angles, the dihedral angle α and the rotational angle β, that relate protomers to each other (Fig. 3.1). If the two protomers are staggered, either when $\beta \neq 0$ or when they are vertically displaced, various helical structures will form depending on the values of α and β. Such associations are called open, because there remain unoccupied binding sets at both ends of the assembly. Accordingly, these structures may be either infinitely long (at least in principle) or prevented from further growth by steric hindrance. The former alternative holds also for a linear association where $\alpha = 0$ and $\beta = 0$ (Fig. 3.2a). Such structures are not found among oligomeric enzymes. It should be mentioned, though, that some enzymes, e.g. jack bean urease [375] and beef liver glutamate dehydrogenase [237], tend to polymerize giving rise to filamentous or sheetlike open associations. In these cases, however, the polymerizing unit is usually itself an oligomer. The fact that this mode of growth sometimes produces aggregates of discrete size can be caused by conformational changes that prevent further (stable) associations, as postulated for protocollagen proline hydroxylase [883].

In contrast, if $\beta = 0$ and $\alpha = 360°/n$, various, ringlike structures are obtained consisting of n protomers (Fig. 3.2b and d). These assemblies will have an n-fold rotational symmetry axis. Such an association is closed, inasmuch as there are no free binding sets in it. Oligomeric enzymes are, as a rule, closed structures. It should be noted that since protomers are genuinely asymmetric (chiral) [158] bodies, rather than perfect spheres, their aggregates may only have axial symmetry [513], if we disregard translational symmetry, which may hold for open oligomers and polymers. Apart from the trivial case when $n = 1$, i.e. a single globular protein without any (strict) symmetry relation, the class where $n = 2$, hence $\alpha = 180°$, deserves special attention (Fig. 3.2c). This dimer has a twofold symmetry axis and only one binding set A which is, however, self-complementary (cf. Chapter 2, p. 6).

SUPRAMOLECULAR ENZYME ORGANIZATION

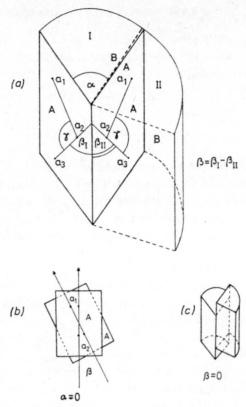

$$\beta = \beta_{\mathrm{I}} - \beta_{\mathrm{II}}$$

Fig. 3.1. Angles characterizing the geometry of association of identical subunits (protomers). (After Hajdu [494].) (*a*) The outlines of two protomers, I and II, are shown, each having two complementary binding sets (contact surfaces) A and B. Three arbitrarily chosen points (centres of given atoms) of binding set A are denoted by a_1, a_2 and a_3, which define plane A. Dihedral angle α is created by the two A planes (drawn in heavy lines) belonging to protomers I and II. The angle $\beta = \beta_{\mathrm{I}} - \beta_{\mathrm{II}}$ gives the degree of rotation of the two protomers relative to each other. In the illustration $\beta = 0$, i.e. the protomers are in register and a ringlike structure would form on binding further protomers. If $\beta \neq 0$, a helical structure ensues which may be an infinite association if β is large enough. If β is small, only a single helical turn can form, further association being prevented by steric hindrance ("hinged helix", [834]). Note that irrespective of the value of β, angles γ on the two A planes are always the same, the subunits being identical. (*b*) At $\alpha = 0$ a linear association occurs, with an element of helicity if $\beta \neq 0$. (On the A planes only two points, a_1 and a_2, are shown for the sake of clarity.) (*c*) A helix may also form if $\alpha \neq 0$ and $\beta = 0$, provided that the protomers are not in register but vertically displaced relative to one another. As in (*a*) above, the helix may be infinite if the vertical displacement is large enough.

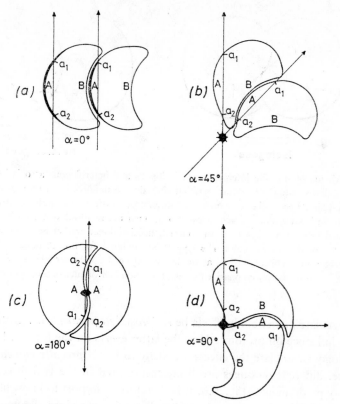

Fig. 3.2. Oligomer associations at $\beta=0$ and various α dihedral angles. In *a*, *b* and *d* each protomer has two binding sets, A (drawn in heavy line) and B, which are complementary. Of the three arbitrary points chosen on contact A only two (a_1 and a_2) are shown and the planes defined by them are perpendicular to the sheet of paper. (*a*) linear (open) heterologous association; (*b*) two protomers of a cyclic (closed) heterologous octamer; (*c*) isologous dimer with a single, self-complementary binding set A; (*d*) half of a cyclic heterologous tetramer.

Isologous and heterologous associations

Monod *et al.* [834] have qualitatively distinguished the structures where $\alpha=180°$ and those where $\alpha<180°$, calling the former *isologous* and the latter *heterologous* associations. The rationale of this distinction is evident from Fig. 3.2. In a heterologous domain of bonding two different (A and B) binding sets stick together, whereas the complementary binding sets in an isologous domain of bonding are identical (type A). In an isologous bonding domain, but not in a heterologous one, each contact $m_{I}n_{II}$ (where n_I is the *n*th residue in protomer I and m_{II} is the *m*th residue in protomer II) has a mirror image contact $m_{II}n_{I}$. These basic features of protomer association are worth remembering when the functional consequences of subunit interactions are being considered. Namely, one would expect *a priori* that in an iso-

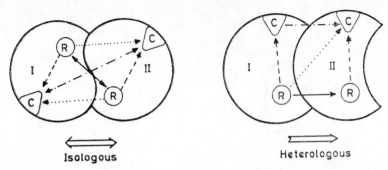

Fig. 3.3. Scheme of site–site interactions in isologous and heterologous associations. C, catalytic site; R, regulatory site. The arrows between sites denote unidirectional (homo- or heterotropic) effects (one-pointed arrows) and bidirectional (homotropic) effects (two-pointed arrows). Note that in the isologous dimer bidirectional arrows go through the two-fold symmetry axis, i.e. identical parts of each subunit are involved. In contrast, unidirectional arrows between subunits involve different parts of the two subunits. The scheme is simplified, for not all possible interactions are shown. Furthermore, effects marked as unidirectional need not be strictly such: the one-pointed arrows merely emphasize the reciprocal non-equivalence of interactions.

logous associate intersubunit effects be *reciprocally equivalent,* i.e. protomer I exerts the same influence on protomer II as the latter exerts on the former. In contrast, in a heterologous associate the effects are likely to be *reciprocally non-equivalent* (vectorial), i.e. different messages are being transmitted in the two directions, one of which may predominate (Fig. 3.3). It is therefore of import to know how many and which type of intersubunit domains of bonding exist in a given oligomer.

Symmetry of oligomeric assemblies

As pointed out above, closed oligomeric assemblies possess various symmetries, depending on the arrangement of protomers, hence they can be classified conveniently on the basis of this symmetry. The reader is referred to special texts for the rigorous treatment of symmetry operations [588, 1312] and for symmetry in proteins in general [515, 666]. In brief, closed oligomers can fall into one of the three classes of Schönfliess' point group symmetry. This classification refers to the collection of symmetry operations, applied about a point, that transpose the object into itself [531]. The three classes are as follows:

Cyclic symmetry (C_n): there is one n-fold rotational symmetry axis and the number of protomers is n;
Dihedral symmetry (D_n): there are n two-fold axes at right angles to any single n-fold axis and the number of protomers is $2n$;
Cubic symmetry is divided into subclasses:

TABLE 3.2.

Symmetry notations, binding set and bonding domain data of some oligomeric protein structures

Number of protomers	Symmetry		Number of binding sets per protomer		Number of types of bonding domain in oligomer		Overall number of bonding domains in oligomer		
	Schönflies symbol	Herman–Mauguin symbol	Heterologous	Isologous	Heterologous	Isologous	Heterologous	Isologous	Total
2	C_2	2	–	1	–	1	–	1	1
3	C_3	3	2	–	1	–	3	–	3
4	C_4	4	2	–	1	–	4	–	4
	D_2	222	–	3	–	3	–	6	6
5	C_5	5	2	–	1	–	5	–	5
6	C_6	6	2	–	1	–	6	–	6
	D_3	32	2	2	1	2	6	3	9
8	C_8	8	2	–	1	–	8	–	8
	D_4	422	2	2	1	2	8	8	16
12	D[a]		2	1	1	1	12	6	18
	T[b]	23	4	1	2	1	24	6	30
24	O[b]	432	2	1	1	1	24	12	36
			4	–	2	–	48	–	48
60	I or Y	532	~	~	~	~	~	~	~

[a] Only the eclipsed case is considered; [b] Only the forms shown in Fig. 3.4h and Fig. 5.5 are considered; ~, several possibilities

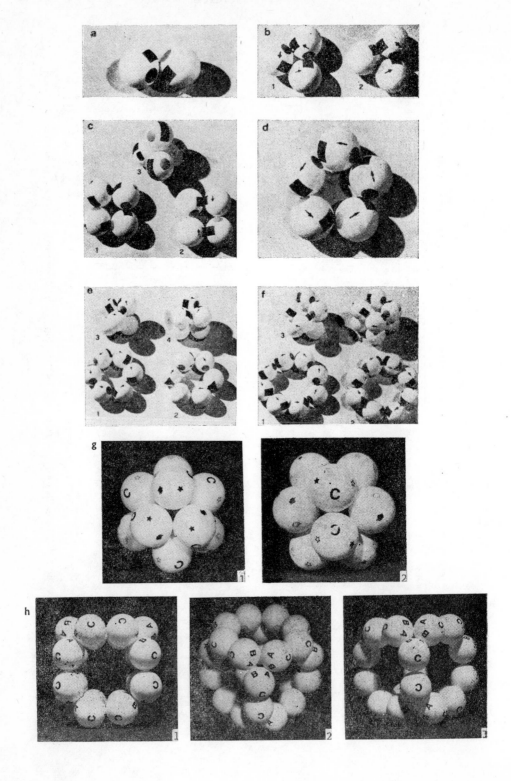

Tetrahedral (T): number of protomers twelve; four three-fold and three two-fold axes; e.g. four cyclic trimers at the vertices of a tetrahedron;

Octahedral (O): number of protomers twenty-four; three four-fold, four three-fold and six two-fold axes; e.g. eight cyclic trimers at the vertices of a cube;

Icosahedral (I or Y): number of protomers sixty; twenty cyclic trimers or twelve pentamers.

All three subclasses of cubic symmetry represent *shell structures,* which are isometric particles whose three coordinate directions are equivalent, i.e. no direction in space can be preferred [172]. All constituent protomers are therefore sterically equivalent.

Although cubic symmetry is rare among enzymes, its characteristic property, *viz.* the steric equivalence of protomers, is also found in D_2 structures [513], exclusively among all other symmetry classes. Tetramers, which are among the most common oligomeric proteins, practically all have D_2 symmetry (cf. below). It appears that the shell-like arrangement of protomers, and hence their steric equivalence, proved to be advantageous not only in virus coats built up from a large number of subunits [172], but also in the much smaller tetrameric enzymes. In the latter the shell is virtual, nothing being hidden "inside", if not the biological function itself of a catalytic–regulatory macromolecule.

In Table 3.2 the symmetries of various oligomeric assemblies, in terms of both the Schönfliess symbols and the Hermann–Mauguin symbols used by crystallographers, together with the number and type of binding sets and bonding domains, are listed. Only structures that have some relevance in nature were included. The decorated ping-pong ball models of the various forms are shown in Fig. 3.4. They will be dealt with in detail in the next section.

Evidence for the occurrence of various oligomeric forms among enzymes

It will be reviewed below how, according to our current knowledge, the oligomeric enzymes found in nature are distributed over the various geometric forms. As we concentrate on enzymes, only occasional mention will be made of non-enzymic

Fig. 3.4. Models for various oligomeric structures. (*a*) Isologous dimer; (*b*) trimers: 1, heterologous (C_3), 2, mixed; (*c*) tetramers: 1, heterologous (C_4), 2, isologous planar, 3, isologous tetrahedral; (*d*) pentamer (C_5); (*e*) hexamers: 1, all-heterologous planar (C_6), 2, all-isologous planar (D_3), 3, spherical staggered (octahedron), 4, spherical eclipsed (trigonal prism); (*f*) octamers: 1, all-heterologous planar (C_8), 2, all-isologous planar (D_4), 3, spherical staggered, 4, spherical eclipsed (cube), (*g*) dodecamer (symmetry class T), viewed along a three-fold (1) and two-fold (2) symmetry axis; (*h*) icosatetramer (24 subunits, octahedral symmetry) viewed along a four-fold (1), three-fold (2) and two-fold (3) symmetry axis.

oligomers, such as transport proteins, hormones, toxins, receptors, lectins, repressors, etc., although several illustrative examples of oligomer symmetry have been found among them.

Dimers whose structure has been examined in sufficient detail (about a dozen enzymes) all proved to fall into class C_2, the isologous dimer (Figs 3.2c and 3.4a). There is one notable exception, yeast hexokinase, which is partly or totally hetero-logous [1177], and further subunit attachment is prevented not by the full occupancy of binding sets but rather by steric hindrance. The asymmetric dimerization has been invoked in the allosteric properties (negative cooperativity) of hexokinase [1176], although it is not clear whether asymmetric dimerization also occurs in solution [1178]. A further doubt is raised by the fact that negative cooperativity toward ATP has been demonstrated also with the succinylated hexokinase monomer [1026]. Solution studies suggest that other kinases, such as creatine kinase and arginine kinase, may also form asymmetric dimers [297, 298].

Trimers can only belong to symmetry class C_3, an all-heterologous ring (Fig. 3.4b, 1). In the case of 2-keto-3-deoxy-6-phosphogluconate aldolase this symmetry has been conclusively established [795], the enzyme has a propeller shape when viewed along the three-fold axis. In Fig. 3.4b, 2 a so-called mixed trimer is seen. In mixed asso-ciations the same type of binding set of the protomer would participate in both isologous and heterologous contacts. With the trimer a mixed structure, comprising two isologous and one heterologous domain, can be produced by turning one of the protomers (the one to the left in Fig. 3.4b) upside down. Mixed structures would imply *facultative* binding set pairings, i.e. a lack of specificity at the subunit contacts. Although Cornish-Bowden and Koshland [253, 254] proposed that mixed oligomers might be thermodynamically viable, there is no evidence for their existence as yet. In fact, it has been claimed [984] that hexameric bovine liver glutamate dehydrogenase consists of two rings of mixed trimers. However, the argument based on crosslinking and subsequent identification of stapled lysyl side chains is not convincing: there is no reason to believe that sequentially identical residues can only be crosslinked across isologous contacts, whereas the crosslinking of sequentially different residues does not even suggest the involvement of a heterologous contact (cf. p. 17).

Tetramers known so far all belong to class D_2, and may assume forms ranging from the planar to the pseudotetrahedral (Fig. 3.4c, 2 and 3). In the former only two, whereas in the latter three binding sets are on each promoter (Table 3.2). The existence of heterologous (C_4) tetramers (Fig. 3.4c, 1) has not yet been substantiated. The earlier assignment [666] that *E. coli* tryptophanase and chicken liver pyruvate carboxylase have C_4 symmetry has been invalidated [972 and 232, 465, respectively]. Likewise, the uncertainty about yeast alcohol dehydrogenase [578] and yeast aldehyde dehydrogenase [1207], has been cleared in favour of D_2 structure [312 and 409, respectively]. The basis of these corrections will be dealt with in some detail later (cf. p. 31).

There seem to be very few *pentameric* enzymes, whose only symmetrical arrangement is the regular pentagon (C_5), an all-heterologous association (Fig. 3.4d). Prohistidine decarboxylase, which is converted into histidine decarboxylase by proteolytic cleavage in each subunit, has been claimed to be a pentamer [986]. Arginine decarboxylase, which has the shape of a regular pentagon under the electron microscope, is in fact a decamer: a heterologous pentamer of isologous dimers [100, 101]. The pentagonal arrangement is altogether rare among proteins. Two further, non-enzymic, cases are clam shrimp haemoglobin [288] and the collar protein of snail α-haemocyanin [1113].

Hexamers can be planar or quasi-spherical assemblies (Fig. 3.4e). The planar hexamer (regular hexagon) can be all-heterologous (C_6) or all-isologous (D_3, "trimer of dimers"). The quasi-spherical forms consist of two layers of heterologous trimers in eclipsed (trigonal prism) or staggered (octahedron) disposition. In the eclipsed case there are two heterologous and one isologous binding sets on each protomer, whereas in the staggered form one further isologous set may be created. It is noteworthy that three out of the four structures in Fig. 3.4e have the same (D_3) symmetry, but the difference between them manifests itself in the type and number of interprotomer binding sets (cf. Table 3.2). As regards occurrence in nature, the few planar hexamers observed usually tend to further aggregate to give stacks of hexagons. Clearly, a closed structure can be made only of two C_6 rings associated in an isologous manner. This seems to be the case for *E. coli* glutamine synthetase, which is then a dodecamer of D_6 symmetry [1253]. Protocollagen proline hydroxylase from chick embryo, which occurs either as a single or a stack of four hexagons [883], in its latter form is not a true closed structure (cf. p. 15). Phosphofructokinase from the alga *Dunaliella salina* has been proposed to have a C_6 form, in addition to others [909]. There is no clear-cut evidence for an all-isologous D_3 structure. Quasi-spherical hexamer arrangements are more common than planar ones: bovine glutamate dehydrogenase [984, 996], Chlorella glutamine synthetase [985], jack bean α-urease [375], and bovine lens leucine aminopeptidase [168, 662, 1282] are all eclipsed or somewhat staggered hexamers. Leucine aminopeptidase is a particularly well-studied example of a distorted trigonal prism with 2 types of isologous and 2 types of heterologous binding sets on each protomer.

Octamers may assume structures analogous to those of hexamers: they may have planar or quasi-spherical (cubic) shape (Fig. 3.4f). The planar octamer can be all-heterologous (C_8) or all-isologous (D_4), whilst the quasi-spherical forms consist of two four-membered heterologous rings in eclipsed (cube) or staggered arrangement (both D_4). The most common and best studied octameric enzyme is ribulose-1,5-bisphosphate carboxylase from chloroplasts. Strictly speaking it is not a true oligomer since it is composed of eight large and eight small subunits, but the L_1S_1 couple can be regarded as the asymmetric protomer. The tobacco leaf enzyme is a cube (D_4) [40] and thus it is structurally analogous to the invertebrate oxygen transport protein, hemerythrin [1283]. The enzyme from *Euglena gracilis* shows up in the electron microscope as a cube, but also as a rosette with apparent eight-fold symmetry [803]. The latter may correspond to either of the two planar octamers in Fig. 3.4f. The other octameric enzymes studied by electron microscopy, namely glutamine synthetase from hamster liver [1221], alcohol oxidase from *Kloeckera* [618] and UDP-glucose pyrophosphorylase from calf liver [732], or by hydrodynamic methods (enolase from *Thermus aquaticus* [1179]), all seemed to have cube-like appearance.

There is a small number of even larger oligomeric enzymes. Among *dodecamers*, *E. coli* glutamine synthetase has already been mentioned to have D_6 symmetry, whereas aspartate-β-decarboxylase, composed of six isologous dimers [116], belongs already to cubic symmetry class T (Fig. 3.4g). In the pyruvate dehydrogenase multi-enzyme complex of *F. coli* the pyruvate decarboxylase "shell" and the lipoamide transacetylase core [281, 306, 413], the latter built up of twenty-four protomers, seem to have octahedral symmetry: they look like truncated cubes, with a trimer at each vertex of the cube (Fig. 3.4h).

Oligomeric enzymes with irregular symmetry

Recently, a few oligomeric enzyme structures have been revealed which, though they exhibit symmetry, are irregular in some respect. We introduce them here not only for their genuine interest, but also to bring out a general facet of living systems: the ability to make "variations to a theme". In the realm of structural symmetry this means that even though the rules of protomer assembly are given, and also obeyed by the overwhelming majority of structures, there may be exceptions where nature has found some other way. Therefore we must not be too preoccupied with our self-constructed categories, rejecting anything that does not fit into the established pigeon-holes.

Yeast pyruvate carboxylase is a tetramer of molecular weight 480,000 and is composed of four apparently identical protomers. The protomers are shaped roughly as prolate ellipsoids and are arranged in such a manner that their centres are located on the corners of a rhombus (rather than a square or a regular or splayed [465] tetrahedron); diagonally opposite pairs of protomers lie in orthogonal planes [233]. The schematic model of this structure is seen in Fig. 3.5, which was derived from electron microscopy aided by computer simulation and biochemical evidence. The peculiarity about this structure is that it forms an open, heterologous association with two types of bonding domain (AD and BC), and half of the binding sets remain unoccupied. Yet polymerization was never observed, probably due to steric hindrance.

Another tenet concerning closed structures has been that the highest possible shell-like aggregate produced purely by isologous associations is the tetramer [513, 834]. However, if protomers are only quasi-equivalent, i.e. are chemically distinguishable but have the same size and shape, a closed octameric, non-ringlike structure may ensue. An example of this seems to be given by yeast phosphofructokinase.

Fig. 3.5. Model of yeast pyruvate carboxylase. (*a*) The protomer has four binding sets, *A, B, C* and *D*. The binding sets need not be located on one side of the protomer as indicated here. (*b*) Two views of the tetramer as the molecule is rotated 180° about the vertical axis. This structure has only a single two-fold symmetry axis (in the centre of the molecule, perpendicular to the paper sheet), thus it does not belong to either class C_2 or D_2 in the sense as interpreted above (Table 3.2). In fact, it has C_2 symmetry with two (identical) protomers making up the asymmetric unit. (From Cohen *et al.* [233].)

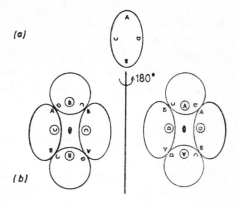

Fig. 3.6. Model of yeast phosphofructokinase. The three two-fold symmetry axes are shown. The two types of protomer, α and β, are represented by grey and white discs. Opposing interfaces are black. It is not decided which of the two types of protomer is in the centre and which on the periphery. Accordingly, the structure may be described as $\alpha_2\beta_4\alpha_2$ or $\beta_2\alpha_4\beta_2$. (From Plietz *et al.* [947].)

The model derived from crosslinking [1222] and small angle X-ray scattering [947] evidence is seen in Fig. 3.6. The structure has D_2 symmetry with an α-β protomer pair forming the asymmetric unit. Although the regular disc-shaped protomers displayed by Plietz et al. [947] are not too helpful as regards symmetry perception, closer scrutiny may convince the reader about the closedness of this hetero-oligomer.

Methods for determining oligomer symmetry

The methods to be described below provide information about molecular shape and neighbourhood relations at different levels of resolution. Therefore they are widely used not only in the structure studies of oligomers proper, but also of monomeric proteins and various protein, and non-protein, aggregates. We are going to discuss mainly those aspects that are pertinent to oligomer symmetry and subunit arrangement.

X-ray crystallography

This technique, with its precise location of most atoms, gives much more than needed to establish the type of oligomer structure. Even at moderate resolution ($\sim$6 Å) the shape of protein is clearly seen, from which oligomer symmetry can be deduced with reasonable confidence. At high resolution ($\sim$2 Å) all interprotomer atomic contacts can be pinpointed, which means that the binding set and bonding domain are identified chemically. This is the ceiling that may be achieved in this respect, at least as regards a static picture. In the case of an isologous tetramer, for example, the three types of interprotomer contact have been visualized by Rossmann and coworkers [738, 839, 1027] in a triangular graph (Fig. 3.7). Comparison of interprotomer contacts in such detail in various dehydrogenases allowed the authors to reveal evolutionary relationships [738]. Thus the Q-axis contact is common in glyceraldehyde-3-phosphate, lactate and malate dehydrogenases, whereas the contacts across the P- and R-axes are quite different. Apparently the three dehydrogenases diverged after Q-axis formation, glyceraldehyde-3-phosphate dehydrogenase branching off first, as judged from the similarity of main chain fold in lactate and malate dehydrogenases. On the other hand, the protomer contact in liver alcohol dehydrogenase, a dimeric enzyme, does not resemble at all any of the contacts in the other three enzymes. Consequently, the separation of alcohol dehydrogenase from the main stem of dehydrogenases took place before Q-axis formation.

The exact three-dimensional picture obtained by X-ray crystallography tells us what these protein "globules" actually look like. As seen in Fig. 3.8, tetrameric glyceraldehyde-3-phosphate dehydrogenase (GAPD) is roughly spherical when viewed along axes Q and R, but is far from being a sphere, rather a dumb-bell, when looked at along axis P [145]. Although the question may seem trivial, one wonders why such a peculiar shape has developed and has been maintained during evolution.

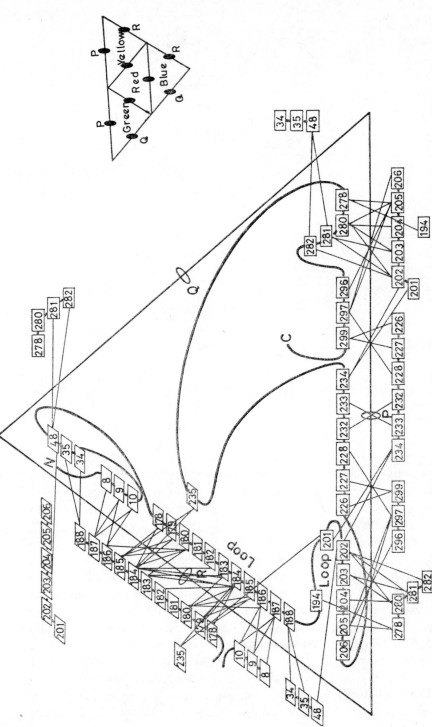

Fig. 3.7. Planar rendering of interprotomer main chain contacts in lobster D-glyceraldehyde-3-phosphate dehydrogenase. Each edge of the triangle corresponds to one interprotomer contact and the triangle itself represents one protomer. Parts of the other three protomers are also seen along the edges. The polypeptide backbone is indicated, in a simplified planar projection, as a continuous heavy line along which residues making up the contacts are marked by their sequence number in boxes. The thin straight lines connecting the boxes show residues whose $C_\alpha - C_\alpha$ distance is <7.5 Å. The three two-fold symmetry axes P, Q and R are in the middle point of triangle edges. Note that the residues on the two sides of each edge are arranged to give two-fold symmetry around the respective axis. The small triangle on the right shows the neighbourhood relations with the other three protomers of the "red" protomer detailed in the main graph. (From Moras *et al.* [839].)

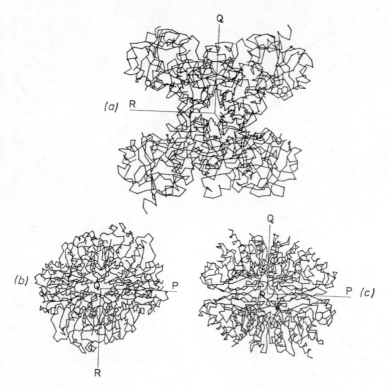

Fig. 3.8. Backbone structure of lobster D-glyceraldehyde-3-phosphate dehydrogenase as viewed down molecular two-fold symmetry axes *P* (*a*), *Q* (*b*) and *R* (*c*). (From Buehner *et al.* [145].)

Of course, there are easy answers of the type "well, it happened like that", which is not so unscientific as it may seem. Nevertheless whoever once tried to assemble a jigsaw puzzle cannot help comparing the P-axis view of GAPD to a piece of this toy. It will be discussed later in this book, whether there is a physical equivalent to this mental association.

X-ray crystallography, being as sensitive as it is, can reveal local or partial asymmetries within a symmetrical oligomer. The example may again be lobster GAPD. For some time this enzyme had been found not to comply with strict 222 symmetry [839], and this inherent asymmetry of association would have neatly served the structural basis for certain functional properties, such as half-of-the-sites reactivity toward some reagents [735, 1088], negative cooperativity in coenzyme (NAD) binding [245, 814, 1059], and partial reduction of bound coenzyme [55]. However, further X-ray investigations disclosed that holo-GAPD (with four NADs bound per tetramer) [881, 882] and apo-GAPD [854] have exact 222 symmetry, and deviations from this are induced by partial saturation with the coenzyme [882] or by abortive ternary complex formation with substrate analogues [420]. Alkylation studies in the crystalline state, as well as in solution, also indicate the identical reactivities of the four active site SH-groups of GAPD [499].

An obvious limitation of X-ray crystallography is that it reflects the molecular form in the crystalline state. Although there is now evidence for some enzymes that they preserve the native (enzymatically active) conformation when in crystals (e.g. [1261]), nevertheless generalizations may be dangerous, also in respect of quaternary structure. We have already referred to α-chymotrypsin, a dimer in the crystal but probably a monomer under physiological conditions (cf. p. 6). Phosphorylase is another, in a sense inverse, example. Although it has been amply proven that phosphorylase a and also the phosphorylase b-AMP complex tend to form tetramers [318, 472, 496], the current most advanced X-ray pictures of these enzymes [595, 1152] tell us nothing about the tetramer, its symmetry or binding set residues. It has been shown by less sophisticated techniques [318, 661] and, incidentally, by a preliminary X-ray work on crystals in which four subunits were contained in the unit cell [47] that the phosphorylase a tetramer has D_2 symmetry. In all fairness, it must be added that an isologous dimer, which phosphorylase normally is, can hardly give a symmetrical tetramer other than D_2.

Finally, X-ray diffraction is a demanding technique in both time and cost, somewhat in proportion to its resolving power. It requires large crystals, which is a practical obstacle to its general applicability [90]. Therefore it is highly probable that, in spite of advances in the speed of X-ray crystallography and the efforts of crystallographers, there will always be many proteins whose structure is unsolved. This circumstance lends significance to more simple and rapid techniques for the determination of subunit arrangement.

Electron microscopy

This technique has been widely used to disclose quaternary structural details. Its main attraction resides, perhaps, in the attitude that "to see is to believe", and electron microscopy is the only means that gives a direct image of an oligomeric protein molecule. It must be remembered, though, that it does so at the expense of somewhat deleterious handling procedures. The latter consist of negative contrasting with the concentrated solution of some electron-dense salt (e.g. sodium phosphotungstate or uranyl acetate) followed by drying, as a result of which the salt forms a dark background around the protein globules, penetrating to some extent between the protomers (for technical details cf. [492]). The picture can be markedly improved through contrast enhancement by image superposition, which already invokes the element of symmetry. Namely, in this process multi-exposure prints are made of the same electron microscopic plate, the printing paper being rotated by $360°/n$ between the exposures each lasting for $1/n$ times the exposure that would be necessary for a direct print. The n-fold symmetry of the object under examination is corroborated if the n-fold pattern is considerably clearer than both the $n-1$ and $n+1$ patterns.

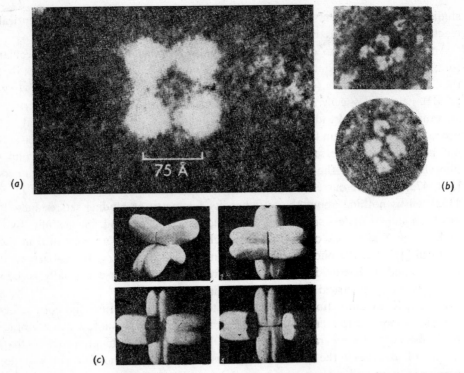

(a)

(b)

(c)

Fig. 3.9. Electron micrographs of "pseudo" and "true" pyruvate carboxylases from chicken liver. (*a*) Picture of a square-shaped tetramer erroneously identified as pyruvate carboxylase (from Valentine *et al.* [1254].) (*b*) Rhombic appearance of two pyruvate carboxylase molecules. The difference in the quality of the pictures taken from the two kinds of proteins (*a* vs. *b*) is due partly to the fact that the unknown protein in (*a*) is highly visible under many conditions of electron microscopy (from Goss *et al.* [465]). (*c*) Tetrahedral model of the enzyme viewed from two different directions (1 and 2). The molecule may become flattened to different degrees (3 and 4) in the absence of the substrate acetyl-coenzyme A, which gives rise to the rhombic appearance on the electron micrographs. (From Mayer *et al.* [796].)

For the successful application of this procedure the pioneering work on tobacco mosaic virus protein rings [779] may serve as an example. A simpler way of contrast enhancement is when the pictures of several individual molecules are superimposed, as in the case of glutamine synthase from *F. coli* [1253].

Even at its best, the limitation of electron microscopy is, apart from the danger of handling artefacts, that only the molecular contours are revealed, which does not always suffice for the unambiguous determination of oligomer symmetry. Thus, the two planar tetramers in Fig. 3.4*c* can hardly be distinguished by electron microscopy. This has been exemplified by the cases of tryptophanase [843] and pyruvate carboxylase [1254]. With tryptophanase the protomers were arranged in a square and the authors could assign the 222 symmetry only with supporting evidence from hydrodynamic and oligomer dissociation properties.

The other enzyme, pyruvate carboxylase from chicken liver mitochondria, has run a strange career under the electron microscope. One of the early accomplishments of protein electron microscopy was the beautiful square-shaped image made by R. Valentine and his coworkers [1254] of a preparation of this enzyme (Fig. 3.9a). The authors cautioned that the picture did not distinguish between C_4 and D_2 symmetry. In spite of this, the image was taken to be a proof of a C_4 tetramer in a comprehensive review [666]. This arbitrary assignment was pointed out in a later survey [409]. A sharp turn came, however, when pyruvate carboxylase was re-examined and showed rhombic and tetrahedral shapes on the electron micrographs (Fig. 3.9b and c), while the square tetramer neatly depicted in the early work [1254] proved to be a contaminating protein [232, 465, 796]. This contaminating protein has now been isolated by the previous authors, but its identity is still a mystery. The moral of this story is that although electron microscopy shows directly the shape of a protein, it does not tell us, as appreciated already by Valentine [480] and now by the re-examiners of pyruvate carboxylase [232, 465], what this protein is.

Hydrodynamic techniques and small angle X-ray scattering

These two methods, though entirely different as regards instrumentation and the underlying physical principles, are both suitable to yield information about the shape of a macromolecule. From the practical point of view, their common featur- is that the behaviour of a complex, but definite structure can be predicted theoreti cally with some accuracy and then compared with the observed behaviour.

The *sedimentation coefficient* of a protein depends upon its size, shape and denf sity, as well as on the density and viscosity of the solvent. Kirkwood [652] derived equations for the calculation of frictional coefficients of structures composed of identical frictional elements (subunits). Bloomfield et al. [98] extended the theory to cover structures containing unlike subunits and used clusters of identical spheres to approximate various protomer shapes (cylindrical rods, ellipsoids of revolution). The closest resemblance to naturally occurring protomer shapes is given by prolate and oblate ellipsoids of revolution, which can be visualized as the products of symmetrical stretching and compressing, respectively, of a sphere. The approximation of these bodies by assemblages of identical spheres is illustrated in Fig. 3.10a. Andrews and Jeffrey [14] applied the Kirkwood–Bloomfield equations to the protomers thus generated and to their various hexameric aggregates (Fig. 3.10b), whereby the sedimentation coefficient of the protomer (s_p) and the oligomer (s_o) could be calculated. Other things being equal, the ratio s_o/s_p will depend solely on the shape of protomer and the mode of aggregation:

$$\frac{s_o}{s_p} = \frac{1+\dfrac{1}{n_o} \displaystyle\sum_{l=1}^{n_o} \sum_{s=1}^{n_o} (R_{ls})^{-1}}{1+\dfrac{1}{n_p} \displaystyle\sum_{l=1}^{n_p} \sum_{s=1}^{n_p} (R_{ls})^{-1}} \qquad l \neq s \tag{3.1}$$

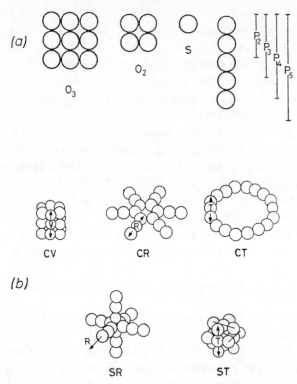

Fig. 3.10. Approximation of oblate and prolate protomer shapes by assemblies of identical spheres (*a*) and some hexameric arrangements made up of prolate protomers (*b*). (*a*) O, S and P designate oblate, sphere and prolate, respectively; (*b*) three cyclic (C) and two spherical (S) hexamers. The second letter under each structure indicates the direction in which the protomer was "stretched": *V*, vertical; *R*, radial; *T*, tangential. The spherical forms actually correspond to the octahedron in Fig. 3.4*e*, 3. (From Andrews and Jeffrey [14].)

where n_p and n_o denote the number of identical spheres (frictional subunits) that make up the protomer and oligomer, respectively, and R_{ls} is the distance between the centres of subunits l and s. The *magnitude* of R_{ls} values depends on the geometry of the structure, whereas the *number* of such distances equals $n(n-1)$, where n is the total number of frictional subunits comprised in the structure.

The theoretical curves of sedimentation ratio values for various hexamers are shown in Fig. 3.11. As an example of application, the case of insulin hexamer may be presented [14]. The experimentally determined s_o/s_p value for this system was 2.4–2.8. Taking into account the measured hydration (0.25 g of water per g of protein) and the frictional ratio, $f/f_o = 1.12$, the insulin protomer can be represented as either an oblate or a prolate ellipsoid of revolution of axial ratio of about 1.5 [885]. Inspection of Fig. 3.11 immediately reveals that oblate CV or CR and prolate CR or CT structures are possible candidates (Fig. 3.12). Spherical arrangements can be excluded. On the other hand, the four closely related structures in Fig. 3.12 cannot

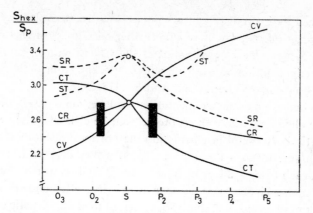

Fig. 3.11. Calculated sedimentation ratios vs. protomer shape for five hexameric structures (cf. Fig. 3.10). Solid lines are for cyclic arrangements, dashed lines for spherical ones. The protomer and hexamer notations are those in Fig. 3.10. Analogous curves can be constructed for oligomers of other protomer number, shape and arrangement. The shaded areas correspond to the ranges of measured values for the insulin hexamer. (From Andrews and Jeffrey [14].)

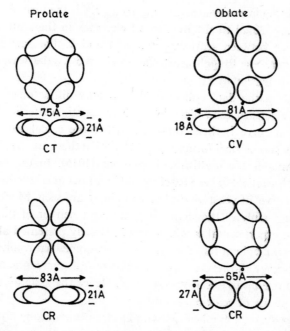

Fig. 3.12. Four cyclic structures for the insulin hexamer that are compatible with hydrodynamic analysis. Structure designations are as in Fig. 3.10. (From Andrews and Jeffrey [14].)

be distinguished by hydrodynamic techniques. In such cases ancillary information, e.g. from electron microscopy, may help further selection. For the insulin hexamer the available X-ray evidence [99] validates the CR structure. Advances in distinguishing by sedimentation various oligomeric models, from dimers to pentamers, have recently been made by Jeffrey and Andrews [15, 592].

An obvious limitation of the sedimentation ratio method is that it is only applicable to dissociable oligomers where the protomer can be characterized separately. Thus it proved particularly useful in elucidating the polymerization of jack bean urease [376]. This enzyme forms a linear array, stabilized by disulphide bond(s), of globular α units, each being a dimer of A_1 subunits [377]. It is worth noting that, as in this case of urease, such heterologous open associations often underlie the appearance of multiple forms of an enzyme, sometimes referred to as non-genetic isozymes.

The *small angle X-ray scattering* technique is based on Debye's theory, which in essence predicts that the pattern of reflections observed will be determined by the size, shape and mass distribution of the scattering particles:

$$I_s = \sum_{j=1}^{n} \sum_{k=1}^{n} f_j \cdot f_k \frac{\sin{(sr_{jk})}}{sr_{jk}} \tag{3.2}$$

where I_s is scattering intensity at scattering vector s, f is the form factor of scattering centres (practically equal to the number of electrons in the scattering atoms) and r_{jk} is the distance between scattering centres j and k. The scattering vector $s = = 4\pi \sin{\Theta}/\lambda$, where $\Theta =$ Bragg's scattering angle and $\lambda =$ the wavelength of radiation; hence the dimension of s is nm^{-1}.

In practice, from the data collected by a Kratky diffractometer [683] plots of log I_s vs. log s are constructed which yield a characteristic scattering curve (Fig. 3.13). This experimental curve is then compared with those calculated on a theoretical basis for various geometrical forms (e.g. [827, 1185]); there are specialized computer programmes available for multisubunit proteins [1016]. In the example given in Fig. 3.13 three theoretical curves based on different octameric models and the measured scattering curve of yeast phosphofructokinase are seen. Models 1 and 2 differ from each other mainly in the shape of protomers; neither of them gives a satisfactory fit to the experimental curve. Model 3, already presented above in Fig. 3.6, differs from model 2 only in that the protomers are somewhat rotated relative to one another. This change, however, made the theoretical curve practically equivalent to the measured one up to about $s = 1.4$ nm^{-1}. Hence the assignment of model 3 as the probable structure of yeast phosphofructokinase.

What has already been emphasized with electron microscopy and hydrodynamic techniques also holds for small angle X-ray scattering: these methods favourably complement each other and thereby increase resolution. Thus, for instance, the trimer structure, built up of tilted ellipsoidal protomers, of histidine decarboxylase was arrived at by small angle X-ray scattering and electron microscopy [462], catalase proved to be a D_2 tetramer by crosslinking [498] corroborated by X-ray scattering

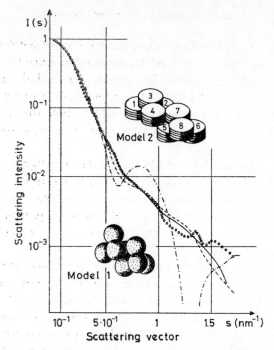

Fig. 3.13. Small angle X-ray scattering curves, log I_s vs. log s, measured for yeast phosphofructokinase
($\circ\circ\circ\circ\circ$) and calculated for three different octameric structures according to [1016]: model 1,
$-\cdot-\cdot-$; model 2, $----$; model 3 (shown in Fig. 3.6), ——. (From Plietz *et al.* [947].)

[693], whereas the C_6 structure assignment to a crosslinked species of phosphofructokinase from a green alga was based upon hydrodynamic and X-ray scattering
measurements [909].

An advantage of small angle scattering, as compared with large angle crystan
diffraction, of X-rays is that it provides information about the protein in solutios
and thus can readily report conformational and quaternary structural changel
induced by specific ligands, for example by NAD in glyceraldehyde-3-phosphate
dehydrogenase [1116].

As an alternative to X-rays, the scattering of neutrons can also be employed in the
study of quaternary structures. Neutrons "see" the atomic nuclei, instead of the
electron cloud, and hydrogen can be easily distinguished by neutron scattering from
all other atomic species, deuterium included. This offers the unique possibility of
recognizing spatial relationships between non-deuterated subunit pairs before a
deuterated background, i.e. the rest of a protein complex [341]. The technique proved
useful in the investigation of various protein–non-protein aggregates, such as ribosomes [1151] and the "bicycle tire" shaped micelles of high density lipoprotein
[1327], but has also been applied with success to oligomeric enzymes, e.g. catalase
[978].

Crosslinking with bifunctional reagents

This technique will be dealt with in detail, because it is the simplest of all requiring no special large instruments, but a gel electrophoretic apparatus, a densitometer and a table-top calculator. It is essentially a chemical method, the only one by which neighbourhood relationships and spatial arrangement in macromolecular aggregates can be studied. By virtue of these facts, it is readily applicable in any biochemical laboratory. Indeed, it is becoming ever more popular, as witnessed by the perusal of current literature.

Although crosslinking can be, and has been, used to analyse a variety of macro-molecular assemblies, such as ribosomes, membranes, receptors, complex enzymes, multienzyme complexes and scaffolded enzyme arrays, our concern here will be confined to the ways in which the symmetry of oligomeric proteins can be deduced by the aid of symmetrical α,ω-bifunctional reagents, i.e. reagents having the same reactive group at both ends.

The basic method has originally been described by Davies and Stark [290] and consists in crosslinking the protein with bifunctional reagent followed by SDS-gel electrophoresis. The previous authors showed that after partial crosslinking there would be as many bands on the gels as there are protomers in the oligomer, since SDS cannot dissociate covalently coupled subunits. The molecular weight of bands corresponds to the multiples of the protomer (polypeptide) molecular weight. Soon after this pioneering work it was realized that not only the number of protomers, but also their spatial arrangement, i.e. oligomer symmetry, might be unravelled from the crosslink pattern [168, 1213]. A major advance to this end was made when the intuitive approach in evaluating crosslink patterns was placed on a quantitative basis backed up by theoretical consideration [495, 578].

Theory

The rationale of the evaluation will be demonstrated for the case of tetramers [495], but it can be extended to higher aggregates as well [409]. On chemical grounds it is expected that different interprotomer bonding domains will have different pro-pensities for crosslinking with a given bifunctional reagent. Kinetically speaking, each bonding domain will have a characteristic rate constant of crosslinking. Clearly, crosslinking occurs in two steps:

$$A + B \xrightarrow{k_1} A{-}B \xrightarrow{k_2} A{=}B$$

where A is the (oligomeric) protein, B the bifunctional reagent and $A{-}B$ and $A{=}B$ denote the protein with monofunctionally and bifunctionally bound reagent, re-spectively. In the overall reaction k_1 is expected to be rate-limiting, since the reaction of the second function is an intramolecular event. Although this need not hold for every case, measurements with several proteins bear out this assumption [495]. Hence the rate constant of crosslinking will be of second order.

Fig. 3.14. "Exploded" scheme of a tetramer with all possible protomer contacts across the three two-fold axes P, Q and R. The contacts (bonding domains) are indicated by various dotted lines. (From Hajdú et al. [495].)

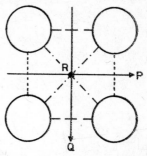

If we take now a tetramer with three types of binding domain and the corresponding three two-fold symmetry axes P, Q and R (Fig. 3.14 and Table 3.2), we can draw up a scheme of crosslinking steps within the oligomer (Fig. 3.15). Differential equations can then be written for the change of each species in time as follows:

$$\frac{\partial[N]}{\partial t} = -2(k_p+k_q+k_r)[N][B] \tag{3.3}$$

$$\frac{\partial P_1}{\partial t} = 2k_p[N]-(k_p+2k_q+2k_r)[P_1][B] \tag{3.4}$$

and analogously for all other species; B is the concentration of bifunctional reagent. If $B \neq 0$, then after division we get for eq. (3.3)

$$\frac{\partial[N]}{\partial \tau} = -2(k_p+k_q+k_r)[N] \tag{3.5}$$

where $\partial\tau=[B]\partial t$ and $\tau=\int_0^t [B]\, dt$.

Integrating (3.5) we obtain

$$[N]=[S]e^{-2(k_p+k_q-k_r)\tau} \tag{3.6}$$

where $[S]$ is the sum total concentration of protein.

In a similar manner one can obtain integrated expressions for all species of Fig. 3.15, from which their amount can be calculated as crosslinking proceeds. However, on SDS-gel electrophoresis not the individual species but their lumped fractions corresponding to 1, 2, 3 and 4 times the protomer molecular weight will be seen, and measured, as discrete bands. Therefore, each species has to be split, as it is expected to dissociate in SDS, and its fraction will contribute to the appropriate n-mer band. For example, the monomer band M will be made up as follows:

$$M=N+0.5(P_1+Q_1+R_1)+ \\ +0.25(P_1Q_1+P_1R_1+Q_1R_1)+ \\ +0.125\,P_1Q_1R_1 \tag{3.7}$$

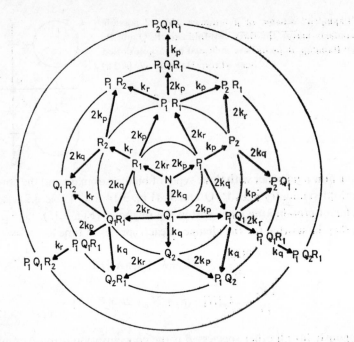

Fig. 3.15. Kinetic scheme of crosslinking a tetramer. N stands for the native protein which can be crosslinked across symmetry axes P, Q and R with rate constants k_p, k_q and k_r, respectively. The rate constants need not refer to a single crosslinkable residue-pair, they may be the resultant of several crosslink combinations. The statistical factor 2 in front of rate constants reflects that in a general (D_2) tetramer each bonding domain type occurs twice (cf. Table 3.2, Fig. 3.14). The various crosslinked species are designated as follows: P_1 and P_2 have one and two crosslinks across axis P, respectively; P_1Q_1 has one crosslink each across axes P and Q, etc. The species are arranged on concentric circles each corresponding to a given number of crosslinks. The scheme is not extended beyond the point where all species already yield tetramers on SDS gel electrophoresis. This stage is reached well before all possible crosslinks are made. (From Hajdú *et al.* [495].)

Figure 3.16 shows theoretical curves for crosslink patterns calculated in this way, the sum of monomer (M), dimer (D), trimer (R) and tetramer (T) bands normalized to unity, by assuming various $k_p : k_q : k_r$ ratios. The crux of the matter is that for a C_4 tetramer the simplified ratio k_p/k_q should be unity (and $k_r = 0$), whereas for D_2 tetramers the k_p/k_q should always differ from unity.

For practical purposes it is useful to convert the above crosslink prediction into symmetry deduction, i.e. to have explicit formulas by which the diagnostic k_p/k_q ratio can be calculated from the measured band intensities [409]. For tetramers there are six such formulas, each containing the intensities of two of the four bands and the sum of band intensities. Four of these formulas are given below, which suffice for practical work. All have the common form

$$\frac{k_p}{k_q} = \frac{\log y_1}{\log y_2} \tag{3.8}$$

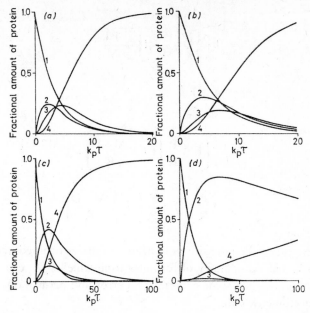

Fig. 3.16. Theoretical curves for the development of crosslinking patterns in tetrameric proteins at various $k_p/k_q/k_r$ ratios. The numbering of curves denotes monomer (1), dimer (2), trimer (3) and tetramer (4). (a) $k_p=k_q=k_r$; (b) $k_p=k_q$, $k_r=0$; (c) $k_p/k_q=4$, (d) $k_p/k_q=50$, $k_r=0$. The abscissa, $k_p\tau$ is linear in time if the concentration of bifunctional reagent is constant. The end points of the curves were chosen arbitrarily; this is why crosslinking is far from completion in D. (From Hajdú *et al.* [495].)

and differ from one another only in the expression for y:

$$y_{1\ 2}=\frac{1}{2}\left(a\pm\sqrt{a+\frac{PD}{M^2}-\frac{2P}{M}}\right) \tag{3.9}$$

where

$$a=\frac{1}{2}\left(1+\sqrt{1+8\frac{P}{M}+4\frac{PD}{M^2}}\right)$$

$$y_{1,2}=b\pm\sqrt{b^2-\frac{P}{M}} \tag{3.10}$$

where

$$b=\frac{1}{2}\left(1+\frac{P}{M}-\frac{PR}{3M^2}\right)$$

$$y_{1,2}=c\pm\sqrt{c^2-\frac{P}{M}} \tag{3.11}$$

where
$$c = 1 + \frac{1}{2}\sqrt{\left(\frac{P}{M} - 1\right)^2 - \frac{PT}{M^2}}$$

$$y_{1,2} = d_1 \pm \sqrt{d_1^2 - d_2}$$
(3.12)

where
$$d_1 = \frac{1}{2}\left(\sqrt{3\frac{P}{R}\left(1 + 3\frac{M+D}{R}\right)} - 3\frac{M+D}{R}\right)$$

$$d_2 = 3\frac{P}{R} - \sqrt{3\frac{P}{R}\left(1 + 3\frac{M+D}{R}\right)}$$

where the letters mean the intensities of various bands as denoted above, and P=
= M+D+R+T. The redundancy of formulas from which k_p/k_q can be calculated
s useful, since it allows for some experimental error and permits one to check the
goodness of densitometric data.

Application to oligomers other than tetramer. Apart from establishing that protomer
number is 2, the crosslinking technique is of no avail to dimers. Occasional deviations from the single symmetric structure, C_2, as in yeast hexokinase [1177], cannot
be detected by it. On the other hand, it may be useful in checking whether a trimer
is C_3 or a mixed structure. As mentioned above (p. 22), the existence of mixed
structures has been considered [253, 254]. Moreover, it has been claimed [984],
though probably ill-foundedly, that bovine glutamate dehydrogenase consists of two
mixed trimers. By the use of suitable reagent(s), it is easy to make the structure
assignment by crosslinking. A scheme analogous to, but much simpler than, that of
tetramers can be solved to give theoretical curves (Fig. 3.17). If the rate constant
ratio $k_a/k_b = 1$, we deal with C_3 symmetry. Significant deviation from unity would
argue for a mixed association.

Complexity markedly increases as we proceed toward higher oligomers. The
theoretical curves of crosslink patterns for two forms of octamers are given
in Fig. 3.18.

Practice

The above theoretical considerations apply to crosslinking with any bifunctional
reagent, if the crosslinks withstand the strongly reducing medium of the SDS-gel-
electrophoresis. There are already a host of reagents that are commercially available
or can be easily synthesized [403, 935], and it may be an individual problem for each
protein which to use. Nonetheless, the α, ω-dimethyl diimidates (or bis-iminoethers)

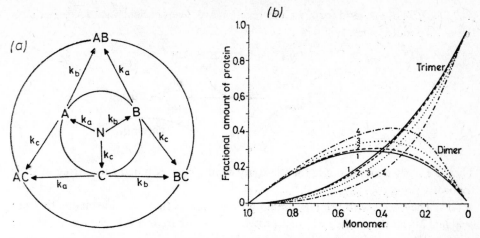

Fig. 3.17. Kinetic scheme and theoretical curves of crosslinking for a trimer. (a) The scheme is analogous to that in Fig. 3.15. In the mixed trimer all three bonding domains (two isologous and one heterologous) are different, thus there may form three types of covalent dimers A, B and C and three types of covalent dimers, A, B and C, and three types of trimer AB, AC and BC, produced as prescribed by the respective rate constants. (b) Theoretical curves: (1) for the C_3 trimer ($k_a = k_b = k_c$), and for mixed trimers, with $k_a : k_b : k_c$ as follows: (2) 3:2:1; (3) 2:1:0; (4) 5:1:0. Instead of the time course, the fractional amounts of dimer and trimer are plotted vs. that of the monomer, for the four selected cases. (From Bartha and Friedrich, unpublished result.)

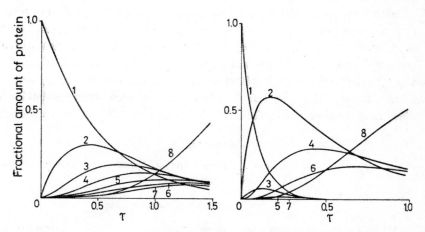

Fig. 3.18. Theoretical curves for the development of crosslinking patterns in octameric proteins. The numbering of curves 1, 2, 3, etc. denotes monomer, dimer, trimer, etc. bands, respectively. The abscissa is linear in time if bifunctional reagent concentration is constant. Octamers: (left) the all-heterologous ($k_p/k_q = 1$); (right) an all-isologous ($k_p/k_q = 10$). (From Friedrich et al. [409].)

[802, 945], whose general formula and reaction are given in Scheme 3.1, are perhaps the most popular ones.

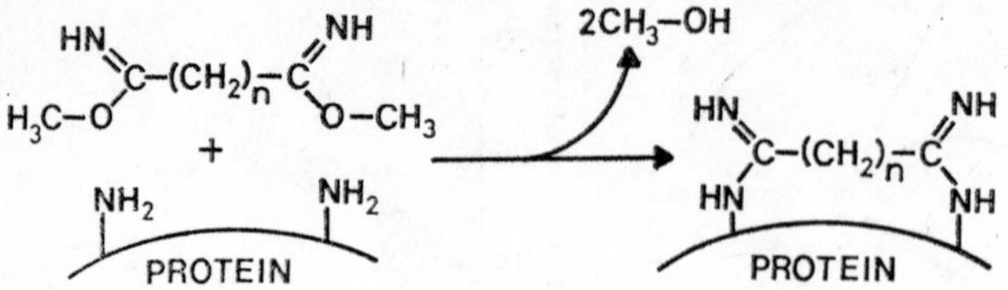

Scheme 3.1

They are particularly suitable for the study of oligomer symmetry by virtue of their following properties:

(a) They react with primary amino groups such as the lysyl side chains that are abundant on the surface of most proteins; hence there is a fair chance to be able to introduce crosslinks across bonding domains.

(b) Their length can be varied by the number (n) of methylene groups connecting the two functions, between the practical limits of 0.37 to 1.45 nm effective reagent length, in 0.12-nm steps. Thus a homologous series of reagents is available, which is most helpful in symmetry analyses. It should be mentioned, however, that the 4- and 5-carbon species (succinic and glutaric diimidates, 0.49 and 0.61 nm, respectively) are inherently poor crosslinkers, presumably because of internal cyclization. Diimidates longer than twelve carbons are not practicable because of their weak solubility in water.

(c) With NH_2-groups diimidates form amidine-groups (Scheme 3.1) whose pK is close to that of primary amino functions, hence no change in surface charge is made. For this reason major conformational alterations due to electrostatic effects are not to be expected, an important criterion for the interpretation of data.

In Fig. 3.19 the SDS gel electrophoretic patterns of two enzymes of D_2 symmetry, skeletal muscle aldolase and lactate dehydrogenase, are seen after crosslinking with four different diimidates and in Table 3.3 the calculated k_p/k_q ratios are given. It is

TABLE 3.3.

The rate constant ratio of crosslinking, k_p/k_q, for aldolase and lactate dehydrogenase obtained with various diimidoesters

Enzyme	k_p/k_q with			
	malonic dimethyl-imidate	glutaric dimethyl-imidate	adipic dimethyl-imidate	suberic dimethyl-imidate
Aldolase	24±6	15±4	2.2±0.5	1.8±0.4
Lactate dehydrogenase	0	0	6.2±1	5.5±0.7

The values were calculated from the dodecylsulphate gel electrophoretic patterns. Number of determinations: 3 to 6. 0=practically no crosslinks formed. (From Hajdu *et al.* [495].)

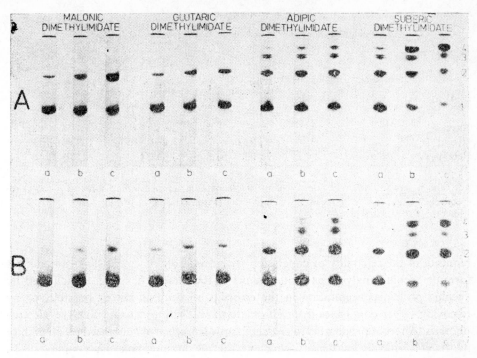

Fig. 3.19. SDS gel electrophoretic patterns of aldolase (*a*) and lactate dehydrogenase (*b*) after different extents of crosslinking with various diimidoesters. (a, b, c) patterns after 20, 50 and 180 min of crosslinking, respectively; 1, 2, 3 and 4 designate monomer, dimer, trimer and tetramer bands, respectively. (From Hajdu *et al.* [495].)

evident that the rate constant ratios differ from unity significantly with *all* reagents tested. The two shorter ones made hardly any crosslinks at all in lactate dehydrogenase, while they could span only one of the interprotomer bonding domains in aldolase producing only dimers (and no trimers and tetramers). The longer reagents were already able to crosslink the other domain(s) as well. A general tendency, neatly obeyed with aldolase, is the gradual decrease of k_p/k_q ratio with increasing reagent length in D_2 tetramers. This is to be expected *a priori,* since the longer the (flexible) reagent, the less it will be able to "sense" small differences, in respect of NH_2-group topologies, between unlike domains. This expectation is also borne out by data obtained with human erythrocyte catalases: the normal enzyme, a mutant and their hybrid (Table 3.4). It may be noted that the k_p/k_q value is a sensitive indicator of small differences in quaternary structure; the values for hybrid catalase fell between those of normal and mutant enzymes with all three reagents tested.

Crosslink patterns become characteristic only at an advanced, but naturally not completed, stage of crosslinking. This can be seen from the triplets of Fig. 3.19, and also from a comparison (Fig. 3.20) based on predicted curves. An important pre-

TABLE 3.4.

Rate constant ratios of crosslinking for human erythrocyte catalases

Catalase species	k_p/k_q		
	Adipic dimethylimidate	Pimelic dimethylimidate	Suberic dimethylimidate
Normal	4.9 ± 0.2	2.6 ± 0.04	1.48 ± 0.08
Hybrid	6.9 ± 0.33	3.0 ± 0.8	1.89 ± 0.13
Heterozygote acatalasemic	9.9 ± 0.8	3.3 ± 0.6	2.05 ± 0.16

k_p and k_q are second-order rate constants of crosslinking across bonding domains p and q, respectively. Figures denote the mean $\pm$ standard error. (From Hajdu *et al.* [498].)

requisite of the reliability of conclusion is that interoligomer crosslinks should not form or become analysed as contaminants of the sample. This can be achieved by keeping protein concentration in the crosslink mixture as low as possible, or by removing larger conjugates by gelchromatography before running SDS gel electrophoresis. A recent proposal is to crosslink a protein when immobilized on an affinity matrix at a critically low density, which would rule out intermolecular coupling [943]. This approach might prove useful in the study of membrane proteins poorly soluble in water. More detailed practical hints can be found in [495].

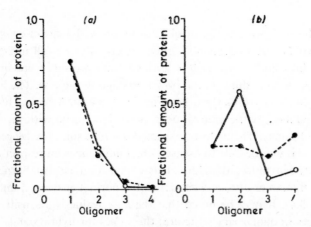

Fig. 3.20. Theoretical distribution of protein in the four dodecylsulphate gel electrophoretic bands at two different degrees of polymerization. (*a*) 25% polymer; (*b*) 75% polymer. (●) $k_p/k_q=1$; (○) $k_p/k_q=10$; $k_r=0$ in both cases. Numbers on the abscissa denote monomer (1), dimer (2), trimer (3) and tetramer (4). Observe that the lines connecting the points only serve better visualization as the abscissa is discontinuous. (From Hajdu *et al.* [495].)

Limitations of the method

Even if all precautions are adhered to, the crosslinking technique may provide no, or erroneous, information about oligomer symmetry for some of the following reasons:

(1) no crosslinks are made. This may happen if reactive side chains are not available around the bonding domains;

(2) different bonding domains appear identical because the distribution and reactivity of relevant side chains are very similar in them. In both cases (1) and (2), the use of a series of bifunctional reagents of different specificity and length is likely to ward off these difficulties;

(3) the flexibility of protein structure distorts the picture, i.e. crosslink formation across one domain influences the reaction at another domain. Although this possibility can never be ruled out completely, experience suggests that such anomalies might be relatively small so that they do not interfere with the conclusion to be drawn. Nevertheless, they are worth bearing in mind.

Affinity crosslinking

A special way of symmetry deduction by a combination of crosslinking and electron microscopy is the approach that might be called affinity crosslinking. It exploits the affinity principle that is already familiar in connection with chromatographic and labelling procedures. In brief, if there is a compound that can bind to the protomers specifically, i.e. at a given site and with reasonably low dissociation constant, then the bis-derivatives of this compound may serve as useful tools in the analysis of oligomer structure. The classical demonstration of the potentialities of affinity crosslinking was the work on avidin structure by Green et al. [480].

Avidin from hen's egg white is a relatively small tetrameric protein (mol.wt: 65,000) [299], each protomer of which binds one molecule of biotin very strongly ($K_d = 10^{-15}$ M) [479], i.e. with great "avidity". Green et al. [480] synthesized various bis-biotinyl derivatives by connecting two biotin rings with spacers of different lengths. After adding these bifunctional biotinyl compounds to avidin, the products were examined by electron microscopy. From a critical chainlength upwards they observed linear polymers of thickness equal to that of an avidin molecule. This form of polymerization unequivocally proved that the four protomers of avidin were arranged with 222 symmetry and that they were in pairs at opposite ends of the short axis of the molecule ($55 \times 55 \times 41$ Å). From the critical reagent length the depth of the biotin-binding cleft could also be estimated as 15 Å.

The elegance and resolving power of affinity crosslinking renders it a highly desirable approach whenever feasible. However, the prerequisite for its application, i.e. a proper ligand whose strong binding is not destroyed by derivatization, is not easily met. The avidin–biotin couple is admittedly a unique example which offered itself. Nevertheless, one may find tightly binding substrates, coenzymes, effectors or their analogues whose derivatives would make this technique amenable to various oligomeric and complex enzymes, as well as multienzyme complexes.

BIOLOGICAL ROLES OF OLIGOMERIC STRUCTURE

Quaternary structure and enzyme activity

A general question that can be raised about oligomeric enzymes is whether quaternary structure is essential for catalytic activity. As we see it today, no general answer to this question can be given, but rather each enzyme poses an individual problem.

With stable oligomers, which do not spontaneously dissociate on dilution even at the lowest practicable protein concentrations, the question proved hard to crack. The inherent difficulty is that such stable oligomers can only be disrupted by drastic influences, such as exposure to extremes of pH, or high concentrations of ions or denaturants (urea, guanidinium hydrochloride and various detergents), which abolish not only the quaternary but also the tertiary and secondary protein structures. Consequently, the protomers produced are unfolded and necessarily devoid of biological activity. In the following we shall survey some experimental approaches developed to obtain meaningful answers to the above question.

Reactivation kinetics

The renaturation of a reversibly denatured oligomer (e.g. by slow removal of the denaturant) can be considered, as a first approximation, to follow the scheme:

$$nP_d \xrightarrow{k_1} nP_f \xrightarrow{k_2} [P_f]_n$$

where P_d and P_f stand for the denatured and folded protomers, respectively, $[P_f]_n$ is the oligomer, k_1 is a first-order-rate constant of folding, and k_2 is a higher-order-rate constant of association. The question is whether the structure of the folded but unassociated protomers sufficiently resembles that of the associated protomers to cause appreciable catalytic activity. In this simple scheme, recovery of activity will be first order if the folded protomers are catalytically competent, but higher order if they are inactive *and* the association step is kinetically significant.

However, experiments have shown that reactivation processes usually follow more complex kinetics. One of the observed cases is when no single rate-determining step exists, but rather both k_1 and k_2 are kinetically significant. Accordingly, reactivation can be described by a two-step (uni- and bi-molecular) consecutive reaction, i.e. enzyme activity is recovered in a sigmoidal manner. This was found to hold by

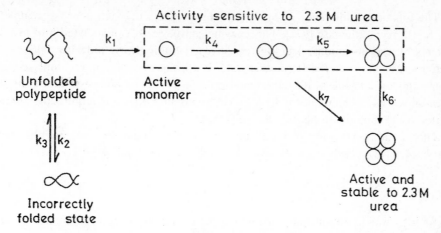

Fig. 3.21. Kinetic model of the renaturation of rabbit muscle aldolase. The enzyme was denatured in 6 M guanidinium-HCl. A small aliquot of denatured aldolase was mixed into a large volume of "renaturation buffer" (50 mM tris-HCl, 1 mM EDTA, 140 mM 2-mercaptoethanol, 1 mg/ml bovine serum albumin, 20% sucrose) at 0°C and the reappearance of enzyme activity was monitored. The measured points conformed to theoretical curves calculated according to the mechanism illustrated in the figure, with the following rate constants: $k_1 = 2.3 \times 10^{-2}$ min^{-1}; $k_2 = 2.0 \times 10^{-2}$ min^{-1}; $k_3 = 1.5 \times 10^{-2}$ min^{-1}; $k_4 = 1.4 \times 10^5$ M^{-1} min^{-1}; $k_5 = 4.0 \times 10^6$ M^{-1} min^{-1}; $k_6 = 1.0 \times 10^6$ M^{-1} min^{-1}; $k_7 = 1.0 \times 10^5$ M^{-1} min^{-1}. The specific activity of active protomers was assumed to be unchanged by association. (From Chan *et al.* [194].)

Jaenicke and coworkers for tetrameric heart lactate dehydrogenase [1035], tetrameric yeast glyceraldehyde-3-phosphate dehydrogenase [1034] and dimeric mitochondrial malate dehydrogenase [587]. The sigmoidal kinetics mean that the separated pro-tomers of these enzymes are inactive or at least less active than the protomers within the oligomer.

Another source of complexity is when folding itself cannot be described by a single first order reaction. The reason for this may be that there are metastable intermediates that are not on the direct pathway of folding. This was claimed by Ikai and Tanford [580] to be a general phenomenon in proteins and supported by the protein analytical studies on the folding of bovine pancreatic trypsin inhibitor by Creighton [265]. The reactivation of guanidinium-HCl-denatured rabbit muscle aldolase seems to proceed through such a mechanism [197]. In these experiments Chan and coworkers were able to show that the folded monomer was sensitive to 2.3 M urea, whereas the associated tetramer was not. Hence, by including 2.3 M urea in the assay mixture, activity coming from subunit and from the tetramer could be clearly distinguished. As expected, the time course of appearance of urea-resistant activity followed a sigmoidal curve, since subunit association is a step subsequent to folding. The model suggested by Chan *et al.* [197] for the renaturation of aldolase is given in Fig. 3.21. Apparently, the kinetics suggest that aldolase subunits possess enzyme activity but are less stable than tetramers.

The functional difference of unlike subunit contacts in a tetramer is exemplified by *E. coli* tryptophanase [971]. This enzyme can be readily dissociated into inactive dimers by various anions. The kinetics of reactivation clearly indicate that enzyme activity is confined to the tetrameric state. It appears that the dimer–dimer contact constrains the enzyme in its active conformation [972]. In contrast, rabbit muscle creatine kinase, which is normally a dimer, apparently does not require the dimerization of its subunits to be fully active [418a].

The assembly pathways of multisubunit proteins, among them several oligomeric enzymes, have recently been reviewed by Friedman and Beychok [402].

Stabilized subunits

Another way of establishing the relation between quaternary structure and enzyme activity is to produce stabilized subunits, usually by chemical modification. A disadvantage of this approach is that side-chain blocking, apart from dissociating the oligomer, may inactivate the enzyme. Therefore only positive results, i.e. when the modified stable subunits are active, can be regarded meaningful. Extensive succinylation [808, 809] or maleylation [1106] of amino groups yields inactive stable subunits of aldolase and glyceraldehyde-3-phosphate dehydrogenase. Careful, partially reverted, two-step acylation of aldolase also failed to reveal active subunits [407]. Reaction of one or two out of the 4 SH-groups per protomer with *p*-hydroxymercuribenzoate in glyceraldehyde-3-phosphate dehydrogenase leads to dissociation into inactive subunits [1127]. The partial recovery of activity that can be monitored in time is due to the disproportionation of the reagent giving fully active tetramers and denatured subunits [1202].

In some instances nevertheless active subunits could be produced by chemical modification. With aldolase this was achieved by arresting the association of folded subunits by iodoacetate, which probably reacted with one of the SH-groups at the subunit contact surface [195]. A crucial point in the strategy of these experiments was that reassociation was slowed down by applying a medium of high viscosity and very low enzyme concentration. Subunits displaying about half of their original enzyme activity could be prepared from yeast hexokinase by succinylating two to three amino groups (out of thirty-seven per subunit) [1026]. It should be mentioned, however, that yeast hexokinase is not a stable dimer, as shown by its behaviour in solution [1101, 1102] and asymmetric dimerization in crystal ([1177]) A somewhat peculiar phenomenon has been reported about tobacco cell phosphodiesterase [1104]. This tetrameric enzyme ($M_r = 280,000$) falls apart giving rise to subunits in the presence of 5 M urea and 2-mercaptoethanol. After removing urea the subunits do not reassociate and exhibit about 80% of the original specific activity.

Oligomeric proteins can also be dissociated by high pressures as proved, for example, with pig muscle lactate dehydrogenase by hybridization [586] and chemical crosslinking [1056]. Dissociation was accompanied by deactivation, both processes being completely reverted upon pressure release. Since the subunits produced by pressure are presumably structured, as judged from the fact that considerably higher pressures are required to denature monomeric proteins [519] than to dissociate oligomers, the above finding would suggest that the quaternary structure is essential for lactate dehydrogenase activity. However, it cannot be ruled out that pressure-deactivation is a volume effect on the catalytic reaction itself [1056].

Immobilized subunits

An elegant approach to the question of protomer activity has been designed by Chan [188]. Its scheme is shown in Fig. 3.22. To an appropriate solid matrix, e.g. Sepharose, the oligomeric enzyme is coupled covalently. Before coupling the matrix is activated only slightly, e.g. with cyanogen bromide, so that each oligomeric molecule would react through but one of its subunits. The immobilized enzyme thus prepared is expected to display full enzymatic activity. A denaturing agent, e.g. concentrated guanidinium-HCl, is then added which unfolds and removes polypeptide chains not linked to the matrix covalently. Upon removal of the denaturant, the matrix-bound polypeptide chains refold. It can then be tested, whether the single, isolated subunit possesses catalytic activity or not. The answer proved to be affirmative for aldolase [188], transaldolase [199] yeast glyceraldehyde-3-phosphate

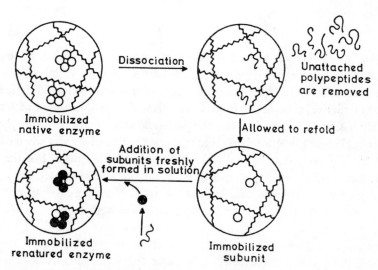

Fig. 3.22. Scheme of experimental procedures in testing immobilized subunits. The large circle with the zigzag lines inside represents a Sepharose bead, which is the polysaccharide matrix. The small circles denote subunits of a tetrameric enzyme held together by non-covalent bonds. Covalent linkage of subunits to the matrix is indicated by a short line. Unfolded subunits are illustrated as irregularly twisted lines. Shaded circles are subunits taken up from solution by the immobilized subunit. (From Chan [191].)

dehydrogenase [33, 34] and human liver arginase [171], and negative for phosphory-lase [367], lactate dehydrogenase [198], phosphoglucose isomerase [138], pyruvate kinase [954] and alkaline phosphatase from *E. coli* [800]. An important check is the last step in Fig. 3.22, which consists in the recovery of subunits missing from the oligomer, upon addition of denatured enzyme followed by dialysis. In the case of a tetramer the amount of matrix-bound protein should increase four-fold; so does the enzyme activity, if the immobilized subunit was active, whereas if it was inactive, catalytic activity reappears on the reformation of the oligomer. Another control, applied in the case of alkaline phosphatase [800], can be made when coupling to the matrix is reversible. The immobilized subunit preparation is then subjected to cross-linking with a bifunctional reagent, and after cleavage of bonds between matrix and protein, the released material is analysed by polyacrylamide gel electrophoresis. Detection of crosslinked material would then indicate that activity exhibited by the immobilized subunit preparation may not be an inherent property of the subunit but may come from incomplete disruption (or partial reformation) of oligomeric structure.

The virtue of this experimental setup resides in its simplicity and quantitative character. The only objection that can be raised is that the isolated subunit may, after all, get some support from the matrix. However, considering the specificity of subunit contacts, the Sepharose matrix can hardly substitute for a complementary protein surface. In fact, the data obtained with the immobilized subunit technique are in accord with other lines of evidence, e.g. for rabbit muscle aldolase as described above.

Complementation by inactive subunits

The relative autonomy of subunit contact surfaces and active sites in some enzymes is indicated by the phenomenon of complementation with inactivated subunits. Namely, in enzymes whose quaternary structure seems essential for catalytic activity, the solitary (inactive) subunit can often be reactivated if complemented with another, irreversibly inactivated subunit. Thus matrix-bound subunits of phosphorylase [367] and phosphoglucose isomerase [138] regained practically full activity when dimerized with soluble inactive subunits modified with pyridoxal phosphate. For the activity of glyceraldehyde-3-phosphate dehydrogenase some degree of quaternary organiza-tion seemed necessary: claims have been put forward confining catalytic competence to the tetrameric form [551, 563, 709, 1263] or detecting activity in dimers [856, 901]. However, recently the monomer also proved to be active [33, 34]. Hybridization experiments revealed catalytic activity even in tetramers containing a single unmodi-fied subunit [809, 1197]. It follows that in these enzymes the activity of a subunit is restored even if complexed to inactive subunit(s).

Extensive complementation studies have been conducted with β-galactosidase from *E. coli* [180, 1241]. Wild type β-galactosidase is a tetramer, whereas its deletion mutant M 15 is an enzymatically inactive dimer lacking residues 11–41 [713]. When a cyanogen bromide peptide comprising residues 3–92 of the normal enzyme was added to this mutant, both activity and tetrameric structure were restored. Immunological evidence suggests that the complementing peptide forms part of the dimer–dimer interface [180]. The experiments do not allow decision whether the recovery of activity is due to the non-covalent binding of complementing peptide or to the concomitant tetramerization.

Quaternary structure and protein stability

The literature abounds in data showing that oligomerization endows protein subunits with enhanced stability toward various deleterious effects, such as heat, denaturing agents and proteolytic attack. A thoroughly studied example is rabbit muscle aldolase. This enzyme is a stable tetramer that does not dissociate into smaller entities without harsh treatments [407, 722, 808, 1106]. While stable, active subunits can be prepared from it, as described above, by immobilization [188] or iodoacetate modification during renaturation [195], these are susceptible to 2.3 M urea in contrast to the reassociated tetramer. The transient formation of urea [197] and heat-labile intermediates [1269], i.e. folded solitary subunits, has also been detected during the renaturation of aldolase denatured by guanidinium-HCl or acid. Another damaging influence, ultraviolet irradiation, also seems to affect subunits more severely than the tetramer, although in an indirect way [410]. Irradiation with ultraviolet light rapidly inactivates aldolase. Between 1 and 20 mg/ml aldolase the dependence of first-order-rate constant of inactivation on enzyme concentration (Fig. 3.23) follows a simple theoretical curve, based on a constant sensitivity parameter and absorption coefficient at the critical wavelength(s). However, below this enzyme concentration range inactivation becomes much faster than predicted by theory, the latter being adhered to only if the irradiation mixture also contains 1% bovine serum albumin (or other inert protein). The explanation offered for this phenomenon ran as follows: UV-induced inactivation of one out of four subunits results in the breakup of quaternary structure producing one inactive (denatured) and three fairly intact subunits. The latter recombine rapidly giving rise to new tetramers. Below a critical enzyme concentration (about 1 mg/ml), however, recombination becomes slow relative to the denaturation of free subunits. This secondary inactivation can be warded off by the well-known, though ill-understood, protective effect of an alien protein [e.g. 197]. One may speculate that quaternary organization might have developed during the course of evolution, among other reasons, to stabilize proteins against the detrimental influence of ionizing radiations.

In mammalian lactate dehydrogenases the quaternary structure is essential for catalytic activity [198]. In a study of the urea sensitivity of the muscle (M) and heart (H) type enzymes, as well as their hybrids, the homologous subunit interactions were found to endow the enzyme with much greater stability than the heterologous ones; the order of decreasing stability was H_4, M_4, H_3M, HM_3 and H_2M_2 [200]. The

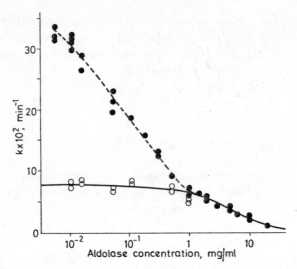

Fig. 3.23. Ultraviolet irradiation-induced inactivation of rabbit muscle aldolase. Dependence of the first-order-rate constant of inactivation on enzyme concentration. Irradiation was carried out in 0.2 M Tris-HCl, pH 7.5 (●), and in the same buffer also containing 1% bovine serum albumin (○). The solid line is a theoretical curve derived from the equation $k = \alpha I_0 \dfrac{1 - e^{-bc}}{bc}$ where k is the first order rate constant of inactivation, α is a sensitivity parameter, I_0 is the incident light intensity, c is protein concentration, and $b = \beta l$, where l is layer thickness and β is the absorption coefficient of aldolase at the critical wave length(s). The broken line is an empirical curve. (From Friedrich *et al.* [410].)

remarkable stability of the tetramer, as compared with that of the subunit, has been clearly shown with both immobilized [198] and transiently produced [1194] subunit. The properties of transient lactate dehydrogenase subunits have been studied in heat denaturation and freeze–thaw experiments. Interestingly, the quaternary structure of this enzyme can be broken up, among other means, by exposure to such extremes of temperature. Dissociation triggered by a primary "hit" followed by recombination of surviving subunits, a mechanism delineated above for the UV-induced inactivation of aldolase, was first proposed by Südi [1193] for the heat inactivation of various lactate dehydrogenase isoenzymes (Fig. 3.24). Freeze–thaw treatment, a procedure applied to produce hybrids from parent lactate dehydrogenase homo-tetramers [211, 778], is based essentially on a combined effect of pH, ionic strength and temperature, since the freezing of salt solutions causes large changes in ionic strength, and often in pH, as various eutectics are approached [211, 1258]. After thawing, the separated subunits recombine fairly slowly, which can be readily monitored by the reappearance of enzyme activity [1194]. Furthermore, while in subunit state the enzyme reacts rapidly with iodoacetamide and is inactivated, this is not observed with the native tetramer. One cannot help remembering aldolase (cf. [195]) where the analogous reaction resulted in the stabilization of *active* subunits

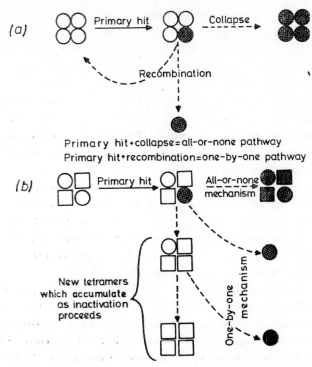

Fig. 3.24. Mechanisms suggested for the heat inactivation of lactate dehydrogenase. (*a*) Primary hit
denatures one subunit, followed either by the collapse of the whole molecule (tetramer) or by the
separation of subunits and recombination of intact ones to new tetramers. Open circles: active subunits; shaded circles: inactive subunits. (*b*) Experiments with the H_2M_2 hybrid tetramer supported
the latter (one-by-one) mechanism: with the progress of heat inactivation the residual native tetramer
population was enriched in subunit M, which is inherently more resistant to heat than the H subunit.
For further details consult the original work by Südi [1193].

by preventing their association. To take the comparison one step further, freeze–thaw
treatment in the presence of iodoacetate or iodoacetamide also inactivates rabbit
muscle aldolase [408]. However, in the case of aldolase the mechanism of this inactivation is fundamentally different from that seen in lactate dehydrogenase: the aldolase subunits apparently do not separate as a result of freezing, the quaternary and
tertiary structures are merely loosened allowing the reagent to attack at several
points. The important stabilizing role of undistorted contact surfaces is probably
reflected by the fact that incorporation of less than four carboxymethyl groups per
tetramer suffices to inactivate a whole molecule [408]. Under the extreme conditions
of freeze–thaw treatment these alkylating agents are apparently much more deleterious to aldolase than under mild conditions, as shown by the catalytic competence
of iodoacetate-stabilized aldolase monomer [195].

In the case of dissociable oligomers the decreased stability of lower aggregational
form(s) is usually manifested in the protein concentration dependence of heat inac-

tivation: the inactivation rate constant increases with decreasing protein concentration. This was observed with pig muscle glyceraldehyde-3-phosphate dehydrogenase [1262] and *F. coli* tryptophanase [455]. It should be borne in mind, however, that specific ligands may shift the dissociation equilibrium, as happens with glyceraldehyde-3-phosphate dehydrogenase and its coenzyme NAD [630].

The greater stability of the oligomer as opposed to the subunit need not be a general phenomenon valid for all kinds of deleterious influence. For example, the active subunit of tetrameric tobacco cell phosphodiesterase is more sensitive to sodium dodecylsulphate than the tetramer is, but the two aggregational forms seem to display the same resistance to heat and urea [1104].

In conclusion, while it can be regarded a general tenet that quaternary structure increases the stability of protein (enzyme) molecules, the stability of quaternary structure may vary over a wide range. Above we referred to dissociable oligomers and will treat them in more detail in connection with enzyme regulation. In contrast, there are oligomers whose protomer association is practically irreversible under *in vivo* conditions. Muscle lactate dehydrogenase and aldolase belong to this category. Recently, dimeric glucose-6-phosphate dehydrogenase was shown not to give the hybrid in a heterokaryon produced by fusing mouse and human fibroblasts, if protein synthesis was arrested prior to cell fusion [461]. Apparently, there is no exchange of subunits between the two types of dimer. It remains to be elucidated what selective pressures operated on various oligomeric enzymes to develop the wide range of protomer interaction energies experienced today.

Quaternary structure and enzyme regulation

The regulation of enzymatic, and other types of biological activities is, perhaps, the field where the quaternary protein structure has been most extensively invoked. This reflects the plain fact that most intracellular enzymes known, or suspected, to have something to do with regulation are oligomers. The basic concepts developed in the area have found their way into biochemistry texts and have been amply reviewed. At all events, the subject matter is so great that here only the outlines can be given, with some comments pertinent to quaternary structure. For a comprehensive description the reader is referred to specialized treatises (for example by Dixon and Webb [314]). We are going to distinguish cases where regulation is exercised through subunit interactions at unchanged degree of aggregation and cases in which regulation involves the reversible dissociation of the oligomer.

Regulation via subunit interactions within the oligomer

The conceptual breakthrough in the regulation of enzyme avtivity occurred in 1963, when Monod, Changeux and Jacob [832] coined the term *allostery*. It was to designate the phenomenon that a compound (effector), chemically different from the substrate, could bind to the enzyme influencing (inhibiting or promoting) its catalytic activity. Thus, an *allosteric site* was assumed to exist on the enzyme with distinct stereospecificity from that of the *active site,* and communication between the two sites occurred through the protein fabric. This model was to interpret the by then well-known phenomenon of feed-back inhibition in biosynthetic pathways, i.e. the frequently competitive inhibition of the enzyme catalysing the first committed step in a reaction sequence by an end-product that does not closely resemble the substrate (see, for example, [1163]).

Thus in its original sense allostery was not related to quaternary structure in any way. However, in a subsequent paper Monod, Wyman and Changeux [834] proposed a mechanistic model for allosteric transitions, a pivotal point of which was that allosteric enzymes were invariably oligomers. The other major features of this *symmetry* or *concerted model* are that the oligomer may exist in at least two conformational states in equilibrium, the enzyme in the R (relaxed) state binding the substrate more strongly than in the T (taught or tensed) state, and that the symmetry of the oligomer is preserved when going from one state to the other. In other words, all protomers undergo transconformation simultaneously, in a concerted manner. The scheme (3.2) of these transitions in a tetramer is as follows:

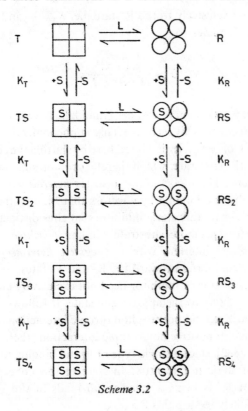

Scheme 3.2

S is the substrate and $L=T/R$, the allosteric constant. K_T and K_R are the intrinsic dissociation constants of the enzyme–substrate complex in the T- and R-states, respectively. An important feature of this model is that K_T and K_R do not change with increasing saturation of the tetramer, they only become modified by the statistical factors to yield the apparent dissociation constants. (For example, for the first substrate molecule binding to the R state and giving RS, the apparent dissociation constant $K_{R1}^{app}=\frac{1}{4}K_R$, because there are four vacant sites on R to bind S, and one occupied site on RS to release S. Conversely, for the fourth step $K_{R4}^{app}=4K_R$. In a tetramer the statistical factor causes a sixteen-fold difference between the apparent dissociation constants of the first and fourth ligand molecules.)

Calculating with the apparent dissociation constants and introducing $\alpha=[S]/K_R$ and $c=K_R/K_T$, one gets the following expression for the fractional saturation, $\bar{Y}_s$, of the tetramer (for the details of derivation see, for example, [314]):

$$\bar{Y}_s=\frac{\alpha(1+\alpha)^3+Lc\alpha(1+c\alpha)^3}{(1+\alpha)^4+L(1+c\alpha)^4}\,. \tag{3.13}$$

If the substrate binds exclusively to the R state, i.e. $K_T=\infty$ and therefore $c=0$, the saturation equation simplifies to

$$\bar{Y}_s=\frac{\alpha(1+\alpha)^3}{(1+\alpha)^4+L}\,. \tag{3.14}$$

If L is finite, the saturation curve will be sigmoidal. (In case of $L=0$ or $L=\infty$, the saturation curve reduces to a hyperbola.) This is phenomenologically indicative of cooperative binding. Cooperativity is caused here by shifting the equilibrium from the T- to the R-state by the substrate, which thereby promotes its own binding *(positive homotropic interaction)*. The model allows also for *heterotropic interactions,* i.e. for the influencing of substrate binding by effectors that bind to the R-state (allosteric activators) or to the T-state (allosteric inhibitors) at sites distinct from the substrate binding site. Activators promote substrate binding *(positive heterotropic interaction),* whereas inhibitors interfere with it *(negative heterotropic interaction),* by shifting the allosteric equilibrium toward the R- and T-states, respectively. Similarly to the substrate, an activator or inhibitor need not exclusively bind to one state, but it has to show higher affinity to the appropriate state. It follows that an activator as well as an inhibitor, will exhibit positive homotropic interaction in its binding.

It may be noted that in positive homotropic interactions there is no "allo-sterism" in the original sense [832], since a single type of ligand binding to a single type of site bears no element of steric non-identity. In spite of this conceptual inconsistency, the adjective "allosteric" is being applied to all kinds of site–site interactions and no puristic effort is likely to stop this usage.

The concerted allostery model, although it proved highly inspiring and useful, had a serious limitation: it could not account for *negative homotropic interactions,* i.e. when the same ligand progressively interferes with its own binding to successive protomers within the oligomer. This phenomenon, often referred to as *negative cooperativity,* is quite common among enzymes [735]. In 1966 Koshland, Némethy and Filmer [679] propounded another allosteric mechanism called the *sequential model,* which allowed for negative cooperativity. The sequential model is an extension of earlier models for cooperative behaviour by Adair [4], Pauling [918] and Wyman [1338], as well as of Koshland's induced-fit theory of enzyme action [678] to regulation. This model proposes that the conformation of each protomer within an oligomer can be changed separately, i.e. the symmetry of the oligomer need not be preserved. The substrate induces or stabilizes a protomer conformation that is not assumed by the neighbouring protomers but which predisposes the other protomers, through the interprotomer contacts, to undergo the similar conformational change on substrate binding (positive cooperativity) or to hinder this conformational transition and thereby further substrate binding (negative cooperativity).

For a dimeric enzyme, in the simplest case, the protomer in state A (square) does not bind the substrate, whereas it does in state B (circle):

The A$\rightleftharpoons$B transition is determined by equilibrium constant *L:*

$$L=\frac{B}{A}.$$

(3.15)

Since this transition in one protomer is influenced by the other protomer, interaction constants were defined, in terms of the following equilibrium constants:

$$K_{AB}=\frac{[AB][A]}{[AA][B]},$$

(3.16)

$$K_{BB}=\frac{[BB][A][A]}{[AA][B][B]}.$$

(3.17)

The third constant, K_{AA}, was taken as unity by convention. These equilibrium constants express the relative stabilities of the different interprotomer contacts. For example, if $K_{AB}>1$, the interaction between A and B is favoured to that between A and A, therefore the AA→AB transition will be facilitated.

Substrate binding is then determined by the apparent dissociation constant of the enzyme–substrate complexes, the *L* equilibrium constant for transition A$\rightleftharpoons$B, and the change in protomer interaction constants. If the intrinsic substrate dissociation constant is K_s, the statistical factor yields the apparent dissociation constants $\frac{1}{2}K_s$

and $2K_s$ for the first and second binding steps in the dimer, respectively. The change in protomer interaction constants for the first substrate molecule is K_{AB}/K_{AA}, and for the second molecule is K_{BB}/K_{AB}. Thus the actual value of the apparent dissociation constant for the first substrate molecule will be

$$K_1^{app} = \frac{K_s K_{AA}}{2LK_{AB}} = \frac{K_s}{2LK_{AB}}, \tag{3.18}$$

and for the second substrate molecule

$$K_2^{app} = \frac{2K_s K_{AB}}{LK_{BB}}. \tag{3.19}$$

Accordingly, there will be positive cooperativity in substrate binding if $2K_{AB} < < K_{BB}/2K_{AB}$ and negative cooperativity if this inequality is inverse.

The fractional saturation of the enzyme, $\bar{Y}_s$, is then

$$\bar{Y}_s = \frac{LK_s K_{AB}[S] + L^2 K_{BB}[S]^2}{K_s^2 + 2LK_s K_{AB}[S] + L^2 K_{BB}[S]^2}. \tag{3.20}$$

The sequential model, like the concerted model, accounts also for positive and negative *heterotropic* effects, i.e. for interactions between substrate and activator or inhibitor binding. These effectors are assumed to have separate, stereospecific binding sites on the protomer, distinct from the substrate binding site, and exert their effect by inducing (or stabilizing) one or another protein conformation. There are a number of mechanistic possibilities, many more than with the concerted model, as detailed by Kirtley and Koshland [660].

The derivation becomes increasingly difficult with oligomers higher than dimer. Koshland *et al.* [679] considered tetramers with "tetrahedral", "square" and "linear" protomer interaction patterns. As practically all tetrameric enzymes known so far in sufficient detail are isologous (D_2 symmetry), the tetrahedral model is of primary import. The apparent substrate dissociation constants for this case were derived by Koshland *et al.* [679] analogously to the dimer above. Thus, for the first substrate

$$K_1^{app} = \frac{K_s}{4LK_{AB}^3}. \tag{3.21}$$

The interaction constant K_{AB} is cubed because there are three AB contacts in the tetrahedral A_3BS species. The other three apparent dissociation constants were as follows:

$$K_2^{app} = \frac{2K_s}{3LK_{AB}K_{BB}}, \tag{3.22}$$

$$K_3^{app} = \frac{3K_s K_{AB}}{2LK_{BB}^2}, \tag{3.23}$$

$$K_4^{app} = \frac{4K_s K_{AB}^3}{LK_{BB}^3}. \tag{3.24}$$

Instead of giving the complex expression for $\overline{Y}_s$ (for the solution see, for example, [314]), we point out a shortcoming of this derivation. Namely, it is an oversimplification that, for example, in the $A_2B_2S_2$ tetramer there are one A–A contact, four A–B contacts and one B–B contact. Since protomers are asymmetric, in a tetrahedral array each protomer may establish three different types of contact with the other protomers (cf. Fig. 3.14). Hence even in an A_4 tetramer three types of contact occur and the contact pattern becomes more complex if ligand binds to the enzyme (Table 3.5). The structural transitions may be described by the following scheme (3.3):

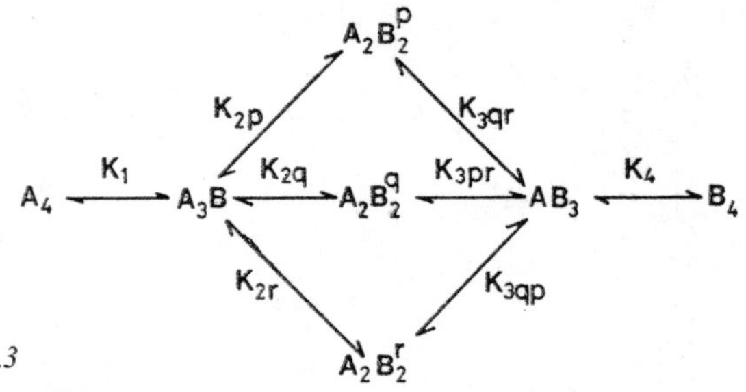

Scheme 3.3

where the designations of molecular forms and equilibrium constants are as in Table 3.5. Theoretical treatment taking into account these isomerizations would certainly be closer to reality than applying a bulk interaction constant, e.g. K_{AB}, for three different vicinity effects. Efforts to this end have earlier been made by Wyman [1338] and Wong and Endrényi [1333], who examined the square tetramer and used phenomenological interaction constants (not defined as equilibrium constants; for example, if K_s is the substrate dissociation constant of the protomer, it becomes $\alpha_1 K_s$ if the "close-neighbour" protomer is occupied, $\alpha_2 K_s$ if the "distant-neighbour" protomer is occupied, and $\alpha_1\alpha_2 K_s$ if both).

All above considerations refer to substrate (or effector) binding and not to the velocity of enzyme reaction. To a first approximation, enzyme saturation may be proportional to velocity, but it remains to be established experimentally whether cooperativity in binding is manifested in kinetic cooperativity. It was argued on theoretical grounds [456] that the concerted model allowed for kinetic *negative* cooperativity. However, this claim was subsequently refuted [921]. Thus, negative cooperativity, whether in binding or in kinetics, can only be explained by a sequential mechanism. Nevertheless, caution should be exercised because heterogeneity in an enzyme preparation may be misinterpreted for negative cooperativity [432].

Several attempts have been made to reconcile the two classical models of allosteric regulation, showing that they are not mutually exclusive but rather the extremes of a continuum of mechanistic possibilities (e.g. [532, 653]). Even more efforts have been made to assign one or the other mechanism to a given enzyme; for yeast glyceraldehyde-3-phosphate dehydrogenase see [337] vs. [247, 1164]. However, evidence is often inconclusive allowing alternative interpretations.

SUPRAMOLECULAR ENZYME ORGANIZATION

TABLE 3.5.

*Protomer contacts and equilibrium constants of enzyme isomerization in an isologous ("tetrahedral")
tetramer at successive stages of substrate saturation according to the sequential allostery model*

Designation	Scheme	Protomer contacts	Equilibrium constants of enzyme isomerization
A_4		$ppqqrr$	
A_3B		$pp'qq'rr'$	$K_1 = \dfrac{[A_3B]}{[A_4]}$
$A_2B_2^p$		$pp''q'q'r'r'$	$K_{2p} = \dfrac{[A_2B_2^p]}{[A_3B]}$
$A_2B_2^q$		$p'p'qq''r'r'$	$K_{2q} = \dfrac{[A_2B_2^q]}{[A_3B]}$
$A_2B_2^r$		$p'p'q'q'rr''$	$K_{2r} = \dfrac{[A_2B_2^r]}{[A_3B]}$
AB_3		$'q''r'r''$	$K_{3qr} = \dfrac{[AB_3]}{[A_2B_2^p]}$ $K_{3pr} = \dfrac{[AB_3]}{[A_2B_2^q]}$ $K_{3pq} = \dfrac{[AB_3]}{[A_2B_2^r]}$
B_4		$p''p''q''q''r''r''$	$K_4 = \dfrac{[B_4]}{[AB_3]}$

On the whole, to understand allosteric regulation the detailed knowledge of the stereochemical events that take place during the course of saturation with ligand, and of their kinetic parameters would be required. So far no single technique can provide us with this information. Structural details are elaborated by X-ray crystallography, but the crystalline state may interfere with the ligand-induced transconformations meant to be unravelled. Clearly, the combination of various techniques can give the fullest picture. It will be illustrated on the example of glycogen phosphorylase, an allosteric enzyme known in considerable detail, that reality is more complex and even less clear-cut than the ideas outlined above.

A fully fledged allosteric enzyme: Glycogen phosphorylase

Glycogen phosphorylase (1,4-α-D-glucan: orthophosphate α-glucosyl transferase; EC 2.4.1.1) catalyses the reaction

$$\text{Glycogen}_{(n)} + P_i \rightleftharpoons \text{Glycogen}_{(n-1)} + \text{Glucose-1-P}$$

where n is the number of glucosyl residues in glycogen. The enzyme from rabbit muscle has been characterized in the greatest detail; the following description refers to this species. Its primary structure is known, and the 841 amino acid residues give a subunit molecular weight of 97,412 [1225]. It is a dimeric enzyme. Its tendency to form tetramers *in vitro* (cf. [472]) seems irrelevant under physiological conditions, because glycogen, with which it coexists in muscle [815], dissociates the tetramer to dimers [496, 1281]. The activity of phosphorylase is regulated in two ways: (1) through reversible phosphorylation at Ser-14 on each subunit; (2) through ligand-induced conformational changes. These two means of control are exercised simultaneously on the enzyme molecule.

Phosphorylation–dephosphorylation is a widespread phenomenon [212, 684] being the main reversible covalent modification that occurs to enzymes *in vivo* (cf. p. 103). Covalent modification usually perturbs catalytic function. Its effects differ from those

◄

The "exploded" scheme of a tetramer with all three possible types of protomer contact is shown (A_4). Protomers can exist in two conformations, A and B. The latter is induced or stabilized by the substrate. Protomer contacts across symmetry axes P, Q and R (cf. Fig. 3.14) are designated by p, q and r, respectively, in the free enzyme (A_4). If one of the protomers is in the B-state, the changed interprotomer contact is distinguished by priming (p', q' and r'). If both protomers are in the B-state, the interprotomer contacts are doubly primed (p'', q'' and r''). The superscript on the A_2B_2 species indicates the type of contact which relates two B protomers. The equilibrium constants of enzyme isomerization reflect the relative stabilities of the different forms without bound substrate. The subscript of the equilibrium constants denote the number of protomers in the B-state and the type(s) of contact (without priming) that relate the newly established B protomer to the other B protomer(s). For example, in the $A_2B_2^p \rightleftharpoons AB_3$ transition the new B protomer is related to the others through contacts q' and r', hence the equilibrium constant is K_{3qr}.

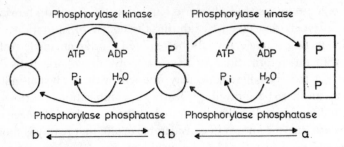

Fig. 3.25. Enzymatic interconversions of glycogen phosphorylase. ○, non-phosphorylated subunit; P|, phosphorylated subunit. Designations: *b*, phosphorylase *b*; *ab*, phosphorylase *ab* (phospho–dephospho hybrid); *a*, phosphorylase *a*. (From Dombrádi [317].)

caused by effector ligands in that the perturbation can persist after removal of the intracellular signal initiating the modification. Also, specific enzymes are required to insert and remove the covalent adduct (e.g. methyl, or, in the present case, phosphate group). The various forms of phosphorylase produced by phosphorylation are shown in Fig. 3.25. Phosphorylase *b,* the enzyme without phosphoryl groups at Ser-14, is converted through a hybrid, phosphorylase *ab,* to the fully phosphorylated *a* form by the enzyme phosphorylase kinase. The reverse reaction is catalysed by phosphorylase phosphatase. The two homodimers markedly differ from each other. Unliganded phosphorylase *b* is inactive, whereas phosphorylase *a* is active. On the other hand, phosphorylase *b* is fully activated by AMP. It follows that the enzyme can be activated in two alternative ways: either by phosphorylation or by AMP. Furthermore, the enzyme function is modulated by half a score of other ligands: the substrates glycogen, glucose-1-P and P_i; the weak activator IMP; and the inhibitors ATP, ADP, UDP-glucose, glucose, glucose-6-P and caffeine. Of all these only caffeine is unphysiological (disregarding the coffee-drinking habit of certain primates), but it is assumed to have a so far unknown counterpart *in vivo* [617]. The question is how all these influences tailor the enzyme to meet the actual catalytic need in the living cell.

The three-dimensional structures of both phosphorylase *b* [1290] and *a* [1152] are known to 0.25–0.3 nm resolution. The gross structures of the two molecules are similar. The schematic picture of phosphorylase *b* is shown in Fig. 3.26. There are four specific binding sites on each subunit: (*i*) the nucleotide or activator site near the subunit interface, (*ii*) the glycogen-storage site, where the enzyme is anchored to its macromolecular substrate, (*iii*) the active site, in a crevice at the far end from the subunit interface, and (*iv*) the nucleoside or inhibitor site, quite close to the active site. The pattern of ligand binding at these sites, superimposed on the effect of covalent modification, adjusts the catalytic performance of the enzyme.

Phosphorylase *b* possesses great conformational plasticity as shown by the marked structural changes detectable upon binding various ligands (cf. [472]). These conformational changes propagate from one protomer to the other resulting in positive

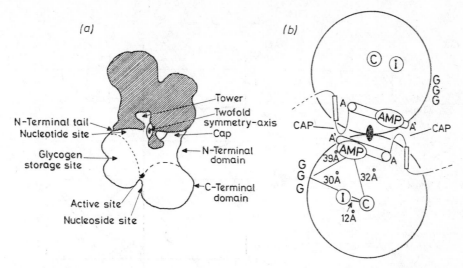

Fig. 3.26. Structure of the phosphorylase dimer, viewed down the two-fold symmetry axis. (*a*) Molecular contours, location of domains and various sites. The symmetry-related subunit is distinguished by hatching. (Adapted from Weber *et al.* [1290].) (*b*) Scheme of the dimer, with inter-site distances indicated. C, catalytic site; AMP, nucleotide (activator) site; I, nucleoside (inhibitor) site; GGG, glycogen storage site; AA', a helical segment of the polypeptide chain. (From Jenkins *et al.* [593].)

homotropic cooperative effect in AMP binding and positive heterotropic effect between AMP and glucose-1-P [57, 151, 496]. It should be recalled, however, that phosphorylase *b* is inactive and activation by AMP may occur *in vivo* only under special conditions [151, 526]. In contrast, phosphorylase *a* is rather rigid, most of the ligands that cause large structural changes in phosphorylase *b* fail to do so in phosphorylase *a* [496]. Nevertheless, conformational changes do occur in phosphorylase *a*: it exists in equilibrium between an inactive T- and an active R-state, thus it is an allosteric enzyme in its own right [527]. Activators (AMP, IMP, glucose-1-P) shift the equilibrium to the R-state, whereas inhibitors (glucose, caffeine) toward the T-state.

The actual changes in three-dimensional structure underlying these transitions are beginning to be known. By chance, X-ray data about rabbit muscle phosphorylase *a* have so far been gathered only from crystals grown in the presence of glucose, which induces the inactive T-state. Ligands favouring the R-state shatter these crystals [765], but phosphorylase *a* from shark muscle has apparently been crystallized in the R-state [381]. Comparison of the two structures may be revealing, even though they originate from different species.

The static picture derived from X-ray crystallography, however, cannot describe the dynamic aspects of the allosteric transitions. Temperature-dependent X-ray diffraction [394], a novel technique which is able to provide information about structural dynamics, has its specific limitations. Therefore solution studies employing a variety of physical and chemical techniques have been conducted (e.g. [151, 318,

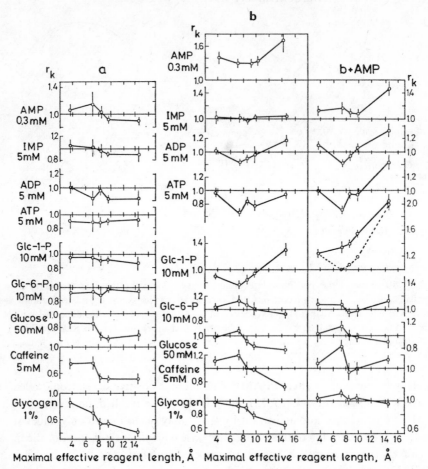

Fig. 3.27. Effect of ligands on the crosslinkability of phosphorylase *b* and *a* with diimidates of different (chain length. Parameter $r_k = k_L/k_O$, where k_L and k_O are the first order rate constants of crosslinking measured by the accumulation of subunit dimers, trimers and tetramers after SDS gel electrophoresis) in the presence and absence of a given ligand. Thus, if $r_k > 1$, the ligand promotes crosslinking, whereas if $r_k < 1$, the ligand hinders crosslink formation. The five reagents used, in the order of increasing maximal effective length, were as follows: malonic, adipic, pimelic, suberic and dodecandeoic diimidates. The lines connecting the points only serve better visualization. Columns: *b*, phosphorylase *b*; *b*+AMP, phosphorylase *b*+0.3 mM AMP; *a*, phosphorylase *a*. The heterotropic effect of glucose-1-P and AMP is reflected in the difference between the measured and calculated (dotted line) diagram; the latter was taken as the average of the r_k values obtained with the two ligands separately. It may be noted that with phosphorylase *a*, only ligands that dissociate the tetrameric species (i.e. glucose, caffeine and glycogen) induce appreciable changes in r_k. (Compiled from Hajdu *et al.* [496] and Dombrádi *et al.* [318].)

496]) and will continue to be useful. From among the host of current questions, we arbitrarily pick out the following one: to what extent can the R- and T-states be regarded as unique structures?

There is one major difference between the structures of phosphorylase *b* and *a*: in the former the N-terminal segment (about twenty residues) is floppy, whereas in the latter this part is immobilized by ionic interaction between the serine phosphate and Arg-69 (in the same subunit), His-36' and Arg-43' (the latter two in the symmetry-related subunit) [1152]. Since AMP activates phosphorylase *b*, the question arose whether it also fixed the N-terminal tail [1290]. Crosslinking studies with dimethyl malonimidate suggest that in AMP-activated phosphorylase *b* tail remains free [485]. In other words, the proper alignment of active site residues can be achieved by different movements in the phosphorylase molecule. Further crosslinking experiments with a homologous series of diimidates on the various phosphorylase forms, with and without ligands, corroborated and extended this conclusion [348, 496]. The rationale of the crosslinking approach for detecting conformational differences is simple: the movement of reactive side chains relative to each other will influence the propensity for crosslinking with a reagent of given length. It is seen in Fig. 3.27 that the various ligands cause pronounced changes in the crosslinkability with the set of reagents in phosphorylase *b*, but much less alteration in phosphorylase *a*. Although the "distance diagrams" of Fig. 3.27 cannot be directly translated into three-dimensional structure, they are characteristic of each enzyme–ligand complex and tell whether conformations are similar or dissimilar. For example, ADP and

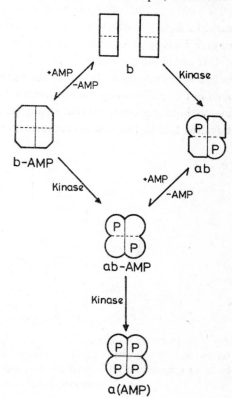

Fig. 3.28. Conformational states detectable by crosslinking in rabbit muscle phosphorylase. Letters *a, b* and *ab* denote the respective forms of phosphorylase. Square, truncated square, and circular protomer symbols designate different polypeptide conformations. Protomer interfaces indicated by dotted line denote the allosterically competent subunit contact *m* (monomer–monomer contact). "Kinase" means the incorporation of phosphate (P) at Ser-14 by phosphorylase kinase. (From Dombrádi *et al.* [318].)

caffeine both inhibit phosphorylase *b*, but, in the absence of AMP, ADP decreases the crosslinkability by the shorter chain diimidates and increases that by the longer reagents, whereas caffeine has the opposite effects. This is consistent with the notion that the inhibited states induced by ADP and caffeine, respectively, are not identical. A generalized conclusion, summarized in Fig. 3.28, from these and other diagrams [318] is that the activators, as well as the inhibitors, establish conformations that are not quite uniform. It follows that both the R- and T-states are heterogeneous if the entire enzyme molecule is looked at.

Glycogen phosphorylase, though a "soluble" enzyme, is part of a multienzyme cascade system organized in isolable particles together with glycogen [815]. Therefore its structural–functional features are to be considered in the wider context of interaction with other components of the system. Knowledge about the details of these interactions is still meagre and is covered in recent reviews on phosphorylase [317, 381].

Regulation via association–dissociation of the oligomer

In 1967, within two years of the birth of the two main allosteric enzyme models described above [679, 834], Kurganov [694, 695], Frieden [397] and Nichol *et al.* [870] independently proposed that cooperativity may result from a change in the state of association of an oligomer without any conformational change being necessary. In fact, the idea was originally advanced by Briehl [132] for interpreting the oxygen binding of lamprey haemoglobin.

One of the simplest cases is that the substrate of a reversibly dissociating enzyme is bound preferentially by either the monomer or the dimer. The latter situation is seen in Fig. 3.29*a*, where each of the two binding sites is built up of elements belonging to both monomers so that strong binding can occur only to the dimer. It should be noted that in this case no conformational change need accompany the "allosteric

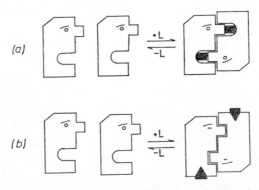

Fig. 3.29. Protomer association promoted by ligand binding. Binding of ligand (*L*, dark object) favours an isologous dimer. (*a*) No conformational change occurs, the ligand binding site is shared by the two protomers; (*b*) ligand binding outside the interprotomer binding domain induces conformational changes that strengthen association.

transition", which is now an association–dissociation process. Alternatively, association (or, by the same token, dissociation) may involve a conformational change in each monomer enforced by quaternary constraint (or its release), which creates the specific binding or catalytic sites (Fig. 3.29b).

The mathematical formulation of such a system is analogous to that of the allosteric symmetry model (p. 55.). In the case of a dimer (E_2) that is in equilibrium with the monomer (E) according to $2E \rightleftharpoons E_2$, the equilibrium constant is

$$L = \frac{[E_2]}{[E]^2}.$$

(3.25)

If both the monomer and dimer bind the substrate (one molecule per monomer) and the two binding sites in the dimer are equivalent and independent, then the intrinsic substrate dissociation constants are as follows:

$$\frac{[E][S]}{[ES]} = K_E$$

(3.26)

and

$$\frac{[E_2][S]}{[E_2S]} = \frac{[E_2S][S]}{[E_2S_2]} = K_{E_2}.$$

(3.27)

By defining $\alpha = [s]/K_E$ and $c = K_E/K_{E_2}$, and taking the statistical factor into account for K_{E_2}, one can work out for the fractional saturation

$$\overline{Y}_s = \frac{\alpha + 2LEc\alpha(1 + c\alpha)}{(1 + \alpha) + 2LE(1 + c\alpha)^2}.$$

(3.28)

In this expression E is the concentration of *free monomeric* enzyme, rather than the total concentration of enzyme, E_0. For the dimeric case, E can be fairly easily expressed in terms of E_0 by solving the enzyme conservation equation, from which

$$E = \frac{-(1 + \alpha) + \sqrt{(1 + \alpha)^2 + 4LE_0(1 + c\alpha)^2}}{2L(1 + c\alpha)^2}.$$

(3.29)

The saturation curve $\overline{Y}_s$ vs. S will range from sigmoidal to hyperbolic depending on the values of L, E_0 and c. It is an important diagnostic feature of this mechanism that the shape of saturation curve changes with enzyme concentration. Whenever this is observed experimentally, one should consider the existence of a reversibly dissociating oligomer.

Phillips [942] gave a critical review of such enzymes up to 1974, including ribonucleoside diphosphate reductase, cytidine triphosphate synthase, aspartokinase–homoserine dehydrogenase, glutamate dehydrogenase and two threonine dehydrases. For each of these enzymes, observations *in vitro* indicated ligand-induced oligomerization, which implied possible regulatory significance. However, the possibility that these observations *in vitro* are relevant to the situation *in vivo* must be examined with particular caution. In any case, for most dissociable enzymes, whose levels in living cells are remarkably constant, the position of the monomer $\rightleftharpoons$ oligomer equilibrium at a given concentration of ligands is also constant *in vivo*. For those inducible enzymes, however, such as biodegradative threonine dehydrase [1313] and glycerol kinase [302], that also show reversible dissociation to an inactive monomer, the equilibrium shifts during induction of the enzyme, leading to an exaggeration of the change in enzyme activity compared to the change in enzyme amount.

Kurganov has worked out equations for a variety of models of dissociating–associating enzymes. These cover the kinetics of the enzyme reaction in rapidly dissociating [705] and associating [697] enzymes, the kinetic criteria for assigning a particular allosteric mechanism [696], the kinetics of the enzyme reaction in slowly dissociating [701] or dissociating and associating [698, 700] enzymes, as well as the kinetics of the dissociation–association process itself [699].

In the following sections a few such enzymes will be introduced briefly, subdivided somewhat arbitrarily into dissociating oligomers and associating oligomers.

Dissociating oligomeric enzymes

Glyceraldehyde-3-phosphate dehydrogenase (GAPD; EC 1.2.1.12) from mammalian muscle is a tetramer that dissociates to dimers and possibly monomers [551, 709]. Substrates and effectors, as well as their various combinations, have been reported to influence dissociation. Thus NAD and P_i favoured the tetramer [551, 709], whereas NAD+glyceraldehyde+P_i promoted dissociation in ultracentrifugation studies [551]. A mixture of the above substrates stabilized the active tetramer in active-enzyme band centrifugation experiments [551, 709], whereas it seemed to produce active monomers and dimers in stopped-flow and gel-permeation studies [898]. Hybridization experiments indicated that NAD "tightened" the tetramer [208, 890, 1058], and NADH "loosened" it [890]. Affinity chromatography on NAD-Sepharose suggested dissociation induced by glyceraldehyde-3-phosphate [607]. On the other hand, GAPD acylated with its substrate 1,3-bisphophoglycerate, which is thought to be the predominant form of enzyme under physiological conditions [96], proved to be a tetramer according to kinetic and gel-chromatographic studies [1263]. The reports about substrate effects are thus controversial. There is agreement among authors that ATP dissociates the tetramer to dimers, but the latter were found inactive when ATP was present in excess [244] and active when ATP was stoichiometric with the enzyme [901]. Other proteins, if present at high concentration, were claimed to shift the association equilibrium toward the tetramer [824], owing to "macromolecular crowding" [823]. In this study [824], however, the conclusion was drawn indirectly from the dependence of catalytic activity on enzyme concentration, which could be interpreted by making the daring assumption that the monomer is about ten times as active as the tetramer.

Glycerol kinase (EC 2.7.1.30) from *E. coli* is a tetramer with one catalytic site and one allosteric site on each subunit [1219]. The allosteric effector is fructose-1,6-bis-phosphate (FBP). De Riel and Paulus have recently provided evidence from kinetic [302], physico-chemical [303] and desensitization [304] experiments for an allosteric regulation of glycerol kinase through dissociation. According to their model the tetramer spontaneously dissociates to dimers; both forms have full catalytic activity, but only the tetramer can bind FBP, which brings the enzyme into an inactive state. Thus the allosteric inhibitor acts by stabilizing an inactive conformation induced solely in the tetramer. Indeed, FBP decreases the apparent dissociation constant for $T \rightleftharpoons 2D$ by two to four orders of magnitude [302, 303]. It is worth noting that in *E. coli* glycerol kinase the allosteric transition occurs at the quaternary rather than tertiary structural level. In fact, it may be a matter of evolutionary chance that in some oligomers the structural transition involves the separation of protomers, while in others it does not. With glycerol kinase the association–dissociation process is fairly slow, which adds to its possible regulatory significance (cf. below). Furthermore, glycerol kinase is an inducible enzyme, its concentration varies widely in the *E. coli* cell, hence activity modulation by enzyme concentration is also conceivable.

Associating oligomeric enzymes

Here we discuss enzymes that are oligomers capable of self-association, i.e. where the basic oligomeric unit usually does not dissociate but can undergo polymerization.

The prototype and most extensively studied example of such enzymes is *glutamate dehydrogenase* (EC 1.4.1.3) from beef liver mitochondria. The polymerizing unit of this enzyme is actually a stable hexamer with $M_r = 340,000$ [378] consisting of six subunits arranged in a quasi-spherical array, two trimeric rings on top of each other (cf. p. 20). In 1961 Tomkins *et al.* [1227] proposed that polymeric glutamate dehydrogenase catalysed the oxidative deamination of glutamate, whereas ligand-induced depolymerization suppressed this activity and evoked alanine dehydrogenase activity. Although this claim was shown later to be wrong [379], the relationship between the state of aggregation and catalytic activity of the enzyme aroused repeated arguments over the past 15 years. Purine nucleotides undoubtedly influence glutamate dehydrogenation, GTP being an inhibitor and ADP an activator, and their effects change with enzyme concentration [400]. However, the data could not be accounted for quantitatively by any of the available models [397, 695, 870] of polymerizing allosteric enzymes, which led in the early seventies to the consensus that the aggregation of glutamate dehydrogenase had little biological significance (cf. [378]). More recently, however, the issue was reopened by a detailed analysis, both theoretical and experimental, made by Cohen and coworkers [235, 237]. They claimed to have reconciled all previous and current experimental data in their model(s) [235]. The salient features of their three, slightly different, models are as follows. The polymerizing unit of the enzyme (i.e. the hexamer) can exist in two conformational states *x* and *y*, the former active and the latter inactive (or less active). The *x* form has a greater

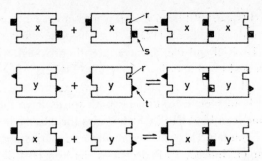

Fig. 3.30. Schematic model of the end-to-end polymerization of glutamate dehydrogenase. Letters x and y denote two conformational states of the hexamer (the basic polymerizing unit); r, s and t represent elements of the contact surfaces. This model illustrates the case when the x–y binding constant is the geometric mean of the x–x and y–y binding constants. (From Cohen and Benedek [235].)

propensity for polymerization than the y form and mixed x-y polymers can be formed. Polymerization does not affect the catalytic activity of either form, but shifts the $y \rightleftharpoons x$ equilibrium toward x, and therefore influences the overall catalytic activity of the enzyme. The parallel measurement of quasi-elastic light scattering and enzyme activity [237] indicated that (1) polymerization is a linear end-to-end association into rigid rods and (2) the best fit was obtained with the model in which the x-y binding constant was the geometric mean of the x-x and y-y binding constants (Fig. 3.30).

This interpretation has been challenged by Thusius [1220], who argues that it is incompatible with the kinetics of polymerization, which indicated a one-state, random self-association mechanism. Such a mechanism was also suggested by pressure perturbation relaxation kinetics [504]. Furthermore, Zeiri and Reisler [1347] presented crosslinking experiments that did not support the existence of two distinct conformational states, and claimed that the effects of regulatory nucleotides on enzyme activity and polymerization were uncoupled. In turn, Cohen and Benedek [236] rejected all the above criticism point by point. At stop press the issue is open.

Phosphofructokinases (PFK; EC 2.7.1.11) from animal tissues such as skeletal muscle [708, 904], heart [773], liver [995] and erythrocytes [613, 1310] undergo open polymerization, the actual mechanism of association depending somewhat on the tissue of origin. With the rabbit muscle enzyme the basic polymerizing unit is the tetramer and association produces sheetlike or filamentous structures [1213]. However, at near-physiological temperature and protein concentration the enzyme is much less aggregated [594] than at lower temperatures (e.g. [904]). Moreover, it has been suggested that the characteristic regulatory behaviour of muscle PFK is due to a rapid equilibrium between an active and an inactive tetrameric form, rather than to polymerization [401]. The human erythrocyte enzyme has been studied in considerable detail: it seems to be a hybrid of two types of subunit [613], analogously to hybrid lactate dehydrogenases. The sigmoidal saturation curve with the substrate fructose-

6-P is probably due to the displacement of the equilibrium inactive dimer $\rightleftharpoons$ active multimer to the right, whereas the inhibitor ATP shifts the equilibrium to the left [1310, 1311]. However, the mechanism of polymerization is sensitive to the reaction conditions (temperature, buffer, effectors, etc.) [1349]. For the sake of comparison, yeast PFK is a stable octamer that neither further associates nor dissociates [675, 999]; its regulation by ATP and fructose-6-P can be described in terms of a mixed concerted-sequential allosteric mechanism [721, 999]. PFKs from bacteria such as *E. coli* [1218] and *B. stearothermophilus* [350] are simple, fairly small tetramers. An interesting feature of the *B. stearothermophilus* enzyme is that the effector site, accommodating both activators and inhibitors, is located at one type of subunit interface, so that the allosteric control is exercised by these ligand bridges between two pairs of protomers [350]. A detailed account of the structure and function of PFKs has recently been given by Goldhammer and Paradies [458].

Ambiguity in the characterization of associating–dissociating oligomers

The brief introduction to associating–dissociating enzymes given above may have suggested to the reader the high degree of uncertainty that exists in this area of enzymology. The problem is partly our as yet inadequate methodology for tackling loose macromolecular associations. Furthermore, the sheer complexity when the enzyme itself is a dissociating system, whose members may possibly each form substrate-, coenzyme- and effector-complexes, renders the clear-cut interpretation of experimental data increasingly difficult. Perhaps this complexity was not at first fully appreciated, and a clearer picture may emerge from very careful, systematic analysis of the basic variables, together with appropriate use of computer modelling, but a reluctance to construct ambitious schemes from limited data. The same holds, possibly to an even greater extent, for weakly interacting multienzyme systems, as will be evident in later chapters.

Hysteretic enzymes

In the allosteric models considered above, with or without changes in the state of aggregation, it was generally assumed that the structural transitions were fast relative to the catalytic reaction. However, steric changes in enzymes may occur during the catalytic reaction [627a]. In 1970 Frieden [398] raised the idea that conformational changes may be slow, resulting in a lag period in the response of the enzyme to an abrupt change in the concentration of substrates of effectors. Frieden coined the term *hysteretic enzymes* and over the years that followed several enzymes with such properties were found (cf. [702]). The hysteretic response is not confined to oligomers, it may be elicited in monomeric enzymes as well. However, hysteresis might be particularly expected in associating enzyme systems, because the second-order reactions involved will be slow at small enough enzyme concentrations. The theoretical work by Kurganov and coworkers on slowly associating–dissociating enzymes was mentioned in the first part of this section. For the detailed treatment of hysteretic behaviour of enzymes, both theoretically and experimentally, the reader should consult the recent review by Frieden [399].

COMPLEX ENZYMES

REGULATION THROUGH NON-CATALYTIC SUBUNITS

The next stage in the structural organization of enzymes after true oligomers is occupied by *complex enzymes*. We use this term to denote enzymes consisting of catalytic and non-catalytic (as a rule, regulatory) subunits. Complex enzymes may be regarded as allosteric enzymes in which the domain building up the binding site for the regulatory effector(s) is separated from the domain accommodating the catalytic site in the sense that it lies in a distinct polypeptide chain. From an evolutionary viewpoint, it is probable that gene fusion did occur in most cases rather than that originally continuous polypeptide chains have separated. In some oligomeric enzymes the allosteric site may have appeared as a result of gene duplication, e.g. in glutamate dehydrogenase [339] and glyceraldehyde-3-phosphate dehydrogenase [340], whereas in others by acquisition of a foreign structural gene, e.g. the NAD-binding domain in glycogen phosphorylase [382]. However, in present day complex enzymes the catalytic and regulatory protein units remained separate, but associable, entities. This arrangement has certain advantages, such as functional versatility (exemplified by lactose synthase) and possibly easier assembly of large and complicated structures, such as the aspartate transcarbamoylase molecule. The regulatory subunits were presumably generated from the catalytic subunit of the same or another enzyme.

Lactose synthase

Lactose synthase catalyses the synthesis of lactose in the lactating mammary gland through the reaction

$$\text{UDP-D-galactose} + \text{D-glucose} \longrightarrow \text{lactose} + \text{UDP}$$

It is the simplest of complex enzymes as it consists of two proteins, one copy of each. One is the enzyme UDP-galactosyl transferase (EC 2.4.1.22) and the other is α-lactalbumin, a non-enzymatic protein abundant in milk. UDP-galactosyl transferase is ubiquitous, certainly not confined to the mammary gland. In other cells it attaches a galactosyl residue to *N*-acetyl glucosamine in the oligosaccharide moieties of glycoproteins, and so participates in the synthesis of cell-envelope constituents. However, when the production of α-lactalbumin is turned on by hormone action at the onset

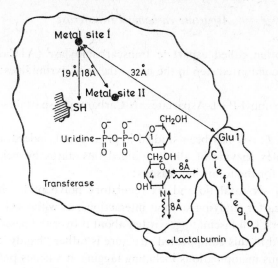

Fig. 3.31. Schematic representation of the active site of bovine galactosyltransferase and its interaction with α-lactalbumin. α-Lactalbumin was uniquely dansylated at the N-terminus (Glu-1), whereas the metal binding sites of galactosyltransferase were liganded with cobalt instead of Mn^{2+}. To stabilize the loose complex, it was crosslinked with pimelic diimidate. The cobalt ion quenched the dansyl fluorescence, which was the basis of distance measurement. Further distances indicated in the figure were derived from other data in the literature. (From O'Keeffe *et al.* [880].)

of lactation, the enzyme found in the membrane of the Golgi apparatus will complex with α-lactalbumin yielding lactose synthase. This modification in substrate specificity is brought about by α-lactalbumin through lowering the K_m for glucose by about three orders of magnitude. This is achieved by the synergistic binding of α-lactalbumin and glucose to the enzyme, which means that α-lactalbumin strengthens the binding of glucose and vice versa. By virtue of its function Hill has named α-lactalbumin a *specifier protein*. Lactose synthase has been reviewed by Hill and Brew [130, 541].

Current research on lactose synthase is focused on the chemistry of galactosyl transferase–α-lactalbumin interaction. Interestingly, the primary structures of α-lactalbumins isolated from several species exhibit extensive homology with that of hen's egg-white lysozyme [129]. In spite of this homology, α-lactalbumin has no enzymatic activity and lysozyme cannot replace α-lactalbumin in forming the lactose synthase complex. Nevertheless, the known three-dimensional structure of lysozyme has permitted the assignment of a putative structure to α-lactalbumin, which helps in interpreting chemical modification data on the α-lactalbumin–galactosyl transferase interaction [879, 959, 1005]. Based upon chemical modification [879] and fluorescence resonance energy transfer measurements [880] Bell and coworkers determined various distances in the lactose synthase complex, as seen in their tentative model (Fig. 3.31). The validity of conclusions can be checked when the current crystallographic studies on baboon α-lactalbumin [31] are completed. The structural analysis of the other party, galactosyl transferase, however, is lagging.

Aspartate carbamoyl transferase

This enzyme, often called aspartate transcarbamoylase (ATCase; EC 2.1.3.2), catalyses the first committed step in the biosynthesis of pyrimidines:

$$\text{Carbamoyl-P} + \text{L-Aspartate} \rightleftharpoons N\text{-Carbamoyl aspartate} + \text{P}_i$$

The enzyme from *E. coli* has been studied in great detail, indeed, it is perhaps the best-known complex enzyme of all. It attained this status by being an unusually attractive enzyme for the following reasons:

1. Consisting of six catalytic and six regulatory polypeptide chains (subunits), it is a bonanza for studying subunit interactions, as witnessed by the incessant flood of papers, still fascinating, written about it over the past 20 years.
2. As chance had it, this complicated structure is rather "handy": it readily lends itself to various manipulations including tagging at various positions, dissociation to single polypeptide chains and subsequent reassembly to yield a product indistinguishable from the native enzyme.
3. Being a microbial enzyme, it can be obtained in mutant forms, an invaluable asset in structure–function studies.

We may start the ATCase story in 1962, when Gerhart and Pardee made the important discovery that the enzyme had a sigmoidal saturation curve with its substrate and was subject to feedback inhibition by the end-product of pyrimidine biosynthesis, cytidine triphosphate (CTP), as well as to activation by ATP [436]. The next major advance was made in 1965 by Gerhart and Schachman [437] who showed that the ATCase molecule could be dissociated by mercurials into two catalytic trimers (C_3) and three regulatory dimers (R_2). The C_3 part was catalytically active and insensitive to CTP, whereas the R_2 part bound the allosteric effectors CTP and ATP. There are three active sites in C_3 and two effector sites in R_2. Further analytical work yielded molecular weights for the catalytic subunit (polypeptide chain*) of 34,000 and for the regulatory subunit of 17,000 [807, 1291], thus the whole complex molecule, C_6R_6, has $M_r \approx 300,000$. Each regulatory subunit contains one Zn^{2+} needed for structural integrity; without Zn^{2+} the native complex cannot be reconstituted from its components [862, 1023]. These and other early developments on ATCase have been reviewed in [584].

The gross molecular structure, according to electron microscopic [1003] and X-ray diffraction data [1284], has D_3 symmetry: the three catalytic subunits are closely associated to form a trimer and the two, nearly eclipsed, trimeric rings are in contact.

---

* Perusal of current literature on ATCase reveals an inconsistency in nomenclature. Some authors refer to the C_3 structure as "catalytic subunit" and to R_2 as "regulatory subunit". Others equate subunits with the individual polypeptide chains and speak of catalytic trimers and regulatory dimers composed of three and two subunits, respectively. We follow the latter usage.

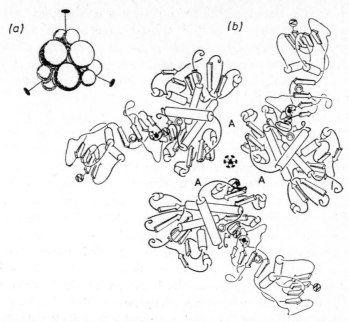

Fig. 3.32. Structure of aspartate carbamoyl transferase. (*a*) Scheme of subunit assembly. Large spheres are catalytic subunits, small spheres are regulatory subunits. The molecule is viewed along the three-fold axis; the three two-fold axes are indicated. (*b*) The "upper half", C_3R_3, of the ATCase molecule as deduced by X-ray crystallography. Cylinders and flat arrows designate α-helices and β-structures, respectively. The catalytic subunit is in heavy, the regulatory subunit in thin line. In the centre the three-fold and three two-fold symmetry axes are shown. (From Monaco *et al.* [831].)

Each C subunit is "crosslinked" by a regulatory dimer (R_2) to another C in the opposite trimer; the two Cs thus connected are 120° apart about the threefold axis (Fig. 3.32*a*). As regards fine structure, unliganded ATCase has been solved to 3 Å resolution and its CTP complex to 2.8 Å [831]. The two structures are practically identical, which suggests that both forms are the inactive T conformation. The schematic diagram of the half molecule C_3R_3 is seen in Fig. 3.32*b*. X-ray crystallographic work on the active isolated catalytic subunit is underway [388]; comparison of the structure to emerge with that of the holoenzyme will markedly contribute to our understanding of the allosteric conformational change.

Subunit interactions in ATCase

The native C_6R_6 enzyme is a very stable assembly. It does not dissociate under mild conditions as witnessed by hydrodynamic measurements at enzyme concentrations as low as 10^{-8} M [1055] and by the lack of subunit exchange with free C or R subunits [1189]. Although the bisubstrate analogue *N*-(phosphonoacetyl)-L-aspartate somewhat enhanced subunit exchange, it was still too slow to be reliably quantitated [1189].

Furthermore, ATCase exhibits both homotropic and heterotropic effects, characteristic of the native enzyme, even at 10^{-10} M enzyme concentration [573]. The dissociation constant of the enzyme, referring to the release of either one C_3 trimer or one R_2 dimer, has recently been estimated to be of the order of 10^{-14} M [115]. Thus we may disregard association–dissociation phenomena in the regulatory transitions of ATCase with confidence.*

As regards the formal mechanism of allosteric transitions there is a consensus in the literature that only an R-and a T-state can be distinguished. There is a body of evidence [343, 573] for the overall validity of the two-state concerted model [834]. On the other hand, more complex mechanisms, claimed to incorporate all available data, have also been proposed [1216], which are specific combinations of the concerted and sequential allostery models. Thus a characteristic feature of Thiry and Hervé's [1216] mechanism is that homotropic cooperative interactions between catalytic sites are concerted transitions, whereas the heterotropic interactions between catalytic and regulatory sites are mediated by sequential conformational changes. Irrespective of the mechanistic subtleties, the substrates carbamoyl-P and aspartate, as well as the activator ATP, favour the R-state, while the inhibitor CTP binds preferentially to the T-state. In the absence of ligands the allosteric constant $L = T/R = 250$, i.e. the T-state is about 14 kJ/mole more stable than the R-state [573]. The latter could be stabilized by crosslinking ATCase with tartaryl diazide in the presence of carbamoyl-P and the substrate–analogue succinate [343]. This covalent "staple" interferes with the structural change underlying the transition. Unfortunately, we do not have as yet a detailed picture of this process. Hydrodynamic [323, 438] and small angle X-ray scattering [835] experiments suggested that the conformational change was large [438] and a model was proposed according to which the rotation of the two catalytic trimers relative to each other was at the heart of molecular rearrangement [435]. However, recent preliminary X-ray diffraction data are consistent rather with a relative elongation of the molecule along the three-fold axis in the R-state [831]. On the other hand, fluorescence energy transfer measurements gave a value of about 30 Å for the distance between catalytic sites in different C_3 trimers and this value did not change upon addition of any of the substrates or effectors [493]. This finding argues against a marked elongation of the molecule, but it does not rule out some degree of shape change if the accuracy of the technique is taken into account.

Although our knowledge about the precise stereochemistry of the allosteric transition is scarce, we know a good deal about the propagation of information within the ATCase molecule, about the strength of subunit interactions and their relative role in regulation. This information came from ingenious experiments on various

---

* However, Cook and Milne [248] observed different binding curves as a function of enzyme concentration and envisaged a polymerizing allosteric model. The source of discrepancy between this report and the bulk of the literature is unknown.

subcomplexes of the enzyme lacking certain subunits, and from studies in which the perturbation by ligand binding of reporter groups attached to different subunits was measured. An important observation of the first type was made in Chan's laboratory [189, 191], by showing that C_3R_6, i.e. the subcomplex lacking one catalytic trimer, exhibited the behaviour of the R-state and showed neither homotropic nor heterotropic interactions. Perhaps not surprisingly, a quaternary constraint present only in the complete C_6R_6 complex is needed for the ligand-induced kinetic effects. The R_2 dimers have to be anchored at both ends, otherwise the local conformational change elicited by the effectors will be "dissipated" instead of transmitted to the catalytic subunits. Incidentally, C_3R_6 was much more resistant than the C_3 trimer to chymotryptic attack [193], showing that the association of C and R subunits affords protection against proteolytic degradation. Studies on the subcomplex C_6R_4 have revealed that the lack of one regulatory dimer considerably loosens the overall structure: C_6R_4 disproportionates into native enzyme C_6R_6 and free catalytic trimers [1189]. This process was accelerated about 300-fold by active site-ligands, which suggests that transition to the relaxed state involves a decrease of interaction energy by 7 kJ/mole per bonding domain between C and R subunits.

A systematic investigation of the communication between various subunits has been conducted by Schachman and coworkers. This "cross-talk"* between subunits was tapped by putting a label on one type of subunit and observing its behaviour after ligand binding to another subunit. By finding the appropriate ways of labelling, the propagation of conformational change could be demonstrated from one catalytic trimer to the other [1339], from C to R subunits [596], and back, from R to C subunits [529].

A promising approach in studying structure–function relationship in bacterial ATCase is to produce mutant enzymes by induced mutagenesis, e.g. by growing the bacteria in the presence of 2-thiouracil [644, 645] or 2-aminopurine [610]. Mutagenesis is superior to chemical modification *in vitro* in that it provides a great number of different modified enzymes each affected at a unique point and with no bulky groups attached to the amino acid side chains. Furthermore, buried residues can be altered, which are inaccessible to modification *in vitro*. Therefore this line of research is now vigorously pursued. So far, mutant ATCases lacking homotropic cooperative interactions but retaining heterotropic interactions [644, 645] or altered in both of these faculties [610] have been isolated and partly characterized. Interestingly, a derivative possessing the former properties could also be produced by crosslinking the catalytic subunits with tartaryl diazide [194]. The mutagenic approach is also used with other enzymes. Its obvious limitation is that only enzymes from organisms amenable to mutagenic manipulations, i.e. mainly microorganisms, can be tackled.

---

* H. K. Schachman's original term instead of "communication", but turned down by journal reviewers.

Subunit assembly in ATCase

The properties of ATCase rendered it a gratifying object for studying pathways of subunit assembly. The polypeptide chains of the enzyme from *E. coli* seem to be synthesized in the order of catalytic chain → regulatory chain [928] from two linked genes, which are part of the same operon, and assembly takes place in the cytoplasm after release from polysomes [927]. The reassembly experiments *in vitro* have therefore physiological relevance.

Work along these lines has been conducted in parallel in the laboratories of Schachman [113, 114, 115, 238] and Chan [189, 192, 196, 844]. Both groups investigated the assembly from the stage of C_3 and R_2, i.e. the route of association of catalytic trimers and regulatory dimers. Although this starting point seems likely in the light of the stability of the two kinds of homo-oligomeric subcomplexes, it cannot be ruled out that *in vivo* the assembly might start with the formation of CR, as cautioned in [115]. Reassembly *in vitro* was monitored by the kinetics of activity changes [115], and the quantitation of various species assembled [114, 115] from differently tagged subunits or as a result of disproportionation of a subcomplex [1190]. Stopped-flow X-ray scattering, a novel combination of the two techniques, has been applied to follow the mercurial-induced dissociation of ATCase [836]; further technical development may make it a valuable tool for assembly studies.

The assembly pathways arrived at by the groups of workers are slightly different. They agree in distinguishing reversible and irreversible steps along the overall route. Bothwell and Schachman's [114] scheme is given in Fig. 3.33. It is seen that only those steps are reversible in which a single C–R bond is to be broken. The sole sub-

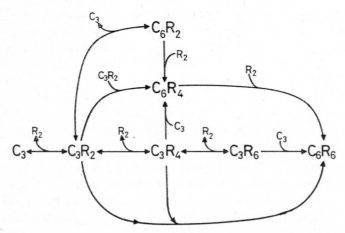

Fig. 3.33. Scheme for the assembly pathways of aspartate carbamoyl transferase from catalytic trimer (C_3) and regulatory dimer (R_2). Only four reactions, those involving the formation or rupture of a single C–R bond, are reversible, as indicated by the arrows. (Adapted from Bothwell and Schachman [114].)

stantial difference in Chan's [192] pathway is that he found "R-redundant" association steps, e.g.

$$C_3R_2 + C_3R_6 \longrightarrow C_6R_6 + R_2$$

to be kinetically significant in the overall assembly. While this may hold for experimental conditions where R subunits are in excess, its relevance *in vivo* is questionable,
because of the coordinate, equimolar synthesis of the two polypeptide chains [928].
An important common conclusion is that the bimolecular reaction of the two types
of subunit is fast, with a rate constant of about 10^6 M^{-1} s^{-1} [115, 192]. Hence
the assembly of the complex is rapidly completed, which may have physiological
significance in *E. coli* [115]. Namely, in the logarithmically growing bacterium culture
a considerable amount of ATCase is to be manufactured every 40 min (the generation time), which meanwhile has to exert its catalytic function. It may be crucial
that incomplete, i.e. unregulated, subcomplexes should not exist even for short
periods of time. The rapidity and virtual irreversibility of the assembly process may
well serve this goal.

A recent concise review on structure–function relationships of *E. coli* ATCase is
given by Kantrowitz *et al.* [611].

CHAPTER 4

INTERMOLECULAR ORGANIZATION: MULTIENZYME SYSTEMS

> "...*Nature has no bottom: its most basic principle is 'organization'. If Nature puts two things together she produces something new with new qualities, which cannot be expressed in terms of qualities of the components. ...Whenever we separate two things, we lose something, something which may have been the most essential feature.*"
>
> Albert Szent-Györgyi [1204]

FUNCTIONAL VS. STRUCTURAL ORGANIZATION

In the preceding chapters we have examined individual enzymes of various degrees of complexity. In doing so we adhered to the practice of classical enzymology, which—ever since its emergence in the late twenties—has been preoccupied with the study of single enzymes, after carefully removing all contaminating proteins. The aim of these endeavours has been clear: to elucidate how a given enzyme works without the interference of other enzymes or proteins.

While these efforts have provided us with a large body of information on a great number of enzymes, they necessarily neglected the *links* or communication *between* different enzymes. In the living cell each enzyme is part of a multienzyme system responsible for running a metabolic pathway which, in turn, is the component of an even larger network, finally building up to what is displayed on the popular and inextricable metabolic maps. Evolution worked at the level of the organism, the viability of which was the resultant of an interplay of the numerous pathways. Hence from the organism's point of view, be it a single cell or a multicellular being, it was not the excellence of any single enzyme but rather the controlled output of the enzyme system that was crucial.

It seems therefore obvious that the multienzyme systems existing in living organisms must possess a high degree of organization. We may distinguish two levels of organization: the purely *functional* and the *structural-functional* ones. Functional organization means that each enzyme has been tailored, evolved in structure, to have functional properties (V_{max}, K_m, K_I, allosteric constants, etc.) that satisfactorily mesh with those of other enzymes in the same and other systems. Thus functional organization involves not only the kinetic complementarity of sequentially consecutive enzymes, but also remote effects underlying feed-back inhibition or forward activation via allosteric transitions. The various, verified and putative, patterns of functional enzyme organization have been the subject of excellent reviews (e.g. [36, 1163]) and will not concern us here. Nevertheless, some aspects of functional organization that have had less emphasis so far, such as those connected with metabolite compartmentation and the systemic approach to metabolic control, will be discussed in later chapters. Our main interest will, however, be focused on *structural organization,* by which is meant all kinds of macromolecular associations involving enzymes and occurring *in vivo.* Structural organization naturally serves functional ends and in this sense it is a stringent way to achieve a functional goal. The importance of precise spatial disposition of enzymes in certain cellular niches, e.g. the membrane-integrated respiratory chain in mitochondria, has long been recognized. In some other cellular compartments structural organization has been not so evident, or even conspicuously absent. The so-called "soluble" enzymes (practically all enzymes discussed in this book up to here) are catalytically competent when isolated in aqueous medium and this prompted the belief that they are subject only to functional organization. However, evidence has accumulated suggesting that structural organization between enzymes is more common than we thought previously. The ever wider appreciation of this notion stimulates experimental work in this field which is eventually to sort out where and how structural enzyme organization is manifest in the living cell. The present, often rather ambiguous, state of affairs will be outlined later in the book through selected examples. In this chapter a brief general survey will be given of the functional corollaries of structural enzyme organization.

FUNCTIONAL CONSEQUENCES OF ENZYME JUXTAPOSITION

PROXIMITY EFFECT

Two or more enzymes can be physically associated in such a manner that their active sites become juxtaposed. The proximity of active sites may then render it possible for the product of the first enzyme to get easy access to the second enzyme. Depending on the physical state of this intermediary metabolite we may distinguish two cases [291]:

Covalently bound intermediate

In certain multienzyme complexes the compound to be transferred from one enzyme to the other is covalently bound to the enzyme complex. This is the case in the pyruvate and α-ketoglutarate dehydrogenase complexes (cf. Chapter 5), where the acyl-group produced from the corresponding oxo-compound is attached, through an *S*-acyl bond, to the lipoamide prosthetic group which, in turn, is covalently linked to the protein. This "swinging arm" is presumed to service the active sites of the three component enzymes of the complex. In the case of such a covalently bound intermediate the efficiency of transfer between enzymes is practically 100%, i.e. there is no leakage from the multienzyme assembly.

Diffusible intermediate

In several instances the transferable metabolite in a multienzyme cluster is not bound covalently to any of the component enzymes, but has to reach the next enzyme in a three-step process: (*i*) dissociation from the active site of enzyme A, (*ii*) diffusion from one enzyme to the other, and (*iii*) association with the active site of enzyme B. Depending on the geometry of the space between the two active sites the intermediate will have different probabilities of reaching the next enzyme without being mixed in the bulk medium. In a stable multienzyme complex, in which the components are permanently associated, this space obviously cannot be completely closed off from the medium. Rather, the space must be open to let in the first metabolite and let out the last metabolite (product) of the complex. It is the appropriate spatial arrangement of active sites relative to each other that ensures that the intermediate will bind to the next enzyme with a high probability instead of leaving the complex. Such *metabolite compartmentation,* often called the *"channelling effect"* is found, for example, in tryptophan synthase (cf. Chapter 5), where indole normally does not leave the complex, but can readily enter it. Let us now consider the consequences and potential advantages of this compartmentation phenomenon.

Protection of metabolite and/or the cell

Some intermediary metabolites are labile because of their susceptibility to hydrolysis or enzymatic attack. For example, most phosphorylated intermediates are attacked by phosphatases. Compartmentation excludes harmful enzymes and, by shortening the transit time (cf. below) of metabolite, lessens the probability of hydrolysis or other routes of decay. On the other hand, some metabolites, e.g. aldehydes, are toxic due to their reactivity toward protein side chains, which makes their segregation from other cellular elements desirable.

Decrease of transient time

The transient time, τ, is a parameter that characterizes the speed of attaining a (new) steady state in a sequence of reactions. In the simplest case there are two consecutive reactions catalysed by enzymes E_1 and E_2 as follows:

$$S \xrightarrow{E_1} I \xrightarrow{E_2} P \tag{4.1}$$

where S is the initial substrate (for enzyme E_1), I is the intermediate metabolite (product of E_1 and substrate of E_2), and P is the final product (of E_2). For ease of kinetic treatment let us assume that both reactions are irreversible and that the concentration of S and hence the velocity of the first reaction, v_1 (the input), is constant and much smaller than the maximum velocity of the second reaction (V_2). Then $[I]$

will be very low and hence the inequality $I \ll K_{m_2}$ is likely to hold. Consequently, the rate equation of the second enzyme simplifies to

$$v_2 = \frac{d[P]}{dt} = \frac{V_2}{K_{m_2}} [I]. \tag{4.2}$$

Apparently, V_2/K_{m_2} is a first order rate constant. The rate of consumption of substrate S is then described by

$$v_1 = \frac{d[I]}{dt} + \frac{V_2}{K_{m_2}} [I], \tag{4.3}$$

and the time course of the reaction will be given by the solution of equation (4.3), with boundary conditions $[I] = 0$, $[P] = 0$ at $t = 0$:

$$[I] = v_1 \tau (1 - e^{-t/\tau}) \tag{4.4}$$

$$[P] = v_1(t + \tau e^{-t/\tau} - \tau), \tag{4.5}$$

where $\tau = K_{m_2}/V_2$ is the *transient time*. An operational definition of transient time can be derived as follows. When $t = \tau$, then

$$[I] = v_1 \tau \left(1 - \frac{1}{e}\right), \tag{4.6}$$

and when $t \rightarrow \infty$, i.e. the steady state is attained, the concentration of I is

$$[I]_{ss} = v_1 \tau \tag{4.7}$$

So

$$\frac{[I]}{[I]_{ss}} = 1 - \frac{1}{e} \approx 0.63 \tag{4.8}$$

i.e. τ is the time required for I to reach $\sim 63\%$ of its steady-state value. Once the steady state is reached

$$[P] = v_1(t - \tau) \tag{4.9}$$

Equation (4.9) gives a straight line if $[P]$ is plotted against t, intersecting the abscissa at τ and the ordinate at $- [I]_{ss}$ (Fig. 4.1). An important conclusion of the above derivation is that the transient time does not depend on the velocity of the first reaction, only on the kinetic parameters of the second, as pointed out by McClure [799] and Hess and Wurster [537]. On the other hand, the rate in the final steady state will be determined by the first reaction only, i.e. it will equal v_1. This follows, of course, from the original condition that in the two-step irreversible reaction the first step is rate-limiting.

The above derivation refers to two enzymes freely in solution. If E_1 and E_2 are physically associated so that I is directly transferred (channelled) from one active

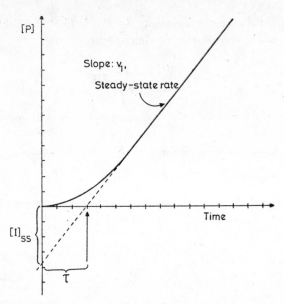

Fig. 4.1. Illustration of transient time τ in an enzymatic reaction sequence. The concentration of the final product of the sequence, *P*, is plotted against time. The intercept with the abscissa gives τ, whereas the intercept with the ordinate yields $-[I]_{ss}$, the concentration of intermediate metabolite, of a two-step reaction, in the steady state. For other details see the text.

site to the other, then τ will be shortened. The reason for this is that *I* does not have to accumulate to secure a v_2 high enough to equal v_1 in the steady state. This point is vividly brought out by Gaertner's [415] illustration (Fig. 4.2).

The prevention of accumulation of intermediate(s) may be of particular economic significance in pathways where the rate-limiting reaction is not the first committed step in the sequence [1256]. Formally, since $v_2 = (V_2/K_{m_2})[I]$, the lowering of $[I]_{ss}$ will be accompanied by a proportionate decrease in K_{m_2}, if we disregard that enzyme association may affect V_2. In this case, then, metabolite channelling manifests itself in a decrease of the apparent Michaelis constant of the second enzyme in a two-enzyme system.

Metabolic pathways usually consist of more than two consecutive enzymatic steps. It can be shown [329, 525] that in a reaction sequence of *n* steps, with the first one being rate-limiting, product formation is described by equation (4.10):

$$[P] = v_1 \left(t + \sum_{i=2}^{n} C_i e^{-t/\tau_i} - \sum_{i=2}^{n} \tau_i \right) \tag{4.10}$$

where

$$C_i = \tau_i^n \prod_{j=2}^{n} \frac{1}{(\tau_i - \tau_j)}, \quad j \neq i \tag{4.11}$$

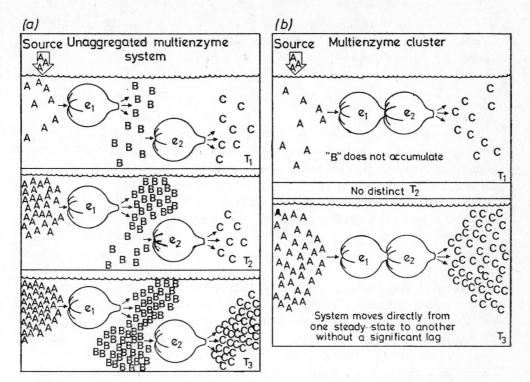

Fig. 4.2. Comparison of the behaviour of a two-enzyme system (e_1, e_2) in the unclustered (*a*) and clustered (*b*) states. The reaction sequence $A \xrightarrow{e_1} B \xrightarrow{e_2} C$ is taken to be irreversible. In (*a*) at time T_1 the system is in steady state at low input of A. At T_2 input of A is increased, but it is not balanced yet by an equivalent output of C, because B has to accumulate for e_2 to work at the appropriate velocity. At T_3 the new steady state is reached, input equals output, the concentration of B is high. In (*b*) the increase of A input is instantaneously followed by an equivalent output of C, because B is compartmented; consequently, there is no distinct transient period. (From Gaertner [415].)

In the steady state ($t \to \infty$):

$$[P] = v_i \left(t - \sum_{i=2}^{n} \tau_i \right) \qquad (4.12)$$

i.e. the transient time for a sequence of enzymes is the sum of the transients of the individual enzymes. It has been calculated [415] that for a sequence of ten enzymes with "average" kinetic parameters the overall transient time may be as large as one hour. An enzyme in the sequence with unusually high transient time can control the switching on of the pathway and thus act as *time-limiting* enzyme [329]. Obviously, the physical juxtaposition of enzymes in a sequence can markedly accelerate the response of the overall reaction by reducing transient time. For more details about transient time, its derivation and limitations in complex systems, the reader may consult a number of papers: on two-enzyme systems [48, 460, 537, 799] three-enzyme systems [51, 537, 799] and in general [329, 1302, 1303].

Decrease of transit time

In an enzymatic reaction sequence transit time is defined as the time required for a metabolite to reach the next enzyme in the pathway. In essence, it is the time of diffusion from one active site to the other. Transit time is expected to be in the order of r_E^2/D_s, where r_E is the average separation distance of enzyme molecules and D_s is the diffusion coefficient of the substrate [1303]. It has been reckoned that even in random, physically unorganized, multienzyme systems the diffusion of metabolites is fast enough not to become rate-limiting [534, 1288]. If so, possible shortening of the transit time by enzyme juxtaposition has no effect on the flux through the pathway.

However, it has also been argued [577] that in the highly viscous intracellular medium diffusion of metabolites may after all be slow enough that changes in transit time will affect the flux. It is to be noted that while the transient time only influences the rate of attaining the steady state but not the final steady-state velocity, transit time may alter the latter. Thus structural enzyme organization can increase the overall velocity of a pathway by reducing transit time. It should be emphasized that this rate enhancement is achieved by the mere proximity of active sites and does not involve any specific molecular interaction increasing individual V_{max}s. Note that in multienzyme clusters where metabolite channelling occurs, the decrease of transient time is, in effect, due to the diminution of transit time to almost zero.

ACTIVATION OF PHYSIOLOGICAL AND SUPPRESSION OF NON-PHYSIOLOGICAL REACTIONS

When two or more enzymes form a cluster, the immediate vicinity (microenvironment) of contact surface residues is inevitably changed. Through these interactions the associating partners may influence each other in a variety of ways. Admittedly, association need not be accompanied by a major conformational change in either of two proteins, especially if their heterologous contact surfaces are "perfectly" complementary. However, the alignment of active site residues in enzymes is extremely precise and therefore apparently small structural alterations, imposed by heterologous constraints, may exercise a profound effect on catalytic efficiency. Thus enzyme–enzyme interactions (associations) may bring about *catalytic facilitation* by modifying the intrinsic kinetic parameters (V_{max}, K_m) of the associating partners. This means that while the separate, unaggregated enzymes may be poor catalysts, association improves them by lowering the K_m and/or increasing the V_{max}. Tryptophan synthase is an example of this, and will be discussed in some detail in Chapter 5. Unfortunately, because of lack of knowledge about three-dimensional structure, in none of the observed cases is the stereochemistry of such enzyme–enzyme interactions clear.

On the other hand, the kinetic and thermodynamic basis of catalytic facilitation in general has been thoroughly analysed by Welch [1303, 1304] for multienzyme systems. Welch set out from the formulation of rate equations for polymeric enzymes by Ricard et al. [1000], in which subunit interactions (oligomer symmetry) were taken into account. According to Ricard et al. [1000] any rate constant can be expressed in terms of an *intrinsic constant* and several *interaction coefficients*:

$$k = k^* \prod_{r=1}^{n} \alpha_r^{i_r} \qquad (4.13)$$

where the intrinsic constant has the form

$$k^* = \frac{k_B T}{h} \exp\left(-\frac{\Delta G_i^{\pm}}{k_B T}\right) \qquad (4.14)$$

where k_B and h are the Boltzmann and Planck constants, respectively, T is the absolute temperature and $\Delta G^{\pm}$ is the intrinsic free energy of activation (per molecule).

The dimensionless interaction coefficient, α_r, is given as

$$\alpha_r = \exp\left(-\frac{G_r^{\pm}}{k_B T}\right) \qquad (4.15)$$

and there is a coefficient for each of the various types of interactions. (The exponent i_r is an integer giving the number of interactions of type r.)

Welch [1303] extended the above derivation by claiming that each enzyme of a clustered system will present an "α-spectrum", which is defined as the sum total of all effects (including protein–protein, protein–matrix and enzyme–ligand interactions) on the rate of the given enzyme reaction (Fig. 4.3). Clearly, the concept of α-spectrum is one way of expressing that individual enzymes in the living cell are bound to be subject to a great number of influences that affect enzyme function, but does not tell us anything about the mode of catalytic facilitation. To this end, Welch borrowed the *molecular enzyme kinetic model* (MEKM) developed by Damjanovich and Somogyi [278, 1139, 1141] for individual enzymes and applied it to multienzyme clusters.

The MEKM intends to give a deeper physical meaning to the phenomenological rate constants, e.g. in the simple scheme:

$$E + S \underset{k_{-1}}{\overset{k_{+1}}{\rightleftharpoons}} ES \overset{k_2}{\rightarrow} E + P. \qquad (4.16)$$

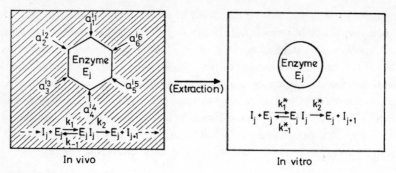

In vitro

Fig. 4.3. Schematic representation of the "α-spectrum" concept. Under conditions *in vivo* the enzyme is influenced by a host of interactions affecting its rate parameters. These influences are abolished by extraction and escape detection in conventional *in vitro* experiments. In the given example enzyme E_j catalyses the reaction $I_j \rightarrow I_{j+1}$. The rate parameters k_s and k_s^* ($s = -1, 1, 2$) for the *in vivo* and in *vitro* conditions, respectively, are related by the expression:

$$k_s = k_s^* \prod_{r=1}^{n} \alpha_r^{i_r} \quad (n=6 \text{ in this example})$$

where i_r is the number of interactions of type r. (From Welch [1303].)

It relies essentially on collision kinetics and certain specific properties of proteins. For example, k_{+1} is given by MEKM as follows [1140]:

$$k_{+1} = \frac{3}{2} \frac{1}{r_s^2} D_s V \left[\frac{1}{2} + \frac{r_0}{\lambda \sqrt{\nu}} \right]^2 \exp\left(-\frac{E_q}{k_B T}\right) \tag{4.17}$$

where r_s = molecular radius of substrate,
   r_0 = radius of a circular area (centred on the binding point) within which *ES* complex formation will be effected by intermolecular forces,
   D_s = diffusion constant of substrate,
   V = recognition volume, within which the substrate will be bound to the enzyme; it is reckoned to be slightly larger than the substrate [1139],
   λ = the distance between two points of a lattice over which the substrate molecule "jumps",
   ν = average number of "jumps" made by the free substrate molecule during its mean lifetime of stay in the recognition volume,
   E_q = threshold energy required for the substrate in the recognition volume to form *ES*.

Then if t' is the mean lifetime of the *ES* complex and L is the probability of transition $ES \rightarrow E + P$,

$$k_2 = \frac{L}{t'} \quad \text{and} \quad k_{-1} = \frac{1-L}{t'}. \tag{4.18}$$

Another major assumption made in MEKM is that the energy required for *ES* formation and *EP* dissociation is gathered from collisions with other (mainly solvent)

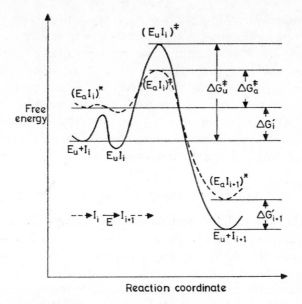

Fig. 4.4. Potential effects of enzyme clustering on the free energy profile of an enzyme reaction. Catalytic facilitation may entail a "smoothing" effect on the overall profile by lowering energy barriers and/or elevating energy valleys. E_a (broken line) and E_u (continuous line) denote clustered and unclustered enzymes (and their free energy profiles), respectively. The reaction catalysed is $I_i \rightarrow I_{i+1}$. The initial and final free energy levels, their difference, and the free energies of activation for the two states, and their difference, are indicated. (From Welch [1303].)

molecules by means of an "energy funnel" conducted by the protein structure as a system of coupled oscillators [278, 1141].

Welch [1303] has pointed out that by applying the formalism of MEKM to multienzyme clusters, catalytic facilitation becomes conceivable. Part of the effect is due to phenomena related to metabolite compartmentation, while others act toward decreasing $\Delta G^{\ddagger}$. Aggregation may also elevate the energy "valleys", smoothing the overall energy profile (Fig. 4.4). For the detailed discussion of these conclusions the reader should consult the original paper [1303] and a recent expansion of it [1304]. It should be clear, however, that MEKM and its extension to multienzyme clusters are theories not yet solidified by experimental evidence. Nevertheless, they have the merit of being inspiring working hypotheses and also of demonstrating that the promotion of reactions in organized enzyme systems can be traced back to physical principles.

Many enzymes are capable of catalysing chemical reactions other than the physiological one. A very common example is the esterolytic activity of proteinases and dehydrogenases. In connection with dehydrogenases Polgár [948] has named such activites *hypocatalytic reactions,* on the grounds that here the enzyme does not utilize its full catalytic sophistication: acyl-enzyme formation with a "hypocatalytic substrate", for example *p*-nitrophenyl acetate, and its subsequent hydrolysis alternate without dehydrogenation. Most of these hypocatalytic reactions are useful only in studying enzyme mechanisms, but are irrelevant *in vivo*, since the corresponding substrates are absent. However, there are certain hypocatalytic reactions that might occur in the living cell. For example, the β_2 enzyme of tryptophan synthase (in detail see Chapter 5) is capable of serine deamination, an undesirable hydrolytic activity destroying the amino acid, but this ability is suppressed when β_2 is bound to the α enzyme. The multienzyme aggregation may serve not only to activate normal metabolic reactions but also to prevent deleterious ones. Again, the stereochemical details of these inhibitory interactions are still unknown.

COORDINATE REGULATORY EFFECTS

As discussed in Chapter 3, the regulation of enzyme activity is, as a rule, exercised through some allosteric mechanism in an oligomeric protein. In a way that we just begin to understand with certain enzymes, the protomers within the oligomer influence each other in respect of turning on or off catalytic activity. In effect, the protomers react in a coordinated manner, whether concertedly or sequentially or both, to the structural impact of effector ligands. It is obvious that the coordinated response requires the physical association of the protomers, even if this association is readily reversible.

This mechanistic principle may hold not only for true oligomers but also for various forms of multienzyme clusters. Messages, i.e. conformational alterations, can be transmitted not only through isologous or heterologous contact domains of identical subunits but through the necessarily heterologous contacts of different proteins as well. We have already seen examples among complex enzymes, e.g. aspartate carbamoyl transferase, where the modifying influence can propagate either way: from the regulatory subunit to the catalytic one and vice versa. Coordinate regulation in a multienzyme cluster means that a single effector ligand affects more than one enzyme, altering their kinetic parameters (V_{max}, K_m) (Fig. 4.5). The economy of such an arrangement is appealing, though the physiological significance is not always evident. Let us take a sequence of enzyme reactions catalysed by the cluster (E_1. $E_2 \ldots E_n$):

$$A \overset{E_1}{\rightarrow} B \overset{E_2}{\rightarrow} C \overset{E_3}{\rightarrow} D \ldots \overset{E_n}{\rightarrow} Z.$$

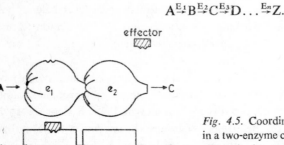

Fig. 4.5. Coordinate conformational change induced in a two-enzyme cluster by an effector acting at a single allosteric site on the first enzyme (e_1). (From Gaertner [415].)

The coordinate *inhibition* of all activities by a ligand acting at enzyme E_1 seems an unnecessary over-insurance of switching the pathway off and saving metabolites, for intermediates B, C, D, etc., are unlikely to accumulate owing to the very probable compartmentation within the cluster (cf. Fig. 4.2). The coordinate *activation* in a sequence has the trivial advantage of allowing metabolite flow at all, unless the rate-limiting step of the sequence can be modulated this way. In the case of the *arom* conjugate (cf. Chapter 5) it has been shown that four of the five enzyme activities of the conjugate are activated by the first substrate. It was proposed that the advantage of such an arrangement would be the relatively greater sensitivity of the *non-activated* enzyme to regulatory effects [1306]. While such a role is conceivable, the idea shares the weakness of most appeals to biological significance about organizational schemes of metabolic networks: logical appearance does not imply physiological reality. Clearly, validity *in vivo* remains to be established.

It is another case when the metabolic pathway is branching:

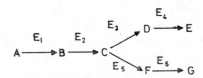

If the synthesis of end-products E and G are to be regulated separately, there must be feedback inhibition by these products on enzymes E_3 and E_5, respectively. Furthermore, intermediates B and C have to provide material for E and G when both limbs are working. However, if only one of the branch-paths is running, economy would dictate a diminished flux through B and C. One way of achieving this is to have isoenzymes for E_1; one isoenzyme is then inhibited by end-product E, the other by G. Now if the appropriate isoenzyme of E_1 is associated with E_3, whereas the other with E_5, then coordinate regulation is feasible in these two-membered clusters through a single allosteric site in either of the associated parties. Such an arrangement seems to be operative in various microorganisms [575, 1238].

Coordinate regulatory effects may include phenomena not strictly related to the control of enzyme activity but pertaining to the control of enzyme level, i.e. enzyme turnover. With the *arom conjugate* it has been demonstrated that the first substrate endows the whole conjugate with enhanced resistance to proteolysis [1272]. Evidently, as long as the first substrate is present, the decay of any activities of the sequence is undesirable. The regulation of enzyme degradation is a complex phenomenon in both microorganisms [566] and animal cells [1080]. The turnover of cellular proteins provides a means to redistribute matter (*viz.* amino acids) to meet the current needs of the cell. In the above example one may conjecture that as long as the first substrate is available, the whole sequence of enzymes will be protected, thereby prolonging the lifetime of enzyme cluster.

MULTIENZYME COMPLEXES AND CONJUGATES

"In nature's infinite book of secrecy
A little can I read."

W. Shakespeare: Antony
and Cleopatra, I, ii, 9.

The structures to be discussed here are the first, in the hierarchy of enzyme organization (cf. Table 1.1), that belong to the supramolecular level, inasmuch as they consist of *two or more enzymes* with different catalytic activities physically clustered either non-covalently (complexes) or covalently (conjugates). Admittedly, multienzyme conjugates, being single polypeptide chains (or their oligomers), may not belong to the supramolecular class in the strict sense of the term. However, we believe that logically they should come under this heading, all the more as multienzyme conjugates have probably evolved from multienzyme complexes by gene fusion.

Multienzyme complexes and conjugates have been widely studied for many years and several of them have become textbook matter. They have been repeatedly reviewed over the past ten years [161, 449, 819, 990, 1303]. Nevertheless, we shall introduce some very typical ones, because without them the overall picture this book is intended to give would be rather incomplete.

STABLE MULTIENZYME COMPLEXES

The adjective "stable" means that these multienzyme complexes can be readily isolated as complexes, i.e. as a rule, they do not fall apart into their constituents during preparation. This operative definition is made to distinguish them from "loose" multienzyme complexes which, in contrast, cannot usually be isolated and the existence of which can be demonstrated only by carefully designed experiments with their constituents. Structures we classify as stable complexes may, however, be dissociable, but their dissociation constants are sufficiently low that these complexes exist in the associated state in the living cell, and are at least partly associated under conditions of protein purification.

TRYPTOPHAN SYNTHASE

The last step of tryptophan biosynthesis in bacteria, yeast, moulds and plants is catalysed by tryptophan synthase (L-serine hydro-lyase [adding indole], EC 4.2.1.20). The enzyme from *E. coli* has been studied in great detail and will be the main object of our attention. It is one of the simplest multienzyme complexes and consists of two enzymes named α and β. The complex contains two copies of each, subject to the dissociation equilibrium

$$\alpha_2\beta_2 \rightleftharpoons 2\alpha + \beta_2$$

In the presence of substrates, L-serine and indoleglycerol-P, and the cofactor pyridoxal-P, the dissociation constant for removal of one α entity from the complex $\alpha_2\beta_2$ or $\alpha\beta_2$ is 4×10^{-10} M, whereas in their absence it is about three orders of magnitude higher [266, 353]. Thus under physiological conditions the structure is in the associated state. Nevertheless, the α and β_2 enzymes can be readily separated from each other, first accomplished in 1958 by the then novel technique of DEAE-cellulose chromatography [263]. The reactions catalysed by tryptophan synthase and its subcomplexes are summarized in Table 5.1. Reaction 1 occurs physiologically, whereas

TABLE 5.1.

Reactions catalysed by tryptophan synthase from E. coli *and its component enzymes*

No.	Reaction — Description	Type	Enzyme species catalysing the reaction
1.	Indole-3-glycerol-P + L-serine $\xrightarrow{PLP}$ L-tryptophan + + D-glyceraldehyde-3-P + H_2O	Physiological	$\alpha_2\beta_2$
2.	Indole-3-glycerol-P $\rightleftharpoons$ indole + D-glyceraldehyde-3-P	Partial	α; $\alpha_2\beta_2$
3.	Indole + L-serine $\xrightarrow{PLP}$ L-tryptophan + H_2O		β_2; $\alpha_2\beta_2$
4.	L-Serine $\xrightarrow{PLP}$ pyruvate + NH_3	Hypocatalytic	β_2
5.	2-Mercaptoethanol + L-serine $\xrightarrow{PLP}$ S-hydroxyethyl-L-cysteine + + H_2O		β_2; $\alpha_2\beta_2$
6.	2-Mercaptoethanol + L-serine + PLP $\rightleftharpoons$ S-pyruvyl-mercaptoethanol + PMP + H_2O		β_2

PLP = pyridoxal phosphate; PMP = pyridoxamine phosphate. (After Miles [819].)

reactions 2 and 3 can be regarded as partial reactions, their sum yielding reaction 1. Reactions 4–6 are hypocatalytic activities of the various species and are irrelevant *in vivo*. Before having a closer look at these various catalytic functions of tryptophan synthase from the point of view of enzyme–enzyme interaction, a brief description of the structural properties of the *E. coli* enzyme complex is given. The reader interested in more details may consult an earlier review, comprising also a historical survey, by Yanofsky and Crawford [1340] or the recent treatise by Miles [819].

Main structural features of tryptophan synthase from E. coli

The α-polypeptide has a molecular weight of 29,000 and consists of 268 amino acid residues of known sequence (cf. [261]). The β_2 enzyme can be reversibly dissociated into two β polypeptide chains by urea [518]; the β-chain has $M_r = 44,200$. The determination of primary structure of β is near completion [262]. Heat inactivation studies indicate strong interdomain interactions in β_2 [1348], indeed, the β–β interface seems to have been conserved during evolution [128]. Each β chain binds one pyridoxal-P through a Schiff-base linkage [384]. The prosthetic group markedly increases the solubility of the β_2 enzyme [3] and it is thought to participate in the catalytic mechanism [1303]. The environment of bound pyridoxal-P is pronouncedly altered on the formation of $\alpha_2\beta_2$ complex, as shown by CD studies [820]. As mentioned above, the dissociation constant of the complex is substantially lowered by the substrates. X-ray diffraction work on the complex and its constituents has so far been hampered by practical difficulties (cf. [1303]). Hydrodynamic measurements [454] indicate that the molecular asymmetry increases in the order $\alpha < \beta_2 < \alpha\beta_2 < \alpha_2\beta_2$

Functional consequences of $\alpha_2\beta_2$ complex formation

Mutual activation of components

Both α and β_2 are enzymes in their own right, as they catalyse the reactions shown in Table 5.1. However, whereas they are rather poor catalysts by themselves, their catalytic efficiency and affinity for the substrates are increased one to two orders of magnitude when complexed in $\alpha_2\beta_2$ [516, 1322]. Kirschner and coworkers [657, 1295] suggested that this activating effect was the result of mutual stabilization of active conformations. Indeed, protein isomerization of both the α [658] and the β_2 [352] component has been detected by fast-reaction techniques and was influenced by complex formation [353]. The binding of two α units by β_2 is strongly cooperative [50]. Further evidence for intermolecular constraint arising from α–β association came from ligand-binding and steady-state kinetic studies [1295]. The binding pattern of pyridoxal-P to β_2 changes from positive cooperativity (Hill coefficient, $n = 1.7$) to non-cooperativity upon $\alpha_2\beta_2$ formation [49, 1240]. The cooperative pyridoxal-P binding of apo-β_2 seems to obey the concerted allosteric model [834]. Recent calorimetric studies revealed a negative enthalpy change on $\alpha_2\beta_2$ complex formation, which was twice as large with β_2 containing the prosthetic group pyridoxal-P than with apo-β_2 [1317]. The above data and other similar findings (cf. [1303]) are all indicative of mutual structural–functional modulation by the two enzymes, although they shed no light on the actual mechanism of activation.

Suppression of serine deaminase activity

As seen in Table 5.1, the β_2 enzyme can catalyse the oxidative deamination of serine. From the organic chemical point of view this is a β-elimination reaction and involves Schiff-base formation between serine-NH_2 and pyridoxal-P [692], a general feature of pyridoxal-P assisted catalysis. This reaction is physiologically undesirable as it degrades L-serine, a precursor of tryptophan biosynthesis. The formation of $\alpha_2\beta_2$ complex prevents L-serine deamination through an as yet ill-understood mechanism (cf. below). The observations mentioned under <i>Mutual activation of components</i> are also pertinent here as symptoms of interaction. One may add that there is experimental evidence for communication between the enzymes: a positive heterotropic interaction has been shown between L-serine binding to β_2 and indole-propanol-P, a substrate analogue, binding to α [657], and the binding of this analogue to α changes the spectrum of pyridoxal-P anchored in β_2 [659].

Indole compartmentation (channelling)

An outstanding feature of the long recognized tryptophan synthase reaction (number 1 in Table 5.1) (cf. [1340]) is that indole is not liberated as an intermediate. Apparently, indole is channelled from the α enzyme to β_2 and does not mix with the bulk medium. The mechanism of this compartmentation effect has been a source of disagreement between authors and its final settlement must probably await the emergence of a detailed three-dimensional picture of the enzyme complex and its various transitions. In fact, the same can be said about phenomena described in the preceding two sections, all the more so, since they are certainly interrelated. The conflicting explanations are (a) diffusion of free indole between active sites; (b) concerted transfer in a composite active site [264]; and (c) covalent attachment throughout, when the condensation of indoleglycerol-P with the serine-Schiff-base moiety precedes the aldol cleavage that liberates D-glyceraldehyde-3-phosphate [813]. In cases (b) and (c) free indole does not actually occur as an intermediate. Whichever mechanism will eventually prove valid, it has to include some basic features as follows: (i) there should be a binding site for indole in both α and β enzymes, as witnessed by the individual catalytic reactions (Table 5.1); (ii) the niche formed by the two active sites and where indole is accommodated should be open to the bulk medium. In fact, there seems now to be a consensus for a two-site model of tryptophan synthases, first set forth by DeMoss [301] for the <i>Neurospora crassa</i> enzyme and confirmed kinetically for both the <i>Neurospora</i> [790] and <i>E. coli</i> [264] enzymes (Fig. 5.1). Isotope tracer experiments with <i>Neurospora</i> enzyme [790] showed that under specially designed

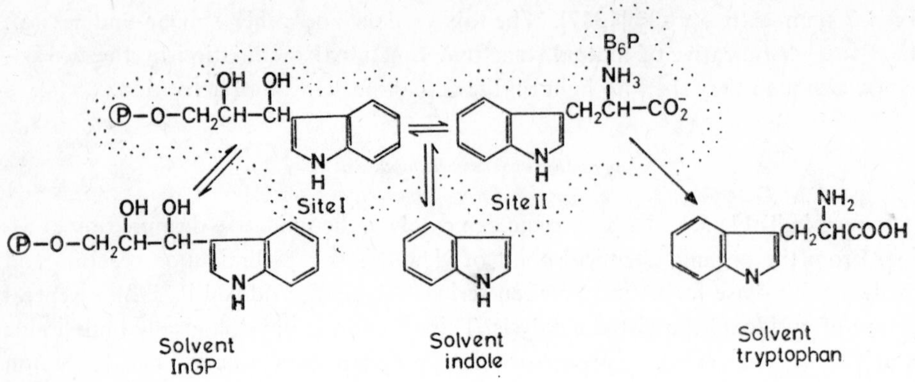

Fig. 5.1. Two-site model for the active site of tryptophan synthase from *Neurospora crassa*. Enzyme-bound indole normally does not communicate with solvent indole, but rather is transferred from site I to site II. (From Matchett [790].)

conditions indole did enter the bulk medium. This finding argues against a covalent attachment of indole and supports a facilitated channelling mechanism. Analogous experiments have not been performed as yet with the *E. coli* enzyme.

In conclusion, the juxtaposition of α and β enzymes results in the compartmentation of indole. Although the underlying mechanism is not clear, we feel that, apart from the possibility of a covalent "handle" [813], the various proposals are only slightly different. After all, indole has to get from site I to site II and it is bordering on semantics whether we call this diffusion or concerted transfer.

Tryptophan synthases from other sources

The enzyme from *Neurospora crassa* is a homodimer and carries the α and β domains on different parts of the same polypeptide chain [790, 791]. As mentioned above, the *Neurospora* enzyme exhibits indole compartmentation just as the *E. coli* enzyme. The other consequences of enzyme juxtaposition cannot be tested with the *Neurospora* enzyme, as there is no mild way to abolish this proximity. The structure of *Neurospora* tryptophan synthase could serve as an example for the concept of Bonner *et al.* [111] according to which physiologically advantageous multienzyme complexes may be converted to multienzyme conjugates by gene fusion.

Yeast, *Saccharomyces cerevisiae,* tryptophan synthase is again a homodimer of polypeptide chain molecular weight 76,000 [307]. An interesting demonstration *in vivo* of the significance of indole channelling has been given in yeast [325, 772]. Two mutant strains were isolated, one of them defective in reaction 2, the other in reaction 3. Neither of them was capable of synthesizing tryptophan. However, when combined in a diploid tryptophan synthesis was resumed, but the cells grew very slowly excreting large amounts of indole into the medium. Indeed, there was a lag-time in growth for the diploid cells during which indole was piled up. The lag-phase could be abolished by adding exogenous indole. It appears that in the diploid organism reactions 2 and 3 operated without indole channelling, instead of reaction 1, because structural complementation was obviated by the coexistence of both α and β enzymes on the same polypeptide chain.

THE α-KETOACID DEHYDROGENASE MULTIENZYME COMPLEXES

The two protagonists of the class are the pyruvate dehydrogenase complex (PDC) and the α-ketoglutarate dehydrogenase complex. The former is the link between gycolysis and the Krebs cycle, whereas the latter is part of the Krebs cycle enzyme system. These giant molecules have been studied in considerable detail over a long period of time and hence a number of excellent reviews are available about them [929, 988, 989, 990]. They are the archetypes of textbook multienzyme complexes, which is attributable to their geometrically appealing, unique subunit arrangement and to their composite active sites where the covalently attached substrate undergoes consecutive transformations by each of three component enzymes. In the following, a description will be given of PDC from *E. coli*, where a novel type of connection between active sites has recently been discovered. The differences of mammalian PDCs from the *E. coli* complex, and some features of the α-ketoglutarate dehydrogenase complex will also be dealt with. We only mention that there are separate α-ketoacid dehydrogenases for branched chain 2-oxo-acids both in mammalian tissues [280, 911, 980] and microorganisms [1009, 1033].

Pyruvate dehydrogenase complex from E. coli

The reaction scheme of PDCs from all sources is given in Fig. 5.2. The net, irreversible reaction—the oxidative decarboxylation of pyruvate—is as follows:

$$Pyruvate + NAD^+ + CoA\text{-}SH \rightarrow Acetyl\text{-}S\text{-}CoA + NADH + H^+ + CO_2$$

This overall reaction is carried out by three different enzymes: pyruvate decarboxylase (EC 1.2.4.1), lipoate acetyltransferase (EC 2.3.1.12) and lipoamide dehydrogenase (EC 1.6.4.3). Thiamine-PP is the prosthetic group of pyruvate decarboxylase, lipoate is bound via a peptide linkage to the ε-NH_2-group of a lysine in acetyltransferase, whereas the third enzyme, lipoamide dehydrogenase, uses FAD and a reversibly reduced $-S\text{-}S-$ as prosthetic group and NAD as coenzyme.

The three-component enzymes are represented in the native complex in several copies, but there is no unanimity in the literature about the precise stoichiometry of chains. There is consensus only in that twenty-four chains of E_2 (acetyl transferase)

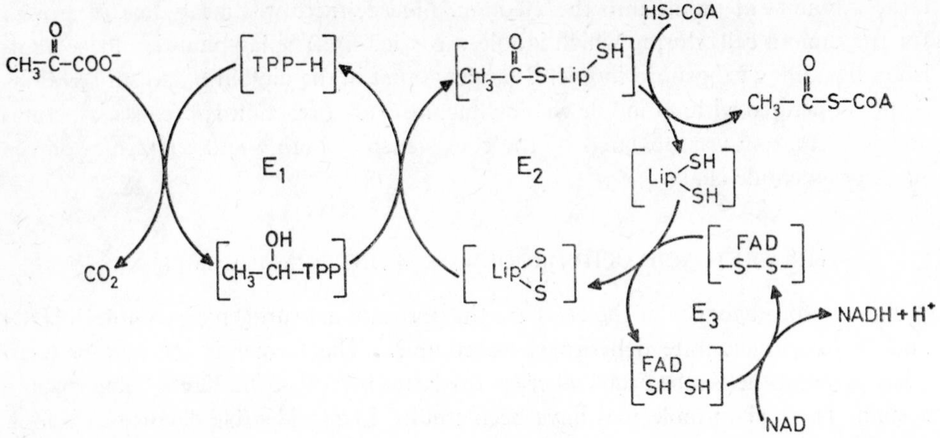

Fig. 5.2. Reaction mechanism of pyruvate dehydrogenase complex. The partial reactions are catalysed by the enzymes indicated as follows: E_1, pyruvate decarboxylase; E_2, lipoate acetyltransferase; E_3, lipoamide dehydrogenase. Symbols: TPP-H, thiamine pyrophosphate; $Lip\langle\begin{smallmatrix}S\\S\end{smallmatrix}$, oxidized lipoate. The species in brackets remain bound to the complex throughout. (After Perham et al. [930].)

form the core of the complex, ordered first as dimers which, in turn, form a cube-like structure of octahedral symmetry [305] (cf. Fig. 3.4h). The preparative molecular weight of the native complex is 4.8×10^6 [17, 989]. The ratio of the three chains $E_1 : E_2 : E_3$ has been claimed to be $2 : 2 : 1$ by Reed and coworkers [17, 992] or $>1 : 1 : 1$ by others [54, 1274] with a limiting value of $2 : 1 : 1$ [931]. A recent re-examination by Perham's group [501] turned out a non-integer ratio of $1.5 : 1 : 1$. The reason for this discrepancy is not yet known.

Another issue has been the number of lipoyl residues per E_2 chain. Early measurements indicated one residue for each E_2 chain [336], which seemed quite logical considering the generally accepted mechanism, found in textbooks, according to which a lipoyl residue attached to a lysine acts as a "pendulum" moving to and fro between E_1 and E_3. However, it was observed later by specific modification of lipoic groups that there were two of them for every E_2 chain ($M_r = 80,000$) [285]. Moreover, quantitation of the lipoic acid/FAD ratio yielded a value of 4 and the unexpected result that only half of the lipoyl moieties were available for reoxidation by E_3-FAD-NAD [581, 1146]. Although ^{35}S incorporation data suggested the existence of three or even more lipoyl residues per acetyltransferase chain [501], more recent investigations [69, 283] seem to settle this value at 2, both lipoic acids being available to oxidation by NAD.

The finding that more than one lipoyl group is on each acetyltransferase chain led the workers in the field to the elucidation of a novel catalytic setup in a giant cluster like PDC. This is active site coupling, via the multiplicity of lipoyl residues, first proposed by Perham and coworkers [53, 284] and subsequently confirmed in other

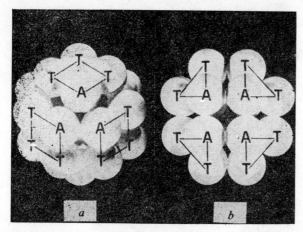

Fig. 5.3. Two models for the mechanism of acetylation of twelve E_2 chains by a single E_1 dimer in the pyruvate dehydrogenase complex. The transacetylase core is viewed along the three-fold (*a*) and four-fold (*b*) symmetry axes. The subunits acetylated through direct contact with an E_1 dimer are marked with *A*, whereas those acetylated via transacylation bear letter *T* and are connected to their "servicing" *A* subunit. The total number of E_2 chains acetylated in either model is twelve. (From Bates *et al.* [53].)

laboratories [7, 18, 243, 396]; it was also found with the α-ketoglutarate dehydro-genase complex from *E. coli* [243]. Bates *et al.* [53] measured the extent of acetylation with 2-[^{14}C]pyruvate of the E_2 component in a series of partly reassembled PDCs and plotted the extent of acetylation of the E_2 core (twenty-four polypeptide chains) against the $E_1 : E_2$ ratio of the subcomplex. They found that the extent of core-acetylation initially rose much steeper than the $E_1 : E_2$ ratio, which meant that more than one E_2 chain was labelled by a given E_1 dimer. (The E_1 dimer was considered because pyruvate decarboxylase is probably dimeric in the free state [336]). On the basis of simple considerations they suggested that any E_1 dimer would acetylate twelve E_2 chains (Fig. 5.3). More direct evidence for active site coupling was obtained by Stanley *et al.* [1165] who used a specific inhibitor for E_1, instead of partially reas-sembled complexes [53], and measured the extent of acetylation of E_2 as a function of residual overall complex activity (Fig. 5.4). The active site coupling in E_2 is indi-cated by the markedly slower decrease in the extent of acetylation than in overall complex activity.

Irrespective of the actual pattern of transacetylation over the E_2 core, the necessity that lipoyl residues must cooperate somehow could be derived from the large distance between catalytic sites that could not have been spanned by a single residue [1096, 1097]. The lipoyl–lysyl swinging arm is 1.4 nm long and the complex is more than 30 nm in diameter [327]. Fluorescence energy transfer measurements indicate that the lipoyl residues are close to each other [19]. They are apparently anchored in elongated cigar-shaped domains sticking out from the surface of the rest of the E_2 core (Fig. 5.5) as discerned recently by limited proteolysis and electron microscopy [93]. Pulsed

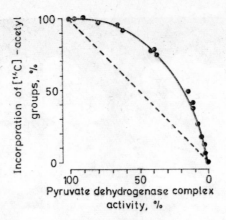

Fig. 5.4. Acetylation with [2-^{14}C] pyruvate of the E_2 component of *E. coli* PDC inhibited to different extents at the E_1 component. Increasing amounts of thiamine thio-thiazolone pyrophosphate, an inhibitor of E_1, were added to the complex, the overall activity was measured, then acetylation was run in the presence of [2-^{14}C] pyruvate for 20 sec. Incorporation of acetyl groups was determined by radioactivity. In the absence of active site coupling within E_2, one would expect the experimental points to lie on the broken line. (From Stanley *et al.* [1165].)

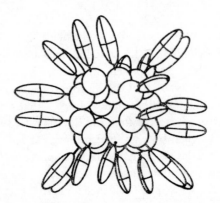

Fig. 5.5. Model of the acetyltransferase core of PDC from *E. coli*, as deduced from electron microscopic images. The spheres represent the subunit binding domains and the ellipsoids represent the extended lipoyl domains. The latter can be removed by tryptic digestion. (From Bleile *et al.* [93].)

quenched-flow experiments evidenced that the transacetylations were fast enough not to be rate-limiting in the overall reaction of the complex [281]. An alternative to transacetylation, the exchange of E_1 dimers in the complex, has been ruled out for it is much too slow to account for the rapid spread of acetyl groups [500]. For the sake of comparison, in the α-ketoglutarate dehydrogenase complex from *E. coli* fluorescence energy transfer measurements yielded distances between active sites that could be spanned by a single lipoyl arm [20], yet the transfer of acyl-groups (here trans-succinylation) among lipoyl groups does occur [243]. Proton NMR studies indicate for both the pyruvate dehydrogenase complex [930] and the α-keto-glutarate dehydrogenase complex [932] from *E. coli* that large segments of the poly-peptide chains carrying the lipoyl residues have remarkable conformational motility, which is probably a prerequisite of active site coupling.

The question of lipoyl residues not available to dehydrogenation is open. These residues have been claimed to amount to about one-half of the total [396]. If they really exist, which has been challenged [283], and are off the main catalytic path [6], they may serve another, so far undefined electron transfer function [396].

The mobility of lipoamide dehydrogenase (E_3) inside the PDC has recently been measured by laser-pulse fluorometry [469]. Surprisingly, the rotational correlation time calculated from the polarized decay of flavin fluorescence was in the same range as that of free lipoamide dehydrogenase. Whether this apparently great mobility of E_3 has anything to do with the catalytic mechanism of the complex, remains to be seen.

Pyruvate dehydrogenase complexes from mammalian tissues

PDC has been isolated from a variety of mammalian sources such as bovine kidney and heart [741, 1014], rat and pig heart [250], and pigeon breast muscle [647]. The bo-vine kidney and pig heart complexes have been studied in the greatest detail. In mam-mals and other eukaryotes the ketoacid dehydrogenase complexes are associated with the mitochondria.

As regards their main features, mammalian PDCs are rather alike. They also resemble the *E. coli* complex inasmuch as their three component enzymes are the same. On the other hand, the morphology of mammalian complexes is different: the acetyl-transferase core has been claimed to consist of sixty subunits arranged to give icosa-hedral symmetry [991]. The actual size of the core is difficult to determine because the preparative molecular weights strongly depend on the experimental conditions [687]. Subunit numbers as low as 30 [762] or 24 [1192] have also been reported. However, recent studies by Reed's group on the bovine heart acetyltransferase, using limited tryptic digestion, sedimentation and electron microscopy, support the 60-subunit model arranged according to icosahedral symmetry (point group 532). Similarly to the *E. coli* complex, the domains anchoring the lipoyl residues are well separated from the subunit-binding domains (Fig. 5.6). The number of attached lipoyl residues is controversial [175, 505, 762]; at present, estimates are one or two for each acetyltransferase subunit. The pyruvate decarboxylase component in mam-mals is a complex protein of structure $\alpha_2\beta_2$ [46] and the catalytic unit is thought to be the $\alpha\beta$ heterodimer (cf. [176]). Indeed, one thiamine-PP has been found per $\alpha\beta$ unit [156]. The molecular weight of the whole complex from bovine kidney is around 7×10^6 [989].

As for the functional properties, the interlipoyl acetyl transfer has also been estab-lished in the mammalian complex [175, 176], which here, too, can be due to the great

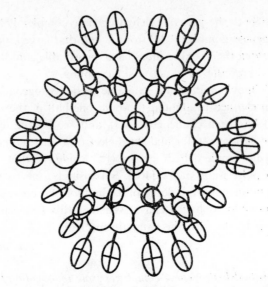

Fig. 5.6. Model of the acetyltransferase core of PDC from bovine heart. Spheres and ellipsoids represent the subunit-binding domains and the lipoyl domains, respectively, as with the *E. coli* complex in Fig. 5.5. The structure has the appearance of a pentagonal dodecahedron with icosahedral, 532, symmetry. The morphological unit is the subunit trimer. (From Bleile *et al.* [92].)

conformational motility of polypeptide segments encompassing the lipoyl–lysine residues [1287]. However, there are distinctive features between microbial and mammalian complexes. In the latter the migration (exchange) of E_1, pyruvate decarboxylase, between complexes is rather fast [175]. Although its rate $(k_{diss} \approx 2 \text{ min}^{-1})$ is too slow to contribute to catalysis, it may play a role in the regulation of activity by enzymatic modifications (cf. below). As a comparison, the propensity for dissociation of the α-ketoglutarate dehydrogenase complex is so great that Severin *et al.* [1087] could interpret the kinetic mechanism of the complex in terms of a multistep dissociation model, where the reverse process (association) is cooperative.

The second and more important attribute of mammalian PDCs is that they are inactivated via phosphorylation by an intrinsic kinase and Mg-ATP, and reactivated by a less strongly bound phosphatase [742]. The kinase is very strongly bound to the acetyltransferase core [174]. During inactivation a particular serine residue on the α-chain of the decarboxylase becomes esterified; a further two serine residues per α-chain can also be phosphorylated but more slowly [643, 1191, 1342]. These other two serines are not as important as the first one in respect of inactivation, yet their phosphorylation is also required for the complete loss of catalytic function [993, 1212]. It has been determined how the various physiological ligands of PDC, such as pyruvate, NADH, acetyl-CoA and ATP, influence the kinase reaction [175, 643]. It has been suggested that the above-mentioned migration of decarboxylase units may provide a mechanism of extensive phosphorylation, and consequential inactivation, inasmuch as the E_1 enzyme can "search" for the rigidly anchored kinase [176].

Another interesting suggestion [174, 1013] is that the kinase attack is elicited by the acetylation of lipoyl residues. It has been conjectured that the effectors of kinase (mainly the substrates of PDC) exert their effect via acetylating the core enzyme. If this idea proves correct, then the multiplicity of lipoyl groups will have a dual role in PDC's function, at least in mammals: a rapid relaying of covalently bound intermediate during catalysis and the regulation of modification by kinase.

Less is known about the control of PDC phosphatase. It has been claimed recently that insulin, which stimulates the enzyme complex, does so by activating phosphatase through a second messenger generated by the plasma membrane [953]. The role of second messenger might be filled by the oligopeptide discovered recently to mimic insulin action by inhibiting protein kinase and activating phosphoprotein phosphatase [716].

As an addition to the complexity of this system, Wieland and coworkers [756, 1314] have purified a protease from rat liver that specifically inactivated the two α-ketoacid dehydrogenase complexes. The phenomenon was not related to phosphorylation. The protease seems to nick the transacylase core at several points, destroying the overall reaction completely but not affecting any of the individual steps.

INTERCONVERTIBLE ENZYMES AND ENZYME CASCADES

In the brief description of mammalian α-ketoacid dehydrogenase complexes, we encountered a very important, rapidly expanding field in enzyme regulation, the realm of interconvertible enzymes. These are enzymes that can exist in two forms distinguished by the presence or absence of a covalently attached group. In fact, interconversion of enzymes forms a subclass of the group of phenomena collectively called *post-translational modification of proteins.* The term indicates that the modification in question, as a rule, the chemical modification of an amino acid side chain, occurs after the biosynthesis of the protein is completed and the native protein is released from the ribosome. We know of about a dozen types of such reactions; the most important ones are phosphorylation, acetylation, methylation, ADP-ribosylation, adenylylation and uridylylation. The modification reactions usually require one equivalent of ATP energy for the introduction of one group. Some of these reactions are irreversible, e.g., the methylation of a number of non-enzymic proteins [905] and cytochrome c [906], whereas others are reversible. Interconvertible enzymes belong to the latter class.

Logically, the interconversion of an enzyme requires two other enzymes, one that puts on and another that takes off the tag. These *interconverting* enzymes may themselves be *interconvertible,* in which case we are faced with a *cascade-like* activation–amplification phenomenon. (Avalanche would perhaps be a better metaphor.) The most well-known multicycle cascades are the glycogen phosphorylase–glycogen synthase system, the glutamine synthase cascade and, above all, the cascade of blood coagulation.

In this book interconvertible enzymes and enzyme cascades will not be reviewed, for two reasons. First, they constitute a vast field each item of which would merit a whole volume. The interested reader may consult recent specialized reviews on post-translational modifications of proteins in general [1329], on interconvertible enzyme cascades [212], on enzyme phosphorylation–dephosphorylation in particular [684] or even more specifically on protein kinases [165], on ADP-ribosylation of proteins [520] and on protease action in the blood coagulation cascade [289]. The second reason for omitting these topics is conceptual rather than practical. In cascade systems the catalysts between two "levels" are in an enzyme–substrate relation with each other. We feel that however wide are the ramifications of supramolecular enzyme organization, it makes sense to draw the borderline where enzymes transform each other instead of working on metabolites in an organized manner.

MULTIENZYME CONJUGATES

In a review of 1976, on multifunctional proteins, Kirschner and Bisswanger [654] gave a selected list comprising fifteen multienzyme conjugates from a variety of organisms. In the five years that have elapsed since then further multienzyme conjugates have been identified, e.g. formyl-methenyl-methylene tetrahydrofolate synthase from sheep liver, a conjugate of three enzymes [917]. A recent selection of papers [88] on "multifunctional proteins" highlights the present state of affairs. No attempt is made here to give a new, comprehensive list. The difficulty of such a compilation resides in the frequent lack of rigorous experimental evidence, which should include clear-cut chemical demonstration of a single polypeptide chain by molecular-weight determination under denaturing conditions, fingerprint analysis together with end-group determination, preferably supplemented by genetic analysis [654].

At all events, it is rather probable that several of our present-day, less characterized multienzyme complexes will eventually turn out to be conjugates. This can be expected simply because during isolation and preparative manipulations peptide bonds can only be cleaved by proteases but not synthesized. Multienzyme conjugates have a conspicuous propensity for being nicked between constituent enzymes, often without impairing individual catalytic activity and overall structure. This suggests that each catalytic activity is more or less confined to a coherent structural domain and that the covalent backbone is not an essential prerequisite for physical association.

In the following a few examples will be given to illustrate how covalent linkage does, or does not in some respects, give rise to the functional consequences of enzyme juxtaposition.

THE *AROM* MULTIENZYME CONJUGATE

Structural properties

The biosynthesis of the three aromatic amino acids, phenylalanine, tyrosine and tryptophan, have a common stretch of reaction sequence, consisting of seven enzymatic steps, that is called the *polyaromatic pathway* (Fig. 5.7). It occurs in microorganisms and higher plants, but is absent in animals. Although the pathway has been studied in a number of organisms (for a recent review see [1308]), attention will be focused here on the case of *Neurospora crassa*, since this system is known at present in the greatest detail. Early genetic work by Giles and coworkers (cf. [446]) has revealed that the structural genes of five enzymes (steps 2 to 6) of the pathway form some kind of a larger unit, the "*arom* gene cluster", which is expressed in a coordinate manner via a polycistronic mRNA and the order of enzymes translated is $2 \rightarrow 6 \rightarrow 5 \rightarrow \rightarrow 3 \rightarrow 4$. Certain anomalies with polarity mutants [447] suggested that the five enzymes remained associated after biosynthesis. Indeed, the isolation and subsequent analysis of the gene products unambiguously proved this association [147, 414]. Thus the picture emerged of an "arom multienzyme complex", thought to be composed of at least four different polypeptide chains.

However, a major improvement by way of affinity chromatography in the preparation procedure of the "arom complex", which considerably reduced the preparation time, fundamentally altered this picture: all five activities were shown to be present on a single polypeptide chain [416]. The earlier finding of several polypeptide chains was due to fragmentation by proteases resident in *Neurospora* [712]. In fact, there are a great number of proteases in this organism [1208] and it has been proposed that the observed proteolytic degradation of the arom polypeptide chain is not only an artefact *in vitro*, but might also be related to enzyme ageing and turnover *in vivo* [419]. Preparation in the presence of various protease inhibitors yielded a single polypeptide chain of molecular weight 150,000 on SDS gel electrophoresis [417]. From the preparative molecular weight ($M_r \approx 300,000$) and from peptide fingerprint analysis [755] it was concluded that the enzyme system is a homodimer consisting of two identical subunits of $M_r = 150,000$. It follows, then, that all five activities must reside in a single polypeptide chain. Accordingly, the arom system is a multienzyme conjugate that further associates to a dimer.

The existence, and by the same token the possible physiological significance, of dimeric structure had been corroborated by earlier work on mutant *Neurospora* strains. Namely, two out of the five enzyme activities of the arom system, DHQ synthase and DHS reductase, exhibited allelic complementation: two strains, each defective in one of the two activities, produced normal growth when their heterokaryon was formed [446, 447]. Figure 5.8 gives a schematic picture of the underlying mechanism. In the example in question the two differently mutated DHS reductase domains are enzymatically inactive when homodimers. However, in their hybrid

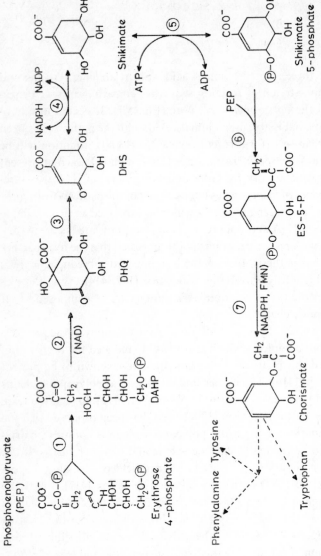

Fig. 5.7. Reaction sequence of the polyaromatic biosynthetic pathway. Abbreviations: DAHP, 3-deoxy-D-arab-inoheptulosonate 7-phosphate; DHQ, 5-dehydroquinate; DHS, 5-dehydroshikimate; ES-5-P, 3-enolpyruvylshiki-mate 5-phosphate. Enzymes: 1, DAHP synthase (EC 4.1.2.15); 2, DHQ synthase; 3, dehydroquinase (DHQ dehydratase, EC 4.2.1.10); 4, DHS reductase (EC 1.1.1.25); 5, shikimate kinase; 6, ES-5-P synthase; 7, choris-mate synthase. The steps after chorismate leading to the three aromatic amino acids are not detailed. Chorismate is precursor also to ubiquinone, folate and vitamin K.

Fig. 5.8. Schematic diagram of allelic complementation in the *Neurospora arom* multienzyme conjugate dimer. The numbers refer to the catalytic steps as indicated in Fig. 5.7. The case illustrated is the complementation of DHS reductase activity. (From Welch and Gaertner [1308].)

Allelic complementation

Wild type (active)

Mutant I
(4 inactive)

Mutant II
(4 inactive)

Heterocaryon hybrid
enzyme (active)

the interaction between the two domains ("quaternary constraint") corrects one (or both) of the active sites. Obviously, such complementation has strict structural criteria and so it may occur only with certain pairs of mutants. It would be a challenging task to unravel the stereochemical details of such an interaction.

The actual shape of the arom conjugate dimer resembles a prolate ellipsoid, as deduced from hydrodynamic measurements [414] and as seen in the electron microscope [419]. More work on the chemistry and three-dimensional structure of the conjugate is needed to understand the catalytic steps proper and the organizational eatures of the system.

Functional properties

The arom conjugate displays several of the properties characteristic of an organized multienzyme cluster. There is a roughly tenfold decrease in *transient time in vitro* over the five reaction steps from DAHP to ES-5-P relative to a hypothetical analogous sequence of unclustered enzymes having the same individual kinetic parameters [1305]. This fact points to the channelling of intermediate metabolites. The same mechanism may underly the phenomenon referred to by Gaertner *et al.* [418] as *catalytic facilitation,* i.e. that the partial reactions of the conjugate, measured by adding one of the intermediate substrates, were markedly slower than the overall reaction starting from DAHP. The same effect can be produced by "interdomain" activation, but this is difficult to assess owing to the covalent linkage of the five enzymes, i.e. the multienzyme *conjugate* character. At any rate, some of the *coordinate regulatory effects* point in this direction. Namely, it was shown that DAHP, the substrate of the first enzyme in the conjugate, enhanced the V_{max} of two reaction steps and decreased the K_m for the respective substrates at four steps [1306]. Furthermore, DAHP protected all five activities from proteolytic attack [1272].

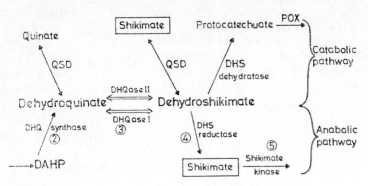

Fig. 5.9. Overlap of the polyaromatic biosynthetic pathway and the quinate catabolic pathway. Symbols: QSD, quinate (shikimate) dehydrogenase; POX, protocatechuate oxygenase; the rest as in Fig. 5.7. QSD is a single enzyme acting on both compounds (44). DHQase I is a constitutive enzyme, DHQase II is an inducible one. In addition to the substrate-product pair of the DHQase reaction, shikimate (framed) is also common to both pathways.

The above findings *in vitro,* however impressive and logically satisfying, nevertheless leave the question of physiological relevance open. Obviously, to demonstrate convincingly that a certain feature is indeed important from the organism's point of view is a hard task and should encompass a wider context of interactions than known to date. Fortunately, in the case of the arom conjugate there is strong evidence that metabolite compartmentation is needed *in vivo,* in order to segregate competing metabolic pathways.

In many microorganisms and plants a *quinate catabolic pathway* coexists with the polyaromatic biosynthetic pathway (Fig. 5.9) and enables these organisms to grow on various alicyclic compounds as carbon source. The two pathways share three metabolites and one reaction step, at DHQase. The catabolic enzymes, including the isoenzymes of DHQase, are coordinately induced by the catabolic intermediates (cf. [446]). It follows that mixing of the intermediates of the two pathways would lead to a special kind of futile cycle, as the compounds generated by the biosynthetic route would trigger the synthesis of catabolic enzymes which, in turn, would degrade the intermediates thus interfering with the production of aromatic amino acids. Early studies by Giles's group [446] led them to propose that the compartmentation of metabolites by the arom cluster served to segregate the two pathways. This contention has been amply corroborated by later work (for a survey see [446]).

While the channelling of "dangerous" metabolites by the arom conjugate is duly fulfilled, this segregation is not absolute. Chaleff [182, 183] has shown that under normal metabolic conditions, in the absence of exogenous inducers, there is a weak flow of biosynthetic metabolites toward protocatechuate, i.e. catabolism. The leakage is apparently permissible as it does not reach the extent where the induction of catabolic enzymes would be unleashed. Moreover, it may be a means of a dynamic regulation of the balance between biosynthesis and degradation as indicated by the kinetic studies of Welch and Gaertner [1305, 1306]. These authors have found that in the presence of DAHP, when the arom conjugate is activated, shikimate kinase (step 5 in Fig. 5.9) becomes rate-limiting for the overall reaction. Since ADP inhibits shikimate kinase, shikimate will not be further processed by the arom conjugate when the *energy charge* [36] of the cell is low. This gives shikimate a chance to escape and fall victim to the catabolic route, which, in turn, will increase the energy charge.

Organization of polyaromatic pathway enzymes in other species

In a recent compilation [1308] thirty species are listed in which the association properties of the polyaromatic pathway enzymes have been examined. Apart from the data for six higher plants, all the work came from Giles's laboratory. This wide phylogenetic screening reveals that, with the exception of bacteria and blue-green algae, the enzymes of the polyaromatic pathway display some kind of physical association in all species examined. Whereas in all fungi the type of association is that found in *Neurospora* (*viz.* enzymes 2 to 6 are clustered), higher plants are characterized by a binary cluster involving enzymes 3 and 4. Interestingly, *Euglena gracilis* has a fungus-type, five-membered cluster [912], whereas *Chlamydomonas reinhardi* possesses a plant-type binary one [67]. Although it is suspected that at least some of these clusters are conjugates, as opposed to complexes, the evidence is not yet as rigorous as that for the *Neurospora* system.

The conspicuous absence of enzyme clustering in bacteria does not mean that here the anabolic and catabolic pathways freely communicate. On the contrary, they seem to be separated but by other, as yet less understood, mechanisms. Evidence has been presented for labile "microcompartments" with the bacterium *Acinetobacter calco-aceticus* [68, 1237] in the physical background of which some protein–protein and protein–membrane interactions may reside.

A remarkable exception among bacteria is *B. subtilis,* which has a three-enzyme cluster including DAHP synthase (enzyme 1), shikimate kinase (enzyme 5) of the polyaromatic pathway and chorismate mutase, the first enzyme of the phenylalanine–tyrosine branch [575, 576]. In this peculiar, non-sequential enzyme cluster chorismate mutase, apart from acting as a catalyst, behaves also as a regulatory subunit of the other two enzymes, inasmuch as these are coordinately inhibited when chorismate mutase binds its substrates.

FATTY ACID SYNTHASES FROM ANIMAL TISSUES

The complex, spiral reaction sequence catalysed by fatty acid synthases is detailed later in this chapter, under yeast fatty acid synthase (p. 113). In contrast to the latter, which is a large molecule ($M_r = 2.4 \times 10^6$), animal fatty acid synthases generally have a molecular weight only around 500,000 and are dimeric in structure, whether isolated from liver, brain or lactating mammary gland (for a review cf. [97]). Because of the large size of the subunits it has been difficult to establish whether they are identical or not, as the usual techniques such as peptide mapping are not applicable. Nevertheless, for the enzyme from mammary gland, recent data obtained by limited tryptic [1130] and elastase [487] digestion, combined with the immunochemical detection of a dissected domain in both subunits [1130], support the existence of a homodimer, i.e. a dimeric *multienzyme conjugate,* Whether the other animal fatty acid synthases have a similar type of structure, remains to be seen. The finding that the (homo?) dimeric enzyme from an insect, *Ceratitis capitata,* requires phospholipid for full activity [428] may be an indication of the possible significance of interactions with membrane components.

Although in bacteria, as a rule, the enzymes of fatty acid synthesis are not physically associated, there is one notable exception. The enzyme from *Mycobacterium smegmatis* has a native $M_r = 2 \times 10^6$, and is an oligomer of a single type of polypeptide chain of $M_r = 290,000$ [1334]. Thus it is a multienzyme conjugate like animal fatty acid synthase, but apparently exists as a high oligomer resembling the yeast synthase (cf. below).

THE INITIATION OF PYRIMIDINE BIOSYNTHESIS:
THE CLUSTER OF CARBAMOYL-P SYNTHASE–ASPARTATE
CARBAMOYL TRANSFERASE–DIHYDROOROTASE

In higher organisms including mammals the first three steps of *de novo* pyrimidine biosynthesis are catalysed by glutamine-dependent carbamoyl-P synthase (EC2.7.2.9), aspartate carbamoyl transferase (EC 2.1.3.2) and dihydroorotase (EC 3.5.2.3). Evidence has been presented, mainly of the co-purification type and coordinate synthesis in mutants, that these three enzymes are clustered, possibly represented on a single polypeptide chain. Such clustering was found in cells spanning a wide evolutionary range, including baker's yeast [753], *Neurospora* [1321], *Ascaris* ovary [23], *Drosophila* [135], *Rana catesbiana* eggs [641], Syrian hamster cells [632] and rat ascites hepatoma [841]. The extreme lability of these enzymes has hampered the analysis of the cluster. Recently, the cluster has been purified to homogeneity from the best source so far available, rat ascites cells, and characterized from the protein chemical point of view [841]. The data are somewhat controversial: the cluster had a $M_r = 870,000$ and in SDS gels gave a single band of $M_r = 210,000$. This would suggest a tetrameric structure, but on sucrose gradient in 1.5 M NaCl the different activities partially separated and the authors found it difficult to reconcile their data with the view that all three activities are borne by the same polypeptide. In contrast, the enzyme purified to homogeneity from an overproducing mutant line of hamster cells had as many aspartate carbamoyl transferase active sites as polypeptide chains, which suggests a true multienzyme conjugate [242]. Clearly, further work is needed to establish the actual type of physical association, and to unravel its functional consequences.

"COMPLEX U": THE TERMINATOR OF PYRIMIDINE BIOSYNTHESIS

The last two steps of the biosynthesis of pyrimidine nucleotides are as follows:

$$\text{Orotate} + \text{Phosphoribosyl-PP} \rightleftharpoons \text{Orotidine-5'-P} + \text{PP} \qquad (1)$$
$$\text{Orotidine-5'-P} \rightarrow \text{Uridyl-5'-P} + CO_2 \qquad (2)$$

Reaction (1) is catalysed by orotate phosphoribosyl transferase (EC 2.4.2.10) and reaction (2) by orotidine-5'-P (OMP) decarboxylase (EC 4.1.1.23). While in prokaryotes and simple eukaryotes such as yeast the two enzymes are distinct and separable [1248], in mammalian cells they occur in a physically associated form named "Complex U" [598]. For the human erythrocyte enzyme a four-chain model of the type $\alpha_2\beta_2$ was proposed on the basis of physicochemical studies [136]. Protein chemical work on Complex U from Ehrlich ascites cells has been hindered by the great lability of both activities [623]. Nevertheless, careful examination of the physical properties of the Ehrlich ascites system, taken together with the lowering of both activities in humans suffering from orotic aciduria [1129], led Traut and Jones [1234] to conclude that "Complex U" was no complex but rather a bienzyme conjugate. The simultaneous induction of the two activities in various mammalian cells supports this view [733, 1196]. However, to preserve something of the complex character, the conjugate can dimerize and also associate with an as yet unidentified protein of $M_r = 29,000$, as shown by detailed sedimentation and gel-filtration studies [1235].

Importantly, Complex U (let us abide with the old name for the sake of brevity) was found to channel the intermediate OMP. This was indicated by two observations. First, OMP synthesized from orotate by phosphoribosyl transferase was preferentially utilized in the second step even in the presence of a large excess of exogenous OMP [1233]. Second, OMP was not available to nucleotidase attack that would have degraded it to orotidine [1232]. As an interesting comparison, Traut [1232] also examined yeast cells, which have two individual enzymes instead of a cluster and cannot compartmentalize OMP. Apparently, yeast can afford this for it contains about 20 times less nucleotidase activity than do Ehrlich ascites cells.

For further details the reader may consult the recent review by Jones [599].

ASPARTOKINASE I: HOMOSERINE DEHYDROGENASE I

In microorganisms L-aspartate is the precursor of a number of other amino acids as shown by the following scheme:

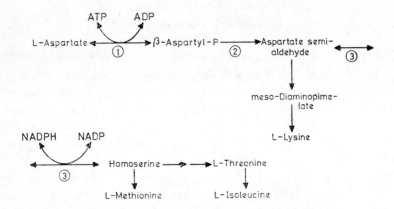

In *E. coli* steps 1 and 3 are catalysed by two different enzyme clusters called aspartokinase homoserine dehydrogenases I and II, and there is also an independent aspartokinase III (cf. [120, 230, 1006]). Much of the fundamental work on this system has been done by Cohen and coworkers [230, 231]. Of these enzyme I has been studied in the greatest detail and proved to be a single polypeptide chain, oligomerized to a tetramer, α_4 (for a review see [1238]). The unusual feature of this conjugate is its non-sequential character; the enzyme catalysing step 2 in the pathway is a separate entity. Aspartokinase I: homoserine dehydrogenase I exhibits coordinate effects often found in multienzyme clusters: both activities are subject to feed-back inhibition by threonine [522, 1265], which is exercised through eight binding sites for threonine on the tetramer. Furthermore, both enzyme activities are protected from heat inactivation by NADPH [230]. A detailed survey of the current knowledge about *E. coli* aspartokinase: homoserine dehydrogenases is given in [231].

OTHER MULTIENZYME CONJUGATES

Chorismate mutase–prephenate dehydrogenase (EC 1.3.1.12) from *E. coli* is the regulatory enzyme in the aromatic biosynthetic pathway leading to tyrosine. The endproduct inhibits the dehydrogenase activity possibly through an allosteric site [671]. The conjugate is basically a dimer of $M_r = 90,000$; its tendency to undergo slow self-association to higher polymers is not observed in a crude cell extract [745].

Phosphoribosyl-anthranylate isomerase–indoleglycerol-P synthase (EC 4.1.1.48) catalyses the two steps of tryptophan biosynthesis that precede the tryptophan synthase reaction. The enzyme from *E. coli* is one of the simplest multienzyme conjugates [267] as it exists as a monomer, arranged into two functional domains corresponding to the two catalytic activities [654]. Equilibrium dialysis [87] and fast-reaction kinetic studies [240] with an analogue of the intermediate substrate revealed two different binding sites for the intermediate, presumably one each on the two functional domains. Although negative cooperativity could not be excluded, there was apparently little interaction between the two sites. This is in keeping with the earlier finding [264] that the intermediate is not channelled in this conjugate but rather accumulates reaching a steady-state level during the overall reaction. This case might serve as a warning that all instances of physical juxtaposition of enzymes do not necessarily have obvious functional repercussions at the level of enzyme reaction. The evolution of a multienzyme conjugate might have been favoured by advantages furnished by other assets, e.g. in terms of the coordinate expression of two, or more, structural genes.

MULTIENZYME COMPLEX-CONJUGATES

FATTY ACID SYNTHASE FROM YEAST

The reaction sequence of the biosynthesis of fatty acids has largely been clarified by the pioneering work of Lynen and his coworkers on baker's yeast, *Saccharomyces cerevisiae* [757, 758, 759]. The long-chain fatty acids are synthesized according to a "recycling" scheme, each cycle consisting of six enzymatic steps and elongating the aliphatic chain by two carbons (Fig. 5.10). During the transformations the intermediates are covalently bound, via *S*-acyl linkages, to two SH-groups on the yeast fatty acid synthase, distinguished arbitrarily by Lynen as "central" and "peripheral". Fatty acid synthases from yeast and animals are sometimes referred to as type I (aggregated or clustered) enzymes, while synthases from bacteria, e.g. *E. coli,* which consist of unclustered, individual enzymes, belong to type II. Various aspects of fatty acid synthases have been the subject of many reviews [97, 449, 757, 759, 1276].

Early work [757, 758] seemed to establish that the eight individual enzymes of yeast synthase were held together by non-covalent bonds, i.e. they formed a multienzyme complex. This contention was supported by finding seven different *N*-terminal amino acids and at least six different polypeptide chains on starch gel electrophoresis. All attempts to isolate the component enzymes in active form have failed, which pointed to the exceptional tightness of the complex. However, Schweizer and coworkers have later demonstrated by protein chemical [1076] and genetic studies [667, 1210] that the enzymes were distributed over only two polypeptide chains of molecular weights around 180,000. The earlier detection of several chains was due to proteolytic fragmentation, just as in the case of the *arom* multienzyme conjugate discussed above. The α (or A) polypeptide comprises the prosthetic group 4'-phosphopantetheine, β-ketoacyl synthase (3) and β-keto reductase (4), whereas the β (or B) polypeptide harbours acetyl transacylase (1), dehydratase (5), enoyl reductase (6), and malonyl (2) (palmitoyl [8]) transacylase.

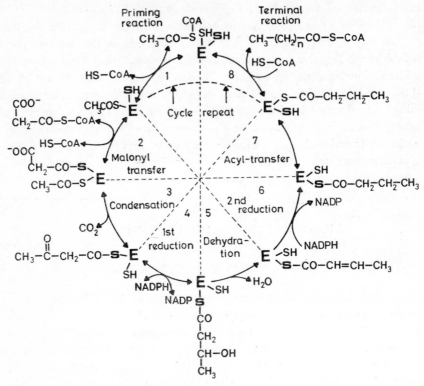

Fig. 5.10. Reaction steps of yeast fatty acid synthase. Reaction (1) (acyl-transfer) initiates the first cycle, whereas reaction (8) (palmitoyl transfer) terminates the cycles. Reaction (8) only occurs if the number of methylene groups $n = 14$, otherwise the cycle (reactions 2 to 7) repeats. The sulphur of the "peripheral" (β-ketoacyl synthase) SH-group is bold face.

Both the "central" and "peripheral" SH-groups were found in the α polypeptide, the former being the SH-group of pantetheine, whereas the latter is located in the active site of β-ketoacyl synthase [688]. More recent SDS polyacrylamide gel electrophoretic measurement yielded somewhat higher molecular weights, 212,000 and 203,000 for the α and β chains, respectively [1184]. As the preparative molecular weight of yeast synthase was found to be $2.3-2.4 \times 10^6$ [1076, 1184], a dodecameric structure $\alpha_6\beta_6$ seemed plausible.

Thus the yeast fatty acid synthase combines the clustering modes of multienzyme conjugates and complexes: it is an oligomeric (hexameric) complex of two multienzyme conjugates, α and β. Electron microscopic studies [1184] revealed a rather unusual shape for the assembly (Fig. 5.11): the α component looks like a plate, whereas the β component resembles an arch, with bead-like domain arrangement. According to recent crosslinking experiments [1186] the two "active" SH-groups on the α component are not farther from each other than 5 Å. This explains the ease of acyl-transfer during the catalytic cycle. In the model proposed the pantetheine- and keto-

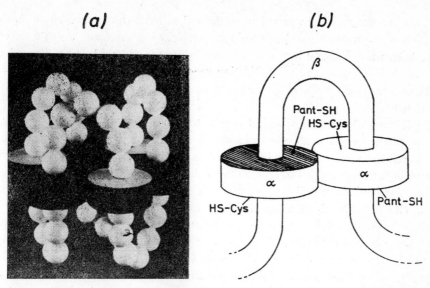

Fig. 5.11. Subunit arrangement in yeast fatty acid synthase. (*a*) Model of the whole molecule, $\alpha_6\beta_6$. The arch-like structures are the β-subunits, each subdivided into 7 bead-like domains; the plate-like structures are the α-subunits, ordered in an all-isologous hexamer (alternating black and white sides on the top). (*b*) Schematic rendering of one active site region. The juxtaposition of pantetheine-SH and cysteine-SH groups in the β-ketoacyl synthase site is tentatively indicated. (From Stoops and Wakil [1186].)

acyl synthase SH-groups are located on opposite sides of the "plates", which are arranged in an all-isologous hexamer (symmetry D_3), so that acyl-transfer occurs between SH-groups belonging to two neighbouring α plates (Fig. 5.12). Thus one catalytic "unit" is composed of three proteins (2+1) to which the intermediate remains covalently bound, though shuttled from one SH to the other, over the seven catalytic cycles that lead to palmitate.

As mentioned at the beginning, in bacteria the enzymes of fatty acid synthesis are unaggregated. In contrast, their inducible *fatty acid β-oxidizing (degradative) enzymes* seem to be clustered. It has been reported that five associated enzymes participate in *E. coli* fatty acid oxidation [83, 877]. Preliminary data suggest that the basic structure is a heterodimer that can undergo further polymerization [58, 957]. It is an interesting, but possibly meaningless, reciprocity that in mammalian mitochondria the β-oxidizing enzymes apparently do not interact with one another.

INTERACTIONS BETWEEN "SOLUBLE" ENZYMES

It has become an ever more appreciated idea over the past 10 years, though its roots date back much earlier, that enzyme–enzyme interactions occur not only between components of well-established multienzyme complexes and conjugates, but also between so-called "soluble" individual enzymes. This operational term only means that the enzyme can be extracted from the cell as a separate entity by an aqueous solution of moderate ionic strength, in enzymatically active form. Soluble enzymes have been the bread-and-butter of enzymologists for some 50 years. The relative ease of their isolation and crystallization rendered them popular objects for study and most of our current knowledge on enzyme structure and function has been derived from them.

While the endeavour in classical enzymology has been to purify each enzyme, painstakingly removing all contaminants through a number of preparative steps, the search for enzyme–enzyme interactions involves a somewhat opposite practice: the mixing of enzymes that have been laboriously separated. In fact, the recognition of such interactions often benefited from the failures in resolving two or more proteins. The potential significance of this *synthetic* trend, as opposed to the earlier *analytic* approach, can hardly be overemphasized. It envisages a much more complex, yet strictly organized structural build-up of metabolic pathways and intracellular space. Authors have not been reluctant to seize this conceptual opportunity and a number of treatises, among them the present book, have been written on the subject, highlighting various aspects of the field [404, 629, 896, 1161, 1308].

It is probably close to truth that concepts are ahead of experimental evidence. By this we do not mean at all that theoretical considerations are moving in a void. On the contrary, we believe that theories have great heuristic value and should be propounded, even if in excess. However, one must clearly see how much is theory and speculation, and how much is solid, experimentally established fact. The "there-must-be" approach [32] is fair if admitted, as done, for example, by Welch and Gaertner [1308], but it is less fair to sell wishful thinking as current knowledge. Supramolecular organization of soluble, individual enzymes is a field particularly prone to such exercises. In the following we try to adopt a sympathetic yet critical attitude toward organizational phenomena and indicate the borderline between facts and fancies.

There are two major compartments in the living cells where "soluble" enzymes are accommodated: the cytoplasm and the mitochondrial matrix. In both, enzyme concentrations are rather high [534, 1153]. The interaction of soluble enzymes thus boils down to the question whether these cellular spaces contain random mixtures of macromolecules, among them enzymes, as implied in most biochemistry texts, or have some kind of subtle substructure of physiological significance that, however, is elusive and does not survive cell disruption.

Since in this chapter we are dealing with enzyme complexes and conjugates, i.e. enzyme clusters that do not need any "third party" to be physically associated, we are going to review evidence of this kind for the soluble enzymes, without going into the role of intracellular structures. However, we do so only to conform to our adopted stepwise advance in enzyme organizational hierarchy. As we shall see in the next chapters, much of the supramolecular organization of enzymes is realized in scaffolded enzyme arrays, which utilize membranes and structural proteins for support.

INTERACTIONS OF GLYCOLYTIC ENZYMES

In the cytoplasm the major set of soluble enzymes belongs to glycolysis. In yeast cells, for example, glycolytic enzymes constitute about 65% of total soluble protein [535]. In fact, glycolysis is generally regarded as the ancient pathway of energy generation catalysed by individual enzymes, as opposed to structure-bound (particulate) terminal oxidation in mitochondria, which was a later acquisition during evolution, possibly by trapping a microorganism in the primordial eukaryotic cell. While this contention may hold true at large, there is no reason to exclude the possibility that some kind of structural organization might have evolved also in glycolysis. Evidently, it must have depended on whether glycolytic organization endowed the cell with properties advantageous in competition (or cooperation) with other cells.

The methodical problems of detecting loose enzyme–enzyme interactions are great. In the following we shall see what approaches have been designed for the various systems studied, and the techniques applied for detecting metabolite compartmentation will be dealt with in more detail in the next chapter. It is pertinent here to refer back to associating–dissociating oligomeric systems (p. 71.), where the experimental difficulties have already been emphasized. These hold more pronouncedly for enzyme–enzyme interactions. Controversial findings may depend upon minor, unnoticed differences in experimental conditions. Furthermore, interaction *per se* does not mean that the *in vitro* phenomenon also occurs *in vivo,* or that it has any physiological significance. Conversely, the working hypothesis of most investigators in this field is that many interactions existing at the high protein concentration, etc. *in vivo,* may escape detection *in vitro* where conditions are inevitably different, unless very careful efforts are made to reproduce them.

The *aldolase-glyceraldehyde-3-phosphate dehydrogenase* (GAPD) interaction, with enzymes from rabbit muscle, has been studied in considerable detail. Ovádi and Keleti [899] provided kinetic evidence for this interaction and for channelling of the intermediate metabolite, glyceraldehyde-3-phosphate (GAP). They measured the coupled enzyme reaction by fast-reaction technique and found a K_m for aldolase-generated GAP with GAPD of about 10^{-5} M. When determined separately with the dehydrogenase, the K_m for GAP supplied from a stock solution was about one order of magnitude higher. The lowering of K_m can be a sign of enzyme–enzyme interaction in general, but in the present case there was a complicating factor. Namely, GAP was known to be predominantly in the hydrated (diol) form when in solution [1236], the ratio diol/aldehyde was approximately 60. The diol form is not a substrate for the dehydrogenase. Hence the simplified scheme of the coupled reaction was as follows:

$$\text{Fructose--}P_2 \overset{E1}{\rightarrow} \text{GAP}_{\text{ald}} \overset{E2}{\rightarrow} \text{NADH}$$
$$k_{-1} \updownarrow k_{+1}$$
$$\text{GAP}_{\text{diol}}$$

where fructose-P_2 is fructose-1,6-bisphosphate, the subscript to GAP designates the aldehyde or diol form and E_1 and E_2 are aldolase and GAPD, respectively. If k_{-1} is small relative to the k_{cat} of E_2, then the low K_m found in the coupled reaction is explained: it refers to the unhydrated form of GAP. However, it was shown that the opposite held, i.e. hydration was faster than dehydrogenation. Therefore the authors assumed that GAP was channelled between the two enzymes, since it could not be "trapped" by bulk water. It should be added that in these experiments enzyme concentrations were rather low (2–12 μM in active site), well below the calculated steady-state GAP concentration. Therefore the trapping of GAP in the aldehyde form by GAPD, which may occur at high enzyme concentration [892], here could not happen.

Interaction must, of course, involve complex formation, however short-lived the complex may be. Ovádi et al. [900] seeking physical evidence for complex formation have measured the polarization of fluorescence of isothiocyanate-coupled fluorophores, covalently attached to either GAPD or aldolase, while increasing the concentration of the other (unlabelled) enzyme. They found an increase in fluorescence polarization in both cases, which suggests complex formation between the two enzymes. The quantitative data could be fitted to the formation of a complex consisting of two aldolase and one GAPD tetramers with an apparent dissociation constant of 3×10^{-7} M at 20°C. A peculiar feature of this phenomenon was the slowness of complex formation: the second-order rate constant of association was 42 M^{-1} s^{-1}, which is several orders of magnitude lower than the average for protein associations [486]. By taking the two experimentally determined values together, one can arrive by simple calculation at a half-life for the complex of about 40 hours. This is a very long time indeed and one wonders how such a durable complex could have escaped detection earlier. In fact, Kwon and Olcott [707] reported the augmentation of tuna aldolase activity by rabbit muscle GAPD, but this effect was instantaneous. On the other hand, Földi et al. [385] did not observe the association of aldolase and GAPD on frontal analysis gel chromatography of a myogen fraction containing both enzymes. In the latter study the apparent molecular weight of aldolase was significantly higher than its actual $M_r = 160,000$, probably due to association with some element of the contractile apparatus [785, 786], but the position of GAPD was not shifted. Comparison of results obtained by different groups is, however, fraught with the difficulty created by using different experimental conditions.

Recently, Grazi and Trombetta [473] reported observing the slow complex formation between aldolase and GAPD, but they were unable to detect the channelling of GAP. The approach adopted relied on their method of quantitating the aldolase.dihydroxyacetone-P intermediate [473] and on the determination of the 3-phosphoglyceroyl-GAPD intermediate. Prolonged preincubation of a mixture of the two enzymes, during which complex formation is expected to take place, did not affect the steady-state concentration of either of these intermediates. Furthermore, the concentration of 3-phosphoglyceroyl-GAPD was the same with and without pre-incubation even in the presence of triose-P isomerase. The latter enzyme was used as a trapping agent for GAP, since it rapidly converts GAP into dihydroxyacetone-P, if the triose phosphate is available in the bulk medium (cf. the enzyme-probe method in Chapter 7). Since pre-incubation did not alter the accessibility of GAP to triose-phosphate isomerase, the authors concluded that the metabolite was not directly transferred from one enzyme to the other. On the other hand, Masters and Winzor [787] failed to detect any interaction between aldolase and GAPD by frontal gel chromatography, as well as by velocity and equilibrium sedimentation analysis. They attribute the positive result obtained by fluorescence polarization [900] to the stickiness introduced by the fluorescent probe. Indeed, it has been demonstrated that prolactin modified with fluorescein isothiocyanate binds to bovine serum albumin, the complex having a dissociation constant of about 10^{-7} M, whereas no interaction was detected with unmodified prolactin [218].

The interaction of aldolase and GAPD was examined by Patthy and Vas [916] by an entirely different method. This was based on the observation that in the presence of electron acceptors, such as $K_3Fe(CN)_6$, aldolase catalyses a suicide reaction when splitting fructose-1,6-bisphosphate [217]. The essence of this suicide reaction is as follows. The carbanion intermediate of aldolase-dihydroxyacetone-P (the same species Grazi and Trombetta [474] detected by acid precipitation) can be oxidized to give hydroxypyruvaldehyde-P, a potent arginine-modifying α-dicarbonyl compound [913]. As there are arginine residues in the substrate phosphate-binding site of aldolase [914, 915], the modification of one of them (Arg-55, [915]) inactivates the enzyme. Importantly, only the nascent oxoform of hydroxypyruvaldehyde-P is an avid arginine modifier; its hydrated form is practically ineffective. The suicide reaction is interpreted as an action of the nascent, active reagent on the parent enzyme molecule [913]. Patthy and Vas [916] added increasing concentrations of GAPD to the aldolase suicide mixture and found that the latter enzyme was also inactivated (Fig. 5.12). The plot of the number of GAPD molecules inactivated per catalytic cycle of aldolase against concentration of GAPD was a rectangular hyperbola. This saturation behaviour and the fact that hydrated hydroxypyruvaldehyde-P is inert suggest that GAPD is attacked by the nascent reagent in the aldolase-GAPD complex. The apparent dissociation constant for a 1 : 1 complex that could be derived from the data was $K_d = 10^{-6}$ M. It is worth mentioning that in this reaction aldolase prefers "murder" to suicide. Namely, while one out of sixty-eight hydroxypyruvaldehyde-P molecules produced inactivates one aldolase subunit, about every sixth inactivates a GAPD subunit. This syncatalytic inactivation is a novel method of detecting enzyme–enzyme interactions. Its scope of application is, of course, limited to enzymes which can be made to catalyse the formation of a highly reactive product in the nascent state.

Aldolase was shown by Batke et al. [56] using the active enzyme centrifugation technique [234] to form a complex with glycerol-3-phosphate dehydrogenase, both enzymes taken from rabbit muscle. The apparent dissociation constant of the complex

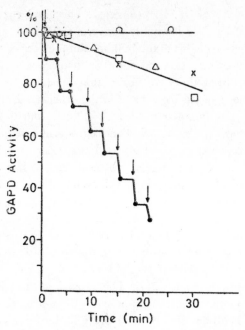

Fig. 5.12. Demonstration of enzyme–enzyme interaction by syncatalytic inactivation. GAPD ($2\times$ $\times 10^{-5}$ M) from rabbit muscle was incubated with rabbit muscle aldolase (10^{-4} M), FBP (5 mM) and repeated doses of hexacyanoferrate (III) at pH 7.5 and 20°C. Hexacyanoferrate doses (1 mM), indicated by the arrows, were reduced immediately, then GAPD activity was assayed ($\bullet$). Control experiments: $\circ$, hexacyanoferrate (III) omitted; $\times$, aldolase omitted; $\triangle$, FBP omitted from the reaction mixture; $\square$, GAPD added *after* the reduction of 8 mM hexacyanoferrate (III). (From Patthy and Vas [916].)

was 10^{-7} M. The dehydrogenase, a dimer, could also dissociate to inactive monomers with a dissociation constant of 7×10^{-7} M. Aldolase only bound to the active dimer, at the same time enhancing the activity of the dehydrogenase. It has been proposed that this complex formation may direct the carbon flow from glycolysis toward lipid synthesis. There is no information as yet how various regulatory metabolites influence complex formation.

Aldolase from rabbit liver has been found by Horecker's group [760, 950, 952] to interact with fructose-1,6-bisphosphatase (FBPase, EC 3.1.3.11). The two enzymes tended to copurify when the livers were obtained from fed rather than fasted rabbits [950]. With the purified enzyme three lines of evidence were provided for complex formation: (*i*) FBPase changed the tryptophan fluorescence emission spectrum of aldolase; (*ii*) aldolase reduced the affinity to Zn^{2+} of the two high-affinity sites of FBPase, and (*iii*) gel-permeation coefficients in batch technique were decreased from about 0.8 for the single enzymes to 0.5 with a mixture of the two enzymes. The quantitative data suggested that two FBPase tetramers and one aldolase tetramer constituted the complex, the formation of which apparently did not require prolonged incubation. The specificity of the interaction was indicated by the fact that replacement of either liver enzyme by its correspondent muscle enzyme abolished the above effects. Pontremoli *et al.* [952] suggest that the formation of such a complex in the liver may promote gluconeogenesis, as in that sequence aldolase reaction is followed by FBPase cleavage. Regulation of FBPase activity by aldolase, through influencing Zn^{2+} binding to the high-affinity sites, has also been envisaged [951, 952]. Aldolase and triosephosphate isomerase from the flight muscle of *Ceratitis capitata* were found by fluorescence and gel-permeation measurements to form a complex [427].

GAPD has been reported to interact also with enzymes other than aldolase: with transketolase (EC 2.2.1.1) in yeast as shown by electrophoresis [672], and with rabbit muscle cytoplasmic aspartate aminotransferase, as detected by fluorescence polarization [900]. The interaction between GAPD and 3-phosphoglycerate kinase, both from rabbit muscle, has been carefully looked for kinetically, by gel-chromatography and fluorescence anisotropy measurements, but all methods gave negative results [1260]. In the case of the yeast enzymes, Hess and coworkers (cf. [534]) made a systematic study of various combinations by measuring transient times over a wide range of enzyme concentration and failed to detect any sign of interaction, i.e. shortening of transient time. (It should be added, however, that the upper limit of enzyme concentration used was still about one order of magnitude below the assumed physiological value.) The only system where some anomaly could be found was the alcohol dehydrogenase–pyruvate dehydrogenase couple.

In conclusion, available evidence on the functionally significant interaction (complex formation) of pure glycolytic enzymes with each other, or with enzymes of other pathways, is meagre. Even the best-studied aldolase-GAPD interaction is controversial. Nevertheless, the observed phenomena may be amplified *in vivo* by the high overall protein concentration, limited amount of free water and, last but not least, through the scaffolding role of intracellular structural elements.

The glycolytic multienzyme complex from E. coli

In contrast to the paucity of yeast and mammalian glycolytic enzymes that have so far been reported to form complexes, in *E. coli* there seems to exist a multienzyme complex accommodating all enzymes of glycolysis. Described first by Mowbray and Moses [851] and recently reinforced by Gorringe and Moses [464], this remarkable structure is able to produce pyruvate from glucose, i.e. it is a complete pathway particle. The fly in the ointment is, however, that there is no solid evidence for its prevalence in the intact bacterial cell. The complex has been isolated by osmotic lysis of *E. coli* spheroplasts followed by differential centrifugation, which yielded membrane fragments in the pellet with overall (glucose→pyruvate) glycolytic activity and a supernatant with no overall glycolytic activity (only individual constituent enzymes were measurable). If the original supernatant and/or the supernatant from

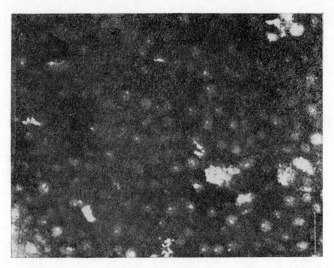

Fig. 5.13. Electron micrograph of particles displaying glycolytic activity. The particles were produced by the concentration of an *E. coli* extract. Negative staining with phosphotungstic acid. Bar represents 100 nM. (From Mowbray and Moses [851].)

washing the membrane pellet was concentrated by ultrafiltration, there appeared in it species of molecular weight at least 1.2×10^6, as tested by Bio-Gel column chromatography. Under the electron microscope these looked like spherical particles of fairly homogeneous size (Fig. 5.13). This glycolytic complex was therefore not isolated, in the strict sense, but rather reassembled from the individual components.

The complex (or particle) has attractive properties. Its size agrees closely with that of a hypothetical complex containing one each of the glycolytic enzymes, which is about 1.3×10^6 daltons. As mentioned above, it exhibits overall glycolytic activity. It establishes fairly tight compartmentation of metabolites, as judged from the weak competition of added unlabelled glycolytic intermediates with intermediates produced endogenously from $[^{14}C_6]$ glucose. Metabolite compartmentation by such a "pathway particle" might be particularly important in a prokaryote, in which membrane-bordered organelles are absent.

Since after lysis the majority of glycolytic enzymes are in the pellet fraction, it is possible that *in vivo* the particles, or their fragments, are membrane-bound. Alternatively, the particle may be a soluble entity in the cell, but becomes attached to the membrane after lysis. (We will be faced with this problem when discussing red blood cell glycolysis in Chapter 6.) Irrespective of these alternatives, the particles produced *in vitro* have no membrane support, therefore they are to be regarded as free complexes. On the other hand, it is not known whether some other (structural) protein is also needed for its assembly.

In conclusion, the present state of affairs is that the glycolytic enzymes of *E. coli* are able to aggregate *in vitro* into a large complex containing all glycolytic enzymes, but this complex may be an artefact. Future research is to tell whether it also exists in the bacterial cell.

The membrane-bounded "glycosome" detected in certain microorganisms will be dealt with in Chapter 6 (p. 172).

INTERACTIONS BETWEEN ENZYMES OF THE MITOCHONDRIAL MATRIX

The possibility of supramolecular organization of enzymes in the mitochondrial matrix has been raised, and pursued ever since, by Srere [1154]. His general arguments, recently summarized in several papers [1156, 1157, 1159] centre around the great probability of organization in such a densely packed "bag" of macromolecules and, on the other hand, how metabolic pathways can be run at all under those conditions. We are going to follow Srere's train of thought in brief.

Electron microscopic studies by Hackenbrock [488] revealed that the matrix does not look like homogeneous matter but rather as an organized reticular network, which undergoes geometric rearrangement depending on the energy status of the mitochondrion. In the low-energy state the protein is tightly packed ("closed phase"), whereas in the high-energy state it forms an expanded lattice ("open phase"). This transformation is accompanied by the extrusion or uptake of water. Hackenbrock estimated the protein concentration to be 56% (w/w) and noted that few proteins had that high solubility. Srere [1156] made a simple calculation which showed that if protein molecules were identical spheres, arranged in the densest possible (cubic) array, then the protein concentration would be about 60% by weight. This value is close to Hackenbrock's estimate. When the matrix contracts (closed phase) to about one-half of its volume at rest (open phase), most of the water not bound to protein will flow out through the inner membrane. This twofold reduction of volume, however, may increase the concentration of metabolites almost *two orders of magnitude*. In this dehydrated state weak interactions between enzymes may become significant.

Garlid's [426] measurements indicate that there are at least two different water compartments in the matrix, one of which is probably free water and the other protein-bound water. The amount of the latter is still a matter of argument, estimates ranging from one [164, 1001] to three [627] layers of H_2O molecules. Whichever is the correct number, it is clear that much of the water present in the matrix is bound. In this respect, the matrix can be compared to a protein crystal, in which the crystal pores serve as communication channels for metabolites [1157]. The pore sizes may be too small for letting through bulky compounds like nucleotides, which then become "locked" between functionally related enzymes, giving rise to much higher local concentrations than calculated from the amount of substance in unit volume.

All the above facts and considerations make it very likely that matrix enzymes interact with each other, whether specifically or not. One would reason, of course, that evolution must have seen to it to exploit these inevitable interactions. Indeed, there are reports of specific, and teleologically "reasonable", complex formations,

though only in a few cases. Thus Halper and Srere [503] detected the interaction
of pig heart citrate synthase and mitochondrial malate dehydrogenase (EC 1.1.1.37)
in concentrated polyethyleneglycol solution, on the basis of their enhanced coprecipitation. Polyethyleneglycol was applied to mimic the macromolecular environment
of the cell and reduce ambient water [818]. The specificity of this interaction was
indicated by the fact that neither cytosolic malate dehydrogenase nor nine other
proteins coprecipitated preferentially with either citrate synthase or mitochondrial
malate dehydrogenase. A possible physiological significance is at hand for a citrate
synthase–malate dehydrogenase complex in the mitochondrial matrix. It would explain, in terms of metabolite channelling, the discrepancy that was found between
the observed high rate of the Krebs cycle and the lower rate of the citrate synthase
reaction calculated by using the apparent intramitochondrial concentrations of acetyl-CoA, oxaloacetate and the enzyme [1155]. Indeed, in model experiments with
artificially immobilized malate dehydrogenase + citrate synthase (+ lactate dehydrogenase to recycle NADH), made by coupling to Sephadex or Sepharose, or by trapping
in polyacrylamide gel, there was a two- to four-fold increase in the rate of citrate
production from malate, as compared with a system of free enzymes [1160]. Citrate
synthase also interacts with the pyruvate dehydrogenase multienzyme complex,
as shown by the kinetics of the coupled reaction (K_m values decreased relative to the
individual enzyme reactions) and by gel chromatography and co-sedimentation
[1195].

The interaction of mitochondrial malate dehydrogenase and aspartate aminotransferase (EC 2.6.1.1), as well as that of their cytoplasmic counterparts, could be
demonstrated by counter-current distribution in an aqueous biphasic dextran-tri-
methylaminopolyethyleneglycol system [39]. Specificity again was witnessed by the
lack of cross-interaction between mitochondrial and cytoplasmic species. At about
the same time, Fasella and coworkers showed by fluorescence polarization and gel
chromatography [1041] that the two enzymes interact physically, and they also
furnished kinetic evidence [141, 1042] for the compartmentation of oxaloacetate,
the intermediate metabolite, in the consecutive reaction of cytoplasmic aspartate
aminotransferase (AAT) and malate dehydrogenase (MDH):

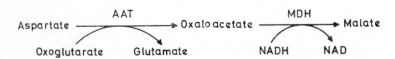

Bryce et al. [141] found that there was no measurable lag phase, i.e. transient time,
in the appearance of malate and when the reaction was run with radioactive aspartate
in the presence of unlabelled oxaloacetate, the specific radioactivity of malate produced was practically the same as that of aspartate. These two observations indicate
a strict channelling of oxaloacetate, still Bryce et al. could not detect any physical
sign of complex formation. They raised an alternative explanation related to the con-

formation of oxaloacetate. Namely, dicarboxylic acids are probably in *trans* conformation when in solution, but in *cis* when bound to the enzymes. Thus nascent, *cis* oxaloacetate would be preferentially utilized by the second enzyme giving rise to the phenomenological criteria of compartmentation. It must be added, however, that recently Manley *et al.* [771] were unable to reproduce the anomalous kinetics and isotope data of Bryce *et al.* [141], and the reason for this discrepancy is not clear (cf. [1042]). At any rate, if oxaloacetate compartmentation between these two enzymes eventually proves correct, then its physiological meaning may be to exclude other reactions of oxaloacetate, which is substrate for a number of enzymes. Furthermore, the two enzymes could cooperate on either side of the mitochondrial membrane in the malate shuttle, which introduces reduced equivalents into the mitochondria.

Malate dehydrogenase has also been shown to form a complex with glutamate dehydrogenase (EC 1.4.1.3) in the presence of palmitoyl-CoA [356]. Fahien and coworkers prepared the enzymes from bovine liver mitochondria. Complex formation was demonstrated by gel chromatography on Sephadex G-200, by chemical crosslinking and coprecipitation in the presence of polyethyleneglycol. In the last type of experiment glutamate dehydrogenase also seemed to interact strongly with mitochondrial aspartate aminotransferase. The calculated tentative stoichiometries in these complexes were three aminotransferase dimers or two malate dehydrogenase dimers for each glutamate dehydrogenase hexamer [356]. In the first complex, palmitoyl-CoA is bound to glutamate dehydrogenase but not to malate dehydrogenase, yet it protects the latter from inhibition by various ligands to which both enzymes are susceptible in the free state [624].

The glutamate dehydrogenase–aminotransferase complex has been studied extensively by Fahien and coworkers using a number of techniques including kinetics [354, 358], fluorescence [362], gel chromatography [358] and crosslinking [357], as well as the coprecipitation mentioned above. Churchich and Lee [219] confirmed the existence of interaction by nanosecond fluorescence emission anisotropy measurements. The peculiarity of this complex is that it enables the dehydrogenation of amino acids, e.g. aspartate and tyrosine, which would not occur with the separate enzymes. In effect, aspartate aminotransferase kindles aspartate dehydrogenase activity in glutamate dehydrogenase, through the following mechanism. Aspartate converts the aminotransferase into the pyridoxamine-P form, which is then dehydrogenated by glutamate dehydrogenase [358]. Thus the two enzymes cooperate in the reaction rather than the aminotransferase acting as an allosteric effector for the dehydrogenase. However, aspartate seems to have an additional role as well, in the control of dehydrogenation of amino acids. Namely, it markedly enhances alanine dehydrogenation by glutamate dehydrogenase+alanine aminotransferase and that of ornithine by glutamate dehydrogenase+ornithine aminotransferase [354]. In the two latter reactions aspartate is not transformed, it only binds to glutamate dehydrogenase increasing the affinity of the dehydrogenase to the various aminotransferases. The metabolic significance of these interactions may be that amino acid dehydrogenation is largely made independent of the presence of α-ketoglutarate. On the other hand, malate weakens the interaction between glutamate dehydrogenase and aspartate aminotransferase thereby decreasing the utilization of α-ketoglutarate [355].

Depending on the metabolic state, glutamate dehydrogenase may be bound to either malate dehydrogenase or some of the various aminotransferases, thereby influencing the flow of metabolites. Alternatively, the complexes may coexist in liver mitochondria, since glutamate dehydrogenase is in excess of the sum of its associating partners [359].

An interesting set of associations has been reported for heart enzymes by Beeck-mans and Kanarek [62] who used immobilized Sepharose-linked enzymes to detect interaction with other enzyme(s). These authors found binding between fumarase and malate dehydrogenase, malate dehydrogenase and aspartate aminotransferase and malate dehydrogenase and citrate synthase. A ternary complex of fumarase, malate dehydrogenase and aspartate aminotransferase was also observed. According to a tentative model four malate dehydrogenase and one fumarase molecules form a core to which either citrate synthase or the aminotransferase would bind depending on the metabolic need, i.e. whether the citric acid cycle or the aspartate–malate shuttle is required to run. Though not strictly relevant to mitochondria, it is worth mentioning that in *Euglena gracilis,* malate dehydrogenase occurs associated with acetyl-CoA carboxylase and phosphoenolpyruvate carboxylase [1331, 1332]. This multienzyme complex is involved in CO_2 fixation.

In conclusion, the interaction of mitochondrial matrix enzymes has been established convincingly, albeit only for a few enzyme couples. On the basis of the general considerations at the beginning of this section one might expect more data of this type. However, the very same arguments can also be used to explain the lack of such data. Namely, if conditions *in vivo* are so much favouring the aggregation of enzymes, with virtually no free water around, then extremely weak affinities between enzymes might suffice to create order. It follows that until we can approximate that milieu experimentally or design meaningful experiments with the intact organelle, there is little hope of finding many enzymes recognizing each other or assembling into a larger entity. This reasoning applies not only to the mitochondrial matrix but also to the cytoplasm and glycolysis. As we shall see in Chapter 7 on compartmentation, even very weak forces between enzymes and short-lived complexes may give rise to supramolecular order. Furthermore, the structure surfaces (mainly membranes) in living cells may serve as support for the organization of soluble enzymes, as will be described in Chapter 6.

ASSOCIATION OF ENZYMES WITH CELLULAR STRUCTURES

"Life is the art of drawing sufficient conclusions from insufficient premises."

Samuel Butler (1612–1680): Note Books
Lord, What is Man? *Life*

When a eukaryotic cell is disrupted and subjected to differential centrifugation, a classical way to obtain various subcellular structures, enzymes are found in the different particulate fractions as well as in the supernatant. The latter, the so-called cytosol, contains the soluble enzymes, i.e. enzymes that are fairly independent of cellular macrostructures. On the other hand, many enzymes are encountered in the particulate fractions. Indeed, it is a characteristic of each fraction which enzymes are present in it, the unique ones being considered as "marker enzymes" for a given particulate fraction.

The intimacy between enzymes and cellular structures varies over a wide range. At one extreme are enzymes inseparable from the structures without fatal consequences to catalytic function. These enzymes are in most cases integrated into lipid bilayers (intrinsic or integral membrane proteins), have considerable hydrophobic portions and can only be solubilized by detergent, which substitutes, in fact, for the original phospholipid environment. Other enzymes, though anchored in some biomembrane, can be released rather easily, because the main bulk of their protein is in the aqueous phase. At the other extreme are enzymes that interact with the cellular structures weakly. The weakest possible interaction is, of course, no interaction at all. It has been a widespread assertion that "soluble enzymes" are at this end of the scale. However, it is almost impossible for these enzymes not to get into contact with one surface or another, with the great abundance of various membranes present in higher cells, and one of the main goals of this chapter is to examine the phenomenon of loose enzyme–structural associations. Our other aim is to give an overview of the different kinds of structure-bound enzymes, through somewhat arbitrarily selected examples. Emphatically, our purpose is to illustrate and no endeavours toward comprehensiveness will be made. This treatment will inevitably result in omissions, but even an attempt at a reasonable coverage would inflate this chapter enormously. Moreover, the examples chosen will be described to the extent required by our context and for further details the reader should consult the references.

ENZYMES INCORPORATED INTO STRUCTURES

INTEGRAL MEMBRANE ENZYMES

These enzymes are embedded, partly or totally, in the lipid bilayer of a membrane. Incorporation into a membrane may serve two ends: (*i*) the establishment of a hydrophobic environment needed for a particular function and (*ii*) connection with transmembrane processes. The latter include not only transport processes in the strict sense, such as Na^+, K^+, phosphate, glucose, etc., crossing the membrane, but also functions coupled to ion gradients that need a barrier to be able to develop. The latter is exemplified by the case of oxidative phosphorylation. On the following pages these systems will be described in brief.

The mitochondrial oxidative phosphorylation system

The term oxidative phosphorylation denotes the chain of events when electrons derived from metabolic carriers (NADH and $FADH_2$) travel through a number of redox proteins to molecular oxygen and the energy provided by this process is utilized for ATP synthesis. The extremely complex macromolecular machinery of this main energy generating system (thirty-two out of the thirty-six ATP molecules produced from the oxidation of one glucose molecule are made here) is located in the mitochondrial inner membrane. It has been, and will remain for a long time, a major challenge to unravel how the various enzymes and non-enzymatic proteins cooperate in the running of oxidative phosphorylation. Since our interest lies in enzyme organization, we are going to touch upon this aspect of this vast field.

The sequence of reactions of the respiratory chain is illustrated in Fig. 6.1. Metabolically generated reducing equivalents from NADH and $FADH_2$ are transferred to CoQ (ubiquinone), a quinone with a hydrophobic tail consisting of three to ten isoprene units. This tail enables CoQ to rapidly diffuse in the lipid bilayer. The electron carriers following CoQ are the various cytochromes equipped with heme group(s). The redox potentials of the members of the chain are such that electron flow is unidirectional. There are three sites in the chain where the free energy of the

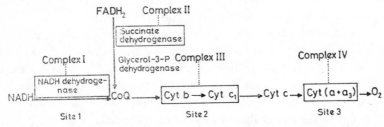

Fig. 6.1. Scheme of the mitochondrial electron transport chain. The various complexes are framed. Sites 1 to 3 designate steps in the electron transfer where free energy sufficient for the synthesis of one ATP molecule per two electrons is conserved. For more details see the text.

redox step is sufficiently large to drive the synthesis of one ATP molecule. These are at the NADH dehydrogenase (site 1), at cytochrome bc_1 (site 2), and at cytochrome c oxidase (site 3). In contrast to earlier notions, energy is conserved at these sites by conversion into a transmembrane proton gradient. The protein machinery that actually performs ATP synthesis is the so-called F_0F_1-ATPase or ATP synthase (not shown in Fig. 6.1).

The protein (enzyme) complexes of oxidative phosphorylation

The mitochondrial oxidative phosphorylation system can be resolved preparatively into a number of components (for a methodological description cf. [517]). One of these is ubiquinone (CoQ), a low molecular weight lipid-like compound; another is cytochrome c, a haem-containing small protein ($M_r = 13,000$) of known tertiary structure (for a review cf. [1040]), whose evolutionary changes are perhaps clarified best among all proteins [309]. The rest of the components are protein complexes (in fact, complex enzymes) designated with Roman numerals I through V and are listed in Table 6.1. In several cases the polypeptide composition is uncertain, hence the figures in Table 6.1 may not be conclusive. Furthermore, the structure of a given complex depends on its origin; most data refer to beef heart mitochondria, but much work has been done also with yeast and *Neurospora* mitochondria, in addition to non-mitochondrial (bacterial) systems [393, 1142]. The stoichiometry in the mitochondrion of complexes I, II, III, IV, cytochrome c and ubiquinone is 1 : 2 : 3 : 7 : 9 : 63 (cf. [489]).

The mechanism of action of oxidative phosphorylation, and of the various complexes therein, is known only in broad outlines. Clear-cut answers can only be expected after the elucidation of the three-dimensional structure of all the constituents, and of the role of the various subunits in the complexes. As regards conceptual framework, the *chemiosmotic theory* put forward by Peter Mitchell [826] seems to prevail over the alternative theories invoking a high-energy intermediate (chemical hypothesis [964, 1124]) or a conformationally activated state of an electron carrier (conformational hypothesis [121, 122, 1125]). The essence of the chemiosmotic theory is that the electron flow generates a proton gradient and membrane potential across the inner mitochondrial membrane, which, in turn, make up the driving force for ATP synthesis in Complex V. Thus the events that are to be understood chemically are how protons are extruded through the proton channel as a result of electron transfer and how the re-entry of protons makes ATP from ADP and P_i.

In spite of a wealth of data no unique picture has so far emerged for either process. It is outside our scope to go into details, for which the reader may consult specialized reviews (e.g. [123, 374, 517, 1257]). We rather add that the mitochondrial inner membrane contains other proteins also that are essential for oxidative phosphorylation. These are the P_i carrier and the ATP/ADP translocator. Their schematic arrangement in the mitochondrial membrane, as recently visualized by Durand *et al.* [326], is seen in Fig. 6.2. The P_i carrier, recognized early as inhibitable by SH-reagents [206, 387];

TABLE 6.1.

Complex enzymes of mitochondrial oxidative phosphorylation

Complex No.	Name	EC No.	Inhibitors	Source	Number of different polypeptide chains	References
I	NADH dehydro-genase (NADH-CoQ reductase)	1.6.99.3	Rotenone, piericidin A, rhein, mercurials, barbiturates	Beef heart	14–17	270, 517
II	Succinate dehydro-genase (succinate-CoQ reductase)	1.3.99.1	Oxaloacetate, malonate, thiol-reagents, 2-thenoyltrifluoro-acetone	Beef heart	8	163, 517
III	Cytochromes b-c_1 (CoQH$_2$-cyto-chrome c reductase)		Antimycin A, 2-alkyl-4-hydroxy-quinoline-N-oxide, SN-5949, polylysine	Beef heart	7	162, 287, 430
IV	Cytochrome c oxidase (a+a_3)	1.9.3.1	CN^-, N_3^-, CO, poly-lysine, sulphide	Beef heart, yeast, *Neuro-spora*	7	162, 348, 517, 956, 1078
V	F_0F_1 ATPase (ATP synthase; ATP-P_i exchange complex)	3.6.1.3	Uncouplers, oligo-mycin, venturicidine, DCCD, triethyltin sulphate, arsenate, mercurials, AMP-PNP	Beef heart	12–13	517, 1198

is a polypeptide of $M_r \approx 32,000$ [11, 490, 1328] and carries out an electroneutra cotransport of phosphate and H^+. Thus it provides P_i for oxidative phosphorylation and participates in the H^+ cycle of the mitochondrion. The nucleotide translocase, itself a protein of about $M_r \approx 30,000$ [326, 1328], catalyses an exchange transport of mitochondrial ATP–cytosolic ADP. Since oxidative phosphorylation is strongly affected by ADP supply, the translocase plays a crucial role in maintaining ATP generation. All the more so, as the kinetic parameters of the translocase seem to be influenced by the energy state of the mitochondrion [43]. The possible orientation of the translocase towards cytosolic hexokinase is discussed later in this chapter.

For the sake of illustration two complexes, cytochrome c oxidase and F_0F_1 ATPase, will be briefly described below. Both complexes have lately been vigorously studied in a great number of laboratories with the main goals to determine architectural detail (i.e. subunit arrangement) and to assign function to the various components. For a current survey of proton-translocating cytochrome complexes see [1318].

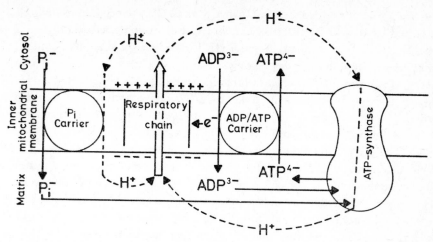

Fig. 6.2. Scheme of H$^+$, P$_i$ and nucleotide transport processes across the mitochondrial inner membrane. (From Durand *et al.* [326].)

Cytochrome c oxidase (EC 1.9.3.1)

The complex IV from beef heart or yeast consists of seven dissimilar polypeptide chains ranging from $M_r \approx 40,000$ (subunit 1) to about $M_r = 5000$ (subunit 7) and the overall molecular weight of the complex is around 2.5×10^5 (for a review cf. [348]). The three larger subunits (1, 2 and 3) are coded by mitochondrial DNA and made on mitochondrial ribosomes [157], whereas the four smaller subunits are coded in the nucleus and synthesized in the cytoplasm [1057]. The holoenzyme spans the mitochondrial inner membrane, as shown in Fig. 6.3, which gives a tentative arrange-

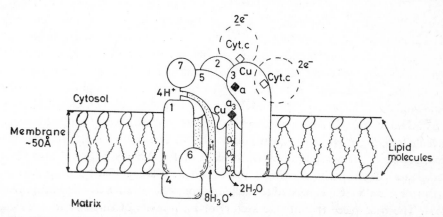

Fig. 6.3. Scheme of subunit arrangement of cytochrome c oxidase in the mitochondrial inner membrane. The numbers denote the seven different polypeptide chains. The drawing also envisages putative channels for the entry and reduction of O$_2$ and the exit of H$^+$ from the matrix space. After Chance *et al.* [202], with the modification that cytochrome c (diameter 34 Å) has been made somewhat larger to be in proportion with membrane thickness ($\sim$50 Å).

ment of the subunits as deduced by Chance *et al.* [202]. The neighbourhood relations in Fig. 6.3 cannot be regarded as conclusive. For example, crosslink data would suggest the vicinity of subunits 4 and 7 [133], as well as of subunits 3 and 6 [677]. Furthermore, subunit 2 as the binding site for cytochrome c [86, 133] has been questioned and some of the smaller subunits were suggested instead [345, 347, 348]. However, it should be borne in mind that crosslink data establishing different contacts need not be mutually exclusive and, more importantly, the mobility of cytochrome c over the functioning complex [1200] may well result in a number of subunit interactions. The orientation of subunits relative to the membrane ("sidedness") is fairly well characterized (cf. [395]), although findings at variance with Fig. 6.3, e.g. the accessibility of subunits 5 and 7 from the matrix side [752], have also been reported.

The seven subunits seem not to be equally important for catalytic activity. The three major subunits containing much α-helix running perpendicular to the membrane plane [528] fulfil an anchoring role in the membrane. Removal of two [940] or all the three [941] of them from the complex had hardly any effect on enzyme activity. The location in the complex of the haem a and a_3 prosthetic groups, and of the two copper atoms that are part of the redox system, is uncertain. The orientation of both hemes is at right angles to the membrane plane [346, 729], just as for the haem group of bound cytochrome c [202]. Fluorescence resonance energy transfer measurements revealed that the latter and the haem in cytochrome oxidase are about 25 Å apart. It appears at present puzzling how electrons can travel that great distance.

Current knowledge about cytochrome oxidase has been compiled in a monograph by Wikström *et al.* [1319].

F_0F_1 ATPase (ATP synthase)

This is the macromolecular structure in which the proton gradient is utilized to drive ATP synthesis. The general outlay of the complex is as follows (Fig. 6.4). It is composed of two separable parts that have different functions: (*i*) the catalytic F_1-ATPase, which corresponds to the projections seen on electron micrographs of the mitochondrial inner membrane [967], and (*ii*) the proton-translocating F_0, divided into a "base piece" and "stalk", which bears the binding sites for phosphorylation inhibitors oligomycin [715] and dicyclohexylcarbodiimide (DCCD) [61]. F_1 can be detached from F_0 in a number of ways, such as treatment with urea [965], organic solvents [60], NaBr [1245], and cardiolipin [1228]. The stalk is built up of two proteins, selectively removable after the release of F_1, namely F_6 [608] and OSCP, the oligomycin sensitivity conferring protein [764]. It seems that F_6 alone is able to fix F_1 to F_0, and OSCP is needed to make the complex sensitive to oligomycin [1036, 1251]. The base piece (membrane sector) of F_0 consists of three or four strongly hydrophobic proteins referred to as "proteolipids" [1082, 1086]; one of them binds DCCD [177, 1115].

The F_1-ATPase is composed of five tightly associated subunits (α to ε) of molecular weights 53,000 (α), 50,000 (β), 33,000 (γ), 17,000 (δ) and 7500 (ε) in case of beef heart

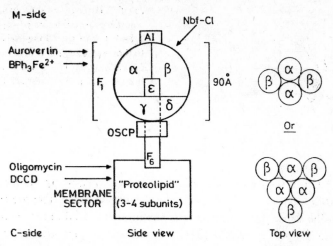

Fig. 6.4. Subunit topology and some inhibitors of mitochondrial F_0F_1 ATPase. In the top view two possible arrangements of α and β subunits are seen, depending on whether these subunits are present in two or three copies in the complex. Orientation in the side view: M (matrix) side and C (cytoplasmic) side. Abbreviations not found in the text: BPh_3Fe^{2+}, tris-bathophenanthroline-Fe^{2+} chelate; Nbf-Cl, 4-chloro-7-nitrobenzofuran. (From Ernster *et al.* [349] with the modification that F_6 is now directly connecting F_1 and the base piece.)

mitochondria (cf. [1082]). In the latter, but not in F_1-ATPases from other sources (non-mammalian mitochondria, chloroplasts and bacteria), a sixth subunit is also found, the ATPase inhibitor, which is readily dissociated from F_1 [922, 1084]. Cross-link experiments have shown the proximity of α-α and α-β, but not of β-β, suggesting the arrangements seen in Fig. 6.4. The catalytic centre is probably on, or involves, the β subunit, which carries the binding site for the inhibitor 4-chloro-7-nitrobenzo-furan [372]. Little is known about the function of the other subunits, though δ has been invoked in coupling F_1 to F_0 [1128]. Progress will probably be speeded up by studies on bacterial ATPases, where genetic work is also feasible (cf. [374]). In *E. coli* all mutants in both F_0 and F_1 map in a single locus called *unc* (cf. [319]), which is transcribed as a single unit, i.e. is within one operon [444]. Several *unc* complementation groups have already been described [444]. Work on mutants, among them also assembly mutants, is likely to shed light on both the function of individual subunits and the details of their topology.

Recent reviews on the mechanism of action and regulation of ATP synthesis by F_1-ATPases were given by Penefsky [925] and Cross [269].

The dynamic organization of oxidative phosphorylation

The sequence of reactions that make up the respiratory chain (cf. Fig. 6.1) can only take place if the components are in appropriate proximity. This can be achieved in two ways: by the build-up of fairly rigid supercomplexes in which the spatial

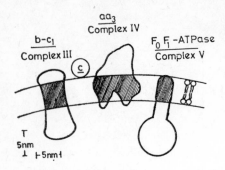

Fig. 6.5. Three major tran smembrane protein complexes of oxidative p hosphorylation are oriented with their long axis perpendicular to the membrane plane an d only a small part (shaded area) is embed ded in the lipid bilayer. (From Hackenbrock [489].)

relations are such to secure the vectorial flow of electrons, or by the free movement and collision of individual components in the membrane plane. The prevailing notion has been that the dense protein packing of the mitochondrial inner membrane (75% protein and 25% lipid by weight [241]) favours the former alternative. However, as pointed out recently by Hackenbrock [489], protein packing in the mitochondrial membrane is not all that dense. Only a minor portion of the major complexes is actually in the membrane, their main bulk protrudes into the aqueous phase (Fig. 6.5). Thus electron microscopy of reconstituted "protein membrane crystals" of cytochrome c oxidase revealed that only 30% of the protein is in the bilayer [412]. The same proportion was found for Complex III [730], whereas with Complex IV the membrane-embedded part is merely 17% of the total mass [1143]. Freeze-fracture pictures testify that only one-third to one-half of total membrane area is occupied by proteins [552], which can also be electrophoresed to one pole of the membrane [1145]. The easy lateral movement of proteins [209] is facilitated by the unusually low microviscosity (0.9 P [365]) in the bilayer for a lipid molecule, which makes the mitochondrial inner membrane one of the most fluid among eukaryotic membranes.

In experiments where liposomes were fused with the inner membrane so that the bilayer lipid was increased up to seven-fold, the distance between intramembrane particles (i.e. protein complexes) visible on freeze-fracture surfaces became greater [1064, 1065]. This enlargement of lipid layer did not inhibit the functioning of the individual complexes, as tested for cytochrome c oxidase and NADH dehydrogenase (in fact, both reactions were somewhat activated by the extra phospholipid), but impaired their coupled reactions proportional to the degree of phospholipid enrichment [1065]. This was found with both the NADH dehydrogenase–cytochrome bc_1 (reaction: NADH→cytochrome c), and the succinate dehydrogenase–cytochrome bc_1 (reaction: succinate→cytochrome c) couples, and less markedly also with the cytochrome bc_1–cytochrome c oxidase (reaction: $CoQH_2 \rightarrow O_2$) couple [489]. The lateral [554] and rotational [626] diffusion of cytochrome c oxidase has indeed been observed. All these kinetic and structural data support the view that at least some elements of the respiratory chain move freely in the membrane plane and electron transfer takes place during the short-lived encounters. The respiratory chain is thus a chain only in the functional but not structural sense. This may then be regarded as an example of dynamic metabolite compartmentation through transient enzyme–enzyme complexes (cf. Chapter 7, p. 189), with the qualifications that here electrons at different energy levels are the "metabolites" and one of the interacting parties, CoQ, is a lipid.

Depending on their size, the various elements of the above system can move with different rates. The average lateral diffusion coefficient of integral protein complexes is 8.3×10^{-10} cm^2/s [1144], that of cytochrome c on the bilayer surface is 1×10^{-8} cm^2/s [902, 1010], whereas CoQ presumably travels in the bilayer as a phospholipid, with about 10^{-8} cm^2/s. Cytochrome c is the most abundant of all the proteins involved and CoQ surpasses it, on a molar basis, about seven-fold. These most mobile parties may determine the overall rate of electron flow by providing the most rapid pathways between Complexes III and IV and Complexes I or II and III, respectively. (It is assumed that electron flow within each complex is much faster than that between complexes.) This suggestion is substantiated by the finding that if CoQ was also included in the liposomes with which the mitochondrial membrane was fused, the decrease in the rate of succinate $\rightarrow$ cytochrome c or NADH $\rightarrow$ cytochrome c reactions was prevented, in spite of the large expansion of the bilayer [1063].

Transport ATPases

These complex oligomeric proteins are functionally reciprocals of the mitochondrial ATPase described above: the latter utilizes an ion gradient to synthesize ATP, the former utilize ATP to generate ionic gradients. They are also called ion pumps, since the characteristic ionic milieu on either side of a biomembrane is created by their pumping activity. In the strict sense they are not to be regarded as enzymes, because they use chemical energy to drive a physical process, the translocation of ions through membranes, rather than a chemical one. Nevertheless, this distinction is not generally made, because transport ATPases have much in common with oligomeric enzymes. They catalyse a process with saturation kinetics so that a great deal of phenomenological enzyme kinetics can be applied to them, they split ATP to ADP and P_i during each turnover, and the mechanism of action presumably involves reversible conformational changes manifested in alternating subunit interactions.

The prototype and best-studied case of these ion pumps is the *$Na^+ + K^+$ activated ATPase* (ATP phosphohydrolase, EC 3.6.1.3). It was discovered by Skou [1122] in 1957, who described an ATPase preparation from crab nerve membrane that was stimulated by Na^+ plus K^+. Since then the $(Na^+ + K^+)$ ATPase has been isolated from a variety of sources, above all from mammalian kidney medulla, shark rectal gland and eel electric organ, and even reconstituted in phospholipid vesicles in the active form. Recent reviews on the structural, enzymatic, transport and energetic aspects of the Na^+–K^+ system are given by Robinson and Flashner [1011], Jørgensen [600] and Schwartz *et al.* [1074].

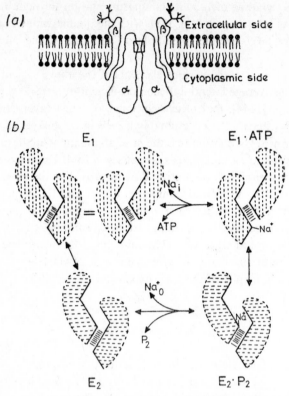

Fig. 6.6. Orientation of subunits relative to the membrane and proposed mechanism of action of (Na$^+$+K$^+$) ATPase. (*a*) Disposition of subunits; the branched structure attached to β-subunit is carbohydrate. The area enclosed by the small rectangle at the α–α contact is assumed to serve as "**gate**" for the ions; (*b*) Functioning of the "gate" at the α–α contact. E$_1$ and E$_2$ designate two conformational states of the enzyme. Na$^+$ transport is effected by the change in subunit contacts. The two structures under E$_1$ are equivalent. (From Guidotti [482].)

The (Na+K$^+$)ATPase consists of two copies each of two different polypeptides. Thus the overall structure is $\alpha_2\beta_2$ (Fig. 6.6*a*). This may just be the basic entity as ligand binding studies [512] suggested the existence of higher oligomers ($\alpha_4\beta_4$ and $\alpha_8\beta_8$). The α subunit ($M_r \approx 100,000$) carries the active site of ATPase, the binding site for the inhibitor ouabain and, as regards topology, it protrudes into the cytoplasmic space. (For recent data on the labelling of α subunit see, for example, [852, 1199 and 1206].) The β subunit is a glycoprotein of $M_r \approx 40,000$ (without the carbohydrate). Its function is unknown. A third type of protein, a proteolipid called γ subunit ($M_r \approx 12,000$) has been identified [389, 1074], but its relation to the major subunits has not yet been clarified. Phospholipids are essential for the ATPase activity [897, 955].

The most probable stoichiometry of the (Na$^+$+K$^+$)ATPase reaction is 3 Na$^+$/2K$^+$/1 ATP, i.e. three Na$^+$ ions are extruded and two K$^+$ ions are taken up at the expense

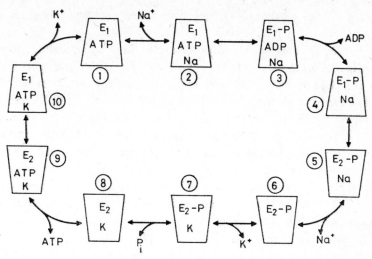

Fig. 6.7. Reaction scheme of the $(Na^+ + K^+)$ ATPase according to Karlish *et al.* [615]. Ten elementary steps and, correspondingly, ten molecular species are distinguished. The two states of the pump-enzyme, E_1 and E_2, are represented by wide-based and wide-topped trapezoids, respectively. (Adapted from Robinson and Flashner [1011].)

of one molecule of ATP (cf. [459]). However, higher Na^+/K^+ ratios cannot be excluded [1038]. A host of models have been proposed for the mechanism of action of this ATPase. The models must account not only for the above ATP-driven coupled Na^+/K^+ exchange, but also for other transport modes (uncoupled Na^+ efflux, Na^+/Na^+ exchange, K^+/K^+ exchange) and enzyme activities (Na^+-dependent ADP/ATP exchange, K^+-dependent enzyme phosphorylation by P_i, K^+-dependent phosphatase activity) observed *in vitro* (cf. [1011]). The reaction scheme deduced from kinetic studies is seen in Fig. 6.7. The main feature of the mechanism is that the ATPase can exist in two different conformational states: E_1 characterized by high affinity for Na^+ and ATP, and E_2 showing lower affinity to these ligands and favouring K^+. In the sequence ATP phosphorylates a specific site (an aspartyl residue) on the α-subunit while the system is in the E_1 state, then ADP is released followed by transconformation into E_2. In the latter state Na^+ is replaced by K^+, which promotes the dephosphorylation of the protein. (A current finding is that if Mg^{2+}-ATP is replaced by Cr^{2+}-ATP, the enzyme becomes inactivated through the formation of an almost non-reactive phosphoprotein [920].) After P_i release the binding of ATP, now with low affinity, induces the recovery of the E_1 state with the consequential discharge of K^+. It should be mentioned that though this reaction mechanism, put forward by Karlish *et al.* [615], appears the most probable one at present, it has not been conclusively proven. Furthermore, the scheme in Fig. 6.7 does not invoke the oligomeric nature of $(Na^+ + K^+)$ATPase. Nevertheless, this can be done by postulating that the two half-complexes are operating in a kind of flip-flop, with 180° phase shift between the two halves [1011]. Such a structural coupling may entail energetic advantages.

Whether the active sites work independently or in a reciprocating manner, the mechanism in Fig. 6.7 throws no light on how vectorial cation transport ensues, i.e. why Na^+ is picked up inside and released outside and the reverse of K^+. The idea of a "mobile carrier", analogous to the ionophore valinomycin [878], which moves with the ion across the membrane, can be abandoned. The highly asymmetric and invariant orientation of the ATPase subunits in the lipid bilayer rules out any kind of rotational movement around an axis parallel to the membrane. The alternative is an "oscillating pore" or "peristaltic channel" through which ions move past, rather than with, the pump proteins. In lack of knowledge about the three-dimensional structure of $(Na^+ + K^+)$ATPase, all mechanisms alluding to protein stereochemistry are at present necessarily hypothetic.

A simple model has been drawn by Guidotti [482] (Fig. 6.6b). Here a small conformational change altering the contacts between the two α-subunits would displace the ion from one compartment to the other. The assumed structural transition resembles the change in the $\alpha_1\beta_2$ contact of haemoglobin on oxygenation, when the interlocking contact clicks from one position to another through a switch in hydrogen bonding [842]. This mechanism attributes immediate significance to the oligomeric nature of transport ATPase, but it fails to explain the $3:2$ ratio of Na^+/K^+.

Beside the above example of a unitary mechanism (i.e. the same channel serves for both Na^+ and K^+), there is a family of binary mechanisms of the "oscillating pore" type, where Na^+ and K^+ pass through different routes. The one set forth by Chandler et al. [203] is quite elaborate, inasmuch as it equates pore oscillations with helical transitions and also offers an explanation for the Na^+/K^+ ratio (Fig. 6.8). According to this proposal Na^+ channels would be formed by helices running perpendicular to the membrane plane, a structural feature indeed characteristic of integral membrane proteins [636]. The K^+ channels would be located between the helices of the same or another subunit. The energy of ATP hydrolysis instigates the sliding of the Na^+ channel from the α-helix to the wider π-helix configuration, then the Na^+ may enter and travel along the whole thickness of the membrane. If the two types of channel are juxtaposed, the dilation and subsequent constriction of the Na^+ channel are transmitted to the K^+ channel, opening the latter for K^+ as the Na^+ channel is closed. It is part of the hypothesis that three intrahelical Na^+ channels flank two interhelical K^+ channels (Fig. 6.8a), consequently, the passage of three Na^+ ions is followed by the counterflow of two K^+ ions. The energy requirements of these helical transitions in a helix-triplet can apparently be met by the hydrolysis of a single ATP molecule [203, 1011]. The mode of using the chemical energy to trigger these transitions is unknown.

An interesting but controversial recent development is the outlining of a "kinase cascade" that eventually results in the phosphorylation of a tyrosine residue on the β subunit impairing the Na^+ pump function [1147]. This cascade, including four en-

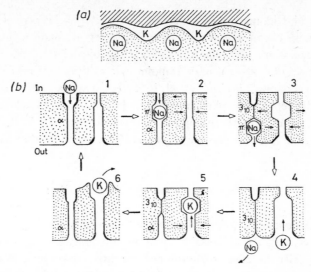

Fig. 6.8. Mechanistic model of Na^+ and K^+ transport: the "gliding edge" hypothesis of Chandler et al. [203] as adapted by Robinson and Flashner [1011]. (a) Two K^+ channels are formed between three Na^+ channels, at the interface of two subunits. The Na^+ channels correspond to intrahelical spaces. (b) Na^+ and K^+ transport. Transitions between three helical forms are assumed: (1) the α-helix (3.6_{13} helix = 3.6 residues per turn and every 13th atom hydrogen bonded), impermeable to cations; (2) the π-helix (4.4_{16} helix = 4.4 residues per turn and every 16th atom hydrogen bonded) having a central cavity large enough to accommodate Na^+ but not K^+; (3) the 3_{10} helix (notation analogous to above), a tight structure that constricts the channel. In the successive steps (1) through (6) Na^+ leaves the cells followed by K^+ uptake. Thin arrows indicate the movements of ions and channel walls.

zymes [1148], was reported to operate only in tumour cells and would explain Otto Warburg's 60-year-old observation of enhanced glycolysis in tumour cells, in terms of an increased production of ADP and P_i by the faulty $(Na^+ + K^+)ATPase$. Regrettably, doubts were cast on the reliability of some of the data and certain already published papers (not cited above) were retracted by the authors [1275]. When writing these lines (September 1981), it is an open issue how much of the kinase cascade will stand the test of time.

The Ca^{2+}-activated ATPase is present in most plasma membranes and in some intracellular membranes, above all in the sarcoplasmic reticulum. The sarcoplasmic Ca^{2+}-ATPase, responsible for maintaining the appropriate Ca^{2+} concentration regulating muscle function, is structurally the best known. It is an oligomer of a single type of polypeptide chain ($M_r \approx 110,000$ [763, 1046]) whose amino acid sequence is partially revealed [9]. The state of oligomerization has been claimed to be tetrameric [767, 853, 1054, 1259], but a dynamic equilibrium between monomers and oligo(tetra?)-mers has also been detected [12, 209, 371, 558]. The mechanism of action of Ca^{2+}-ATPase may be analogous to that of the $(Na^+ + K^+)ATPase$ [482], but a mechanism with oligomerization-induced Ca^{2+}-release has been invoked, too [1259]. From the protein chemical point of view, the latter mechanism is an adaptation of enzyme regulation via oligomer association–dissociation (cf. p. 66) to the membrane phase. Our present understanding is insufficient to decide which of the two types of mechanism, if either at all, is correct. The Ca^{2+}-ATPase is stimulated by cyclic AMP-dependent phosphorylation [1075], and by calmodulin either through direct binding to the pump, as in erythrocytes [591, 717], or via protein kinase action, as in the sarcoplasmic reticulum [569]. For details see Sarkadi's review [1045].

Enzymes of cyclic nucleotide metabolism

The cyclic nucleotides, cyclic AMP (cAMP, adenosine-3',5'-monophosphate) and cyclic GMP (cGMP, guanosine-3',5'-monophosphate) are *second messengers* mediating to the cells the effect of hormones, the first messengers. Both cAMP and cGMP are generated from the corresponding nucleoside triphosphates by complex enzymes called adenylate cyclase and guanylate cyclase, respectively. They are degraded by a variety of cyclic nucleotide phosphodiesterases. In this system of enzymes only adenylate cyclase is a typical integral membrane enzyme, the others have been found in both particulate and soluble fractions of cells. (For a review see Daly [275].)

The *adenylate cyclase system* of many mammalian cell membranes consists of three components: the catalytic subunit C, the hormone receptor protein R, and the guanine nucleotide-binding protein G. Three types of subunit C have been described. The first behaves as a true intrinsic membrane protein on the basis of detergent binding; it was found in bovine cerebral cortex [860] and cultured lymphoma cells [491]. The second type does not bind detergent, even though it requires detergent to be dislodged from the membrane; this enzyme was encountered in rat renal medulla [859]. The third type of catalytic subunit is water soluble (60 to 80% found in the supernatant after simple buffer extraction and centrifugation); this was detected in rat testis [127] and bovine sperm [531]. It is possible, however, that the second and third types are preparational artefacts that lost their hydrophobic anchoring domain during isolation [860].

Subunit R is a hormone receptor protein. The pioneering work of Sutherland and coworkers (cf. [1012]) revealed that catecholamines stimulated some adenylate cyclases with the concomitant increase of intracellular cAMP level. The receptor was then identified by affinity labelling with radioactive agonists (reviewed in [726] and [766]) and antagonists [727, 1249]. It was isolated by digitonin solubilization and affinity chromatography [167]. Though its molecular details are not known yet, agonists increase the apparent molecular weight of the receptor either through a conformational change or by association with another protein [167, 739]. This partner might be the guanine nucleotide-binding protein, G. The latter is also an intrinsic membrane protein located on the cytoplasmic face of the plasma membrane [616].

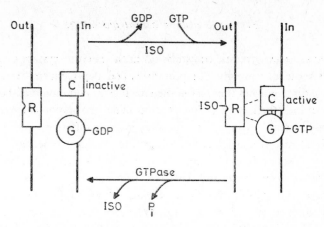

Fig. 6.9. The guanine nucleotide exchange cycle in the regulation of adenylate cyclase. C, Catalytic subunit; R, hormone receptor; G, guanine nucleotide-binding protein; ISO, isoproterenol (an agonist). Membrane sidedness is indicated. (Adapted from Spiegel *et al.* [1150].)

The recent discovery of protein G was anticipated by the demonstration of GTP requirement for the glucagon activation of adenylate cyclase more than 10 years ago [1015]. The mode of participation of guanine nucleotides in influencing adenylate cyclase activity has recently been dealt with by Spiegel *et al.* [1150]. The system studied by these authors was the beta-receptor of turkey erythrocytes. Here GTP binding to protein G activates the enzyme possibly through heterologous interaction between proteins G and C (Fig. 6.9). Enzymatic cleavage of GTP to GDP and P_i brings the system to the inactive state. Nothing is known about the protein chemistry of these transitions. In most cell types cAMP exerts its effect by activating protein kinase(s). Interestingly, in turkey erythrocytes the increased cAMP level induced by agonists regulates Na^+ and K^+ transport [421], by phosphorylating a membrane protein ($M_r=240,000$) named goblin [59].

Guanylate cyclases differ from adenylate cyclase in that they can be found in both soluble and particulate fractions. The two types of enzyme are apparently different entities [649, 858]. Little is known about their protein chemistry. In the nervous system guanylate cyclases have a presynaptic localization (cf. [275]).

cAMP phosphodiesterases and *cGMP phosphodiesterases* cleave the respective cyclic nucleotides to the 5'-phosphate. Both have soluble and particulate forms, moreover, some isoenzymes hydrolyse both cAMP and cGMP (e.g. [606, 946]). In brain, heart and some other tissues there is a Ca^{2+}-dependent activator protein for phosphodiesterase, which is also found in both soluble and particulate fractions [452]. This is the ubiquitous calmodulin. In yeast, the particulate enzyme is bound to the membrane(s) through ionic interactions rather than hydrophobic embedding [748]. Furthermore, one of the several isoenzymes seems to be associated with ribosomes [749].

Enzymes degrading neurotransmitters

These enzymes arrest synaptic transmission in the nervous system by transforming various small molecular weight compounds called neurotransmitters. This highly specialized role distinguishes them from the enzymes of common metabolic pathways and also determines their cellular localization. The two major enzymes briefly described here achieve anchoring at specific sites in quite different ways.

Acetylcholinesterase (EC 3.1.1.7) hydrolyses the cholinergic neurotransmitter acetylcholine in synapses and motor endplates in muscle. Its richest source is the electric organ of the fish *Electrophorus* and *Torpedo*. Enzymes from all origins seem to be similarly tailored and exhibit multiple molecular forms (cf. [8, 108, 110]). When extracted at low salt concentration the G (globular) forms can be obtained, which are the monomer (G_1), dimer (G_2) and tetramer (G_4) of the catalytic subunit. *In vivo* the tetramer seems to prevail, with all subunits being catalytically active [1267]. At high salt concentration the A (asymmetric or aggregating) forms are extracted from tissues. These forms have a collagen-like tail and carry one (A_4), two (A_8) and three (A_{12}) catalytic tetramers. The molecular weight of A_{12} is about 10^6 [110]. The shape of the molecule is seen in Fig. 6.10. The mode of immobilization of acetylcholinesterase forms at the postsynaptic membrane (the site of occurrence of acetylcholine receptor protein) is intriguing. The collagenous tail has long been thought to serve as an anchor by analogy with brush-border enzymes (cf. below), the tail is apparently not embedded in the lipid bilayer, but rather fixed in the synaptic cleft by ionic interactions [21, 109, 110]. Furthermore, the tail is not a prerequisite of binding, as the A forms seem to be absent from the central nervous system of birds and mammals [774, 1008]. In the latter case it is assumed that the G forms are bound through hydrophobic forces.

Monoamine oxidase (MAO; monoamine: O_2 oxidoreductase, EC 1.4.3.4) is located in the outer mitochondrial membrane and degrades amine neurotransmitters in the nervous system and biogenic amines in most tissues. Two types of MAO, A and B, can be distinguished on the basis of substrate specificity and inhibitor sensitivity [668, 861], as well as susceptibility to neuramidinase [571]. The question whether these forms correspond to distinct polypeptide chains or are just conformers stabilized by the particular lipid environment seems now be settled in favour of the first alternative. Thus, MAO can be specifically inhibited by pargyline, by forming a stable adduct with its flavin cofactor [1320]. After labelling MAO A and B with [³H]pargyline, two radioactive polypeptides were obtained by SDS gel electrophoresis with molecular weights around 60,000, the A form being larger than B by about 3000–5000 daltons. This distinction has recently been made for human [137, 179] and rat [160, 178] enzymes, and non-identity was also supported by limited proteolysis and peptide mapping. The two types of MAO are bona-fide integral membrane proteins, in contrast to acetylcholinesterase. At present it is not clear whether the ~60,000 dalton subunits oligomerize or associate with other subunit(s) not binding pargyline, in the mitochondrial outer membrane.

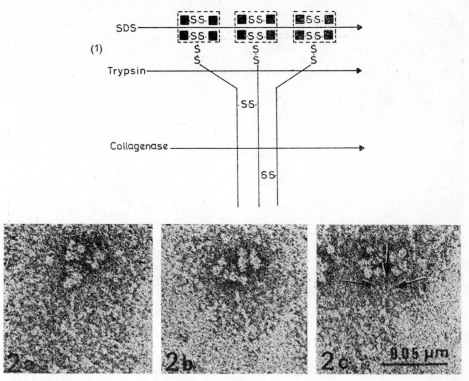

Fig. 6.10. The structure of acetylcholinesterase. (1) Model of eel 18S acetylcholinesterase molecule showing disulphide bridges and points of cleavage by sodium dodecyl sulphate, trypsin and collagenase. (From Anglister and Silman [21].) (2) Electron micrographs of eel acetylcholinesterase molecules negatively stained with uranyl formate. In (*a*) and (*b*) the head is formed by the assembly of three tetramers and a 30–40 nm long tail is discernible. In (*c*) the arrows point to the three filaments that make up the tail and are separated close to the head. (From Cartaud *et al.* [170].)

Brush-border enzymes

The brush border is a collective name for mammalian cell surfaces that carry cylindrical projections, called microvilli, about 0.1 μm in diameter and 1 to 2 μm long. Whenever a mammalian cell surface is not in contact with another cell but rather is turned toward a liquid-filled space, it tends to develop such microvilli. This trend can be observed with many cell types but it is most conspicuous in the epithelial cells of the intestine, the kidney tubules and the placenta. Here the thick carpet of microvilli bordering the lumen of the respective organ gives the impression of a "brush" on micrographs; hence the name. The microvillus consists of a *core*, which is a bundle of microfilaments and ensures mechanical stability, and the *membrane* wraps the core. The latter is the anchoring layer of brush-border enzymes (Fig. 6.11). It has been claimed that the villus is to be regarded as a separate compartment,

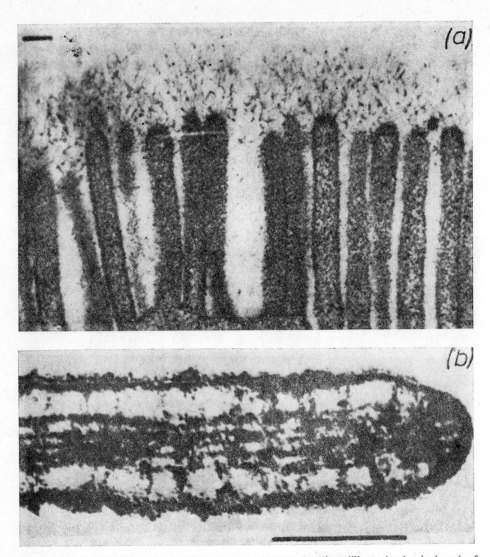

Fig. 6.11. Microvilli of intestinal brush-border membrane (*a*) Microvilli at the luminal end of intestinal epithelial cells. The polysaccharide glycocalyx is seen as the "fuzz" at the tip of microvilli. (From Ito [582].) (*b*) Ultrastructure of the microvillar core. In the middle the parallel actin filaments are seen, which are cross-bridged to the membrane by α-actinin. Bar = 100 nm. (From Mooseker and Tilney [838].)

as it seems to be separated from the cytoplasm as witnessed by the apparent absence of glycolytic enzymes in it [153]. A review on the ultrastructural and molecular organization of microvilli was given by Kenny and Booth [638].

Table 6.2 lists the enzymes found associated with the brush-border membrane. The three major classes of enzymes are peptidases, glycosidases and phosphatases. The role of these hydrolytic enzymes in the gut is quite clear: they complete the digestion of food by producing the readily absorbable monomeric units such as amino acids and monosaccharides. It is noteworthy that all carbohydrate hydrolysing enzymes in the brush border are active on disaccharides. Glucoamylase also attacks amylose or amylopectin at the non-reducing end [650], but prefers oligosaccharides of about nine residues [631]. Phlorizin hydrolase splits phlorizin into glucose and phloretin; its natural substrates are probably complex glycolipids [725]. The presence of disaccharidases in the brush-border membrane may not only serve the breakdown of substrates but may also contribute to the transport of products. At least this is suggested by the phenomenon called "kinetic advantage", by which it is meant that hexoses when combined in disaccharides are taken up by the intestine more readily than are free hexoses [976]. This has been confirmed for glucose uptake from phlorizin [510]. In fact, a similar relation has been found between the absorption of dipeptides and amino acids [792]. The γ-glutamyl transpeptidase has been particularly invoked by Meister [810] as a member of a cycle running amino acid uptake; this enzyme catalyses the transfer of γ-glutamyl residues from glutathione to amino acids, the product entering the cell and being cleaved by another enzyme. However, the transport role of γ-glutamyl transferase is still ambiguous, because in a patient who had a defect in this enzyme amino acid transport was normal [924, 1071]. Likewise, the involvement of alkaline phosphatase in phosphate transport through the brush border has recently been refuted [1215].

It is a common feature of brush-border enzymes that they are anchored in the membrane by a fairly short polypeptide segment (30 to 50 residues) at the N-terminus. They can be extracted from the membrane by either detergent (Triton X-100 or Emulphogen BC 720) treatment or proteolysis with papain, trypsin or elastase. The "proteolytic form" is shorter than the "detergent form" by a segment of about 3000–10,000 molecular weight. Thus brush-border enzymes are integrated in the lipid bilayer only through a short stalk, while the main bulk of the protein globule is in the aqueous luminal phase. For this reason Semenza and coworkers [139] suggested the term "stalked intrinsic membrane protein" for them. The proteolytic and detergent forms are equally active, but the latter tends to polymerize in the absence of detergent, unlike the former, probably due to its hydrophobic extension.

A typical and fairly well-studied example of brush-border enzymes is the sucrase–isomaltase complex (cf. [725]). It is a glycoprotein of $M_r \approx 220,000$, the isomaltase component being slightly larger than the sucrase component. The complex is rather tight and can be dissociated only by boiling in SDS or extensive chemical modification (citraconylation). Of the two enzymes only isomaltase interacts with the mem-

TABLE 6.2.

Brush-border enzymes

Class	Name	EC number	Tissue	References
Glycosidases	Sucrase[a]	3.2.1.48	Intestine ⎫	10, 126, 139,
	Isomaltase[a]	3.2.1.10	Intestine ⎬	279
	Lactase[a]	3.2.1.23	Intestine ⎭	84, 279, 478, 1241
	Phlorizin hydrolase[a]	3.2.1.62	Intestine	84, 725, 769
	Maltase[a]	3.2.1.20	Intestine	1241
	Trehalase	3.2.1.28	Intestine	857, 1047
	Glucoamylase[a] (γ-amylase)	3.2.1.3	Intestine	279, 631, 1061
Peptidases	Neutral endopeptidase	3.4.24.–	Kidney	646
	Aminopeptidase A	3.4.11.7	Kidney, intestine	16, 451
	Aminopeptidase M	3.4.11.2	Kidney, intestine	781, 979, 1277
	Aminopeptidase P	3.4.11.9	Kidney	639
	Aminopeptidases I and II		Intestine	475
	Dipeptidyl peptidase IV	3.4.14.–	Kidney, intestine	279, 875
	γ-Glutamyl transpeptidase	2.3.2.2	Kidney, intestine	471, 568, 875
	Peptidyl dipeptidase (angiotensin I converting enzyme)	3.4.15.1	Kidney	344
	Carboxypeptidase P (prolyl)	3.4.12.–	Kidney	639
Phosphatases	Alkaline phosphatase	3.1.3.1	Intestine, kidney, placenta	166, 239 279, 637
	5′-Nucleotidase	3.1.3.5	Kidney	434
	Phosphodiesterase I	3.1.4.1	Kidney	293
	D-Inositol 1-2 cyclic phosphate 2-phospho-hydrolase	3.1.4.36	Kidney	293
Other enzymes	$(Na^+ + K^+)$ ATPase		Kidney	963
	Mg^{2+} ATPase (azide insensitive)		Kidney, intestine	152, 651
	Adenylate cyclase		Kidney	736
	Cyclic AMP dependent and independent protein kinases		Kidney	433

[a] The following complexes of these enzymes have been detected: sucrase–isomaltase [126, 139]; maltase–sucrase–glucoamylase, maltase–sucrase and maltase–glucoamylase [477]; and lactase–phlorizin hydrolase [84, 977].

brane, carrying the hydrophobic stalk which probably consists of residues 12 to 31 near the N-terminus:

$$(CHO)$$
$$|$$

Ala-Val-Asn-Ala-Phe-Ser-Gly-Leu-Glu-Ile-Thr-
1   5   10

Leu-Ile-Val-Leu-Phe-Val-Ile-Val-Phe-Ile-Ile-Ala-Ile-Ala-Leu-Ile-Ala-Val-Leu-Ala-
15   20   25   30

x-x-x-Pro-Ala-Val-

The underlined segment is one of the most hydrophobic ones found so far, which explains its strong anchoring function. The complex does not seem to be part of the Na^+, D-glucose transporter, as shown by their selective inhibition [511]. The clinical significance of this phenomenon, and of brush-border enzymes in general, will be treated in Chapter 9.

The enzymes listed in Table 6.2 other than glycosidases and peptidases are, with the further exception of alkaline phosphatase, less characteristic brush-border enzymes, but have all been detected in the brush border with reasonable confidence. Their disposition relative to the membrane is scarcely known. For more detail the reader is referred to the references in Table 6.2.

RIBOSOMAL AND NUCLEAR ENZYMES

These enzymes distinguish themselves by having an intimate connection with nucleic acids. This relationship is not necessarily of the enzyme–substrate nature, though it may include this, but rather the nucleic acids act as a scaffolding for these enzymes. Some other enzymes, e.g. the aminoacyl-tRNA synthases [1060], that are active on nucleic acids, are not considered here. In the ribosome and cell nucleus the enzymes are harboured by DNA or RNA, the tightness of binding varying over a wide range.

Proteins are manufactured on *ribosomes,* which are the only organelles, beside nucleosomes, that do not have a boundary membrane of their own. They translate the genetic message that arrives to them in the form of messenger RNA, into a polypeptide of given amino acid sequence. In a manner of speaking, chemically the ribosomes are able to catalyse the synthesis of peptide bonds and, physically, they see to it that the appropriate amino acid be incorporated. Ribosomes are mainly found in the cytoplasm, either free or attached to the endoplasmic reticulum membrane, but they also occur in other, partially autonomous entities such as the mitochondria and chloroplasts.

Investigations on the structure of ribosomes have been vigorously pursued ever since their discovery in the early 1950s. A bacterial cell contains, on the average, 10^4 ribosomes, whereas cells of higher organisms may have many times more. Ribosomes from prokaryotes are smaller (70S) than those of eukaryotes (80S), but the

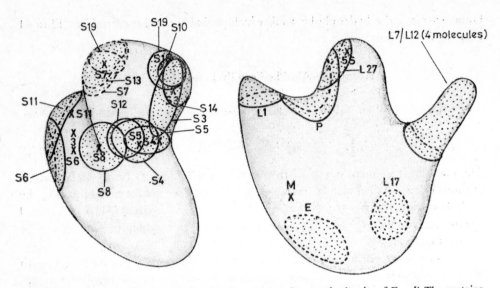

Fig. 6.12. Artist's view of the small (left) and large (right) ribosomal subunits of *E. coli*. The proteins located are numbered in the small (S) and large (L) subunits and are shown as encircled areas, the centre of masses designated by crosses. Peptidyl transferase activity is located on the left-hand ridge of the large subunit (area P). Further sites: M, point of attachment to the membrane; 5S, location of 5S RNA; E, site where the newly synthesized polypeptide chain emerges from the ribosome. (From Lake [710].)

gross structures are similar. The ribosome from *E. coli* has been studied in the greatest detail (for reviews cf. [134, 706, 710]).

Ribosomes have an overall molecular weight of about 2.5×10^6. They consist of two asymmetrical subunits, each a complex nucleoprotein. The smaller subunit (30S) is composed of an RNA molecule of $M_r \approx 600,000$ and one molecule each of twenty-one different proteins. The total molecular weight of the small subunit is $\sim 900,000$. The large subunit (50S) is built up of two RNAs and thirty-four different proteins; its molecular weight is 1.6×10^6, out of which two-thirds are again RNA. Electron microscopy has revealed the overall shape of both subunits (Fig. 6.12) and various other techniques like immune electron microscopy [711, 949], neutron-diffraction analysis of partially deuterated ribosomes [342, 837], crosslinking with bifunctional reagents [351, 640, 1138] and fluorescence energy-transfer measurements [502, 574, 714] have led to the localization of some of the proteins and functional sites in the ribosome. It may appear surprising, but there is but a single enzymic activity tightly associated with ribosomes: that of peptidyl transferase, which catalyses the transfer of the growing polypeptide chain from one tRNA to the amino group of the next amino acyl-tRNA, as dictated by reading the mRNA. This long-sought peptidyl transferase is located on the large ribosomal subunit, near protein L-27. However, it seems that peptidyl transferase activity is not confined to a single polypeptide but comes about as the interplay of several proteins. This conclusion has been reached in experiments with partially assembled large subunits, where the lack of any one of several proteins slowed down but did not arrest chain elongation. Thus peptidyl transferase would correspond to a complex enzyme in our classification, with the addition that it also uses a nucleoprotein, or simply RNA, support for scaffolding.

In fact, there are further steps in ribosomal protein synthesis where enzymatic activity must be involved: these are associated with the hydrolysis of GTP to GDP and P_i. As we know it today, GTP hydrolysis occurs at chain initiation when the three initiation protein factors (IF-1, -2, -3) bind and get released, in the elongation of polypeptide chain (2GTP molecules per residue) promoted by elongation factor EF-Tu and EF-G, and at chain termination along with the binding of two protein termination factors.

The function of these GTP hydrolyses is not clear, but they probably supply the energy for moving the mRNA relative to the ribosome by one codon (base triplet) and possibly for certain structural changes, such as the flip-over of bound amino-acyl-tRNA from the R site to the A site on the ribosome (cf. [710]). Hence these "GTP hydrolases" are akin to myosin ATPases: they are macromolecular machines of motion fuelled by nucleotide triphosphates. We do not know yet the physical carrier of these activities. Presumably they also stem from the cooperation of several proteins, which are likely to include some of the soluble initiation, elongation and termination factors.

Enzymes with no immediate relevance to protein synthesis may also be harboured by ribosomes. Thus it has been found that in yeast the low K_m cyclic AMP phosphodiesterase was largely associated with ribosomes attached to the nuclear membrane [749]. The physiological meaning of this interaction is not known.

The *cell nucleus* of eukaryotic cells contains most of the DNA, the histones, a family of strongly basic proteins that serve as scaffolding for the organization of chromatin into nucleosome structure [804], and a number of non-histone proteins. The latter include the enzymes and other proteins that work on DNA and the proteins that build up a network called the nuclear matrix [621]. The two major processes that take place in the nucleus are DNA synthesis (replication) and the transcription of genetic information from DNA to RNA (messenger RNA synthesis, transcription), supplemented with a number of ancillary processes. Replication is catalysed by *DNA polymerases* (deoxynucleosidetriphosphate: DNA deoxynucleotidyltransferase, EC 2.7.7.7, for a review see [1299]). In eukaryotes there are three distinct DNA polymerases α, β and γ. Enzyme β is rather small ($M_r=30,000$ to 50,000), the other two, α and γ, are larger (M_r ranging from 120,000 to 300,000). Their substructure is as yet ambiguous, but they probably interact with other proteins involved in DNA replication.

Transcription is carried out by *DNA-dependent RNA polymerases* (ribonucleoside-triphosphate: RNA nucleotidyltransferase, EC 2.7.7.6). These are complex enzymes

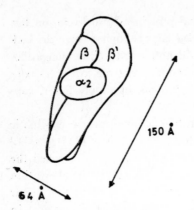

Fig. 6.13. Subunit arrangement and overall shape of DNA-dependent RNA polymerase from *T. thermophilus.* The drawing was deduced by electron microscopy of the images of negatively stained enzyme crystals, by the optical filtering method. (From Tsuji *et al.* [1242].)

composed of several dissimilar subunits (for reviews cf. [185, 186, 682, 751]). The prokaryotic DNA-dependent RNA polymerases have the general structure of $\alpha_2\beta\beta'\sigma$ (Fig. 6.13); there is only one kind of enzyme in each cell [184, 316]. The best known of all is the RNA polymerase from *E. coli,* but even here the role of the different subunits is not unambiguously established (cf. [1181]). What has been clearly demonstrated is that the core enzyme ($\alpha_2\beta\beta'$) is sufficient for RNA synthesis and is needed as an initiation factor. On the other hand, there are multiple nuclear DNA-dependent RNA polymerases in eukaryotic cells, which differ from each other in structure, localization and function (cf. [186, 583]). Three classes have been distinguished: class A (or I) enzymes are insensitive to amanitin, a highly toxic mushroom poison; class B (or II) enzymes are very sensitive to α-amanitin (10^{-9}–10^{-8} M is inhibitory); class C (or III) enzymes are moderately sensitive to α-amanitin (10^{-5}–10^{-4} M inhibits). Their molecular weight is around 5×10^5 and they consist of six different polypeptide chains (not counting tightly bound protein kinase, cf. below). The three types of enzyme carry out different functions: class A enzyme is located in the nucleolus and is involved in ribosomal RNA synthesis; class B enzyme synthesizes heterogeneous nuclear RNA, the precursor of mRNA, whereas class C enzyme produces 5S and pre-4S RNA.

One of the fundamental questions about these RNA polymerases is how the DNA segment to be transcribed is selected. It is generally held that RNA polymerase I or II alone is insufficient for high-fidelity transcription [869, 1293], and a number of protein factors have been described that have no clear-cut function but seem to stimulate RNA polymerase(s) in a quasi-specific manner (cf. [1246] and references therein). Eukaryotic polymerases also display protein kinase activity, carried possibly by one subunit of polymerase II in maize seedlings [590] or by two subunits out of the eight of RNA polymerase I from rat hepatoma cells [1021]. Protein phosphorylation is generally assumed to be instrumental in the specificity of eukaryotic gene transcription [663, 691]. In yeast, a protein factor P_{37} stimulates transcription by binding to the 23,000 dalton subunit of both RNA polymerases I and II [1053]. The latter enzyme, which is responsible for mRNA synthesis, requires divalent cations (first of all Mn^{2+}) for activity and it has been suggested [1252] that the accumulation of metal ions demonstrated in the cell nucleus [1123, 1255] may affect transcription. Control may also be exercised through RNA polymerase levels, as suggested by the changes of the concentration of enzyme II during the life cycle of myxomycetes [680].

Certain enzyme–enzyme interactions, resembling those in the cytoplasm and mito-chondrial matrix (cf. Chapter 5, p. 115), have been claimed to occur in the nucleo-plasm. Thus ornithine decarboxylase (EC 4.1.1.17), the first enzyme in polyamine biosynthesis, when rapidly induced in rat liver led to enhanced RNA synthesis [338]. If this cytoplasmic enzyme was added to isolated nuclei RNA polymerase I activity was enhanced, from which the inference was made that ornithine decarboxylase migrated into the nucleus and acted as a positive regulator of polymerase I [770]. However, this model has recently been challenged by finding no correlation between the changes of nuclear ornithine decarboxylase and RNA polymerase I activities [89]. It may be worth mentioning that prolonged higher activity of ornithine decar-boxylase has been associated with the development of cancerous loci in rat liver [884].

There are enzymes in the cell nucleus that act on DNA conformation, like the DNA unwinding enzymes (ATPases) [227, 570] and DNA topoisomerases type I [187] and II [255, 744], catalysing the concerted breakage and rejoining of DNA backbone bonds (for a review cf. [431]). A current survey of enzymes and proteins affecting the helical structure of DNA was made by Geider and Hoffmann-Berling [429].

The cell nucleus is a peculiar cellular compartment. It is surrounded by a double sheet of nuclear envelope, the outer layer of which continues into the endoplasmic reticulum membrane, a network that pervades the cytoplasm. In spite of the apparent continuity the two membranes can be distinguished on a compositional and organi-zational basis [1004]. The nuclear envelope does not constitute a major barrier to solutes even of macromolecular range, except perhaps only for DNA [1107]. In this respect it is quite different from the mitochondrial inner membrane or the chloroplast membrane. Its almost unrestricted permeability is due to the so-called *pore complexes,* which are protein-bounded holes in the nuclear envelope of about 100 Å diameter [366]. The function of pore complexes is related to the transfer, and processing, of mRNA-protein complexes from the nucleoplasm to the cyto-plasm and the entry of nuclear enzymes and proteins from their cytoplasmic site in synthesis. The upper size limit of RNA-protein particles that can pass was esti-mated to be 40–50S [907]. The transfer is apparently an energy-requiring, controlled activity [1049, 1073], but other molecules including fairly large enzymes can freely go in and out. The ubiquitous dehydrogenase coenzyme, NAD, is synthesized in the nucleus by NAD pyrophosphorylase, an enzyme anchored inside the nucleus, from nicotinamide mononucleotide and ATP that diffuses into the nucleus, whereas much of the NAD produced migrates to the cytoplasm. The confluence of cytoplasm and nucleoplasm (cf. also Chapter 7) in respect of many solutes is also indicated by the fact that glycolytic enzymes can be found equally in the two locales, with the exception of the nucleolus, where glycolytic enzymes seem to be absent [1092].

In conclusion, we are left with a picture of nuclear enzymes, those typically occurring there and briefly described above, as proteins that are confined within the boundaries of the nuclear envelope, though most of them could leave that space through the pore complexes. What keeps them still there? Although we are very far from knowing, let alone understanding at the chemical level, all the macromolecular interactions prevailing in the nucleus, we may not be far from the truth if we attribute the practically exclusive localization of nuclear enzymes to the abundance of appropriate binding sites. Nuclear enzymes stay where they are because they interact with DNA and other proteins, i.e. with chromatin and the nuclear matrix. (For the participation of matrix in various nuclear functions, see, for example, [45, 445, 530].) This is, in fact, a macromolecular variety of compartmentation via binding sites, a phenomenon that will be scrutinized in connection with metabolites in Chapter 7.

ENZYMES LOOSELY BOUND TO STRUCTURES

The embedding or very strong anchoring of enzymes in cellular structures is only one extreme of the wide range of modes for the interaction of enzymes with these structures. In a great number of cases the enzyme is "loosely" bound to a membrane or structural protein, by which it is meant that the enzyme can be removed from its binding site without drastic treatment, such as extraction with detergents or organic solvents. As already mentioned in the previous section, the so-called *peripheral membrane proteins,* as opposed to the integral ones, belong to this class. In most cases the binding forces are electrostatic in nature and hence the enzyme is eluted from the scaffolding structure by solutions of high ionic strength. Often this "high" ionic strength does not surpass isotonicity, i.e. the equivalent of 150 mM NaCl.

Enzymes that are loosely bound to one or another cellular surface are, in most cases, typically "soluble" enzymes and can be extracted from the disrupted cell by aqueous solutions. Although from the enzymological point of view they can be, and in fact are, studied as individual entities, in the cellular context their binding (adsorptive) properties must not be disregarded. We have reiterated in previous chapters that intracellular protein concentration is high and available water is limited. Here it may be added that structural surfaces are abundant. This holds for all higher cells. Even in prokaryotes, which have no intracellular organelles, the cell membrane is thought to be involved in many a binding phenomenon. In eukaryotes the greatest macro-compartment is the cytoplasm, which is full of membraneous septa that belong to a great variety of organelles, as the nucleus, mitochondria, Golgi-apparatus, a host of different "somes", and last but not least the endoplasmic, or in muscle the sarcoplasmic, reticulum. The latter comes from the nuclear membrane and forms an intricate network throughout the cytoplasm. In muscle cells there are all the proteins of the contractile apparatus, but even in non-muscle cells there are the ubiquitous cytoskeletal proteins [228, 229]. In a manner of speaking, it would be as hard for the "soluble" enzymes not to interact with some of these as it is impossible to walk in a thick forest without touching the trees.

The question is, of course, whether these interactions are indifferent or are exploited to serve some physiological cause. In the former case we should reasonably expect haphazard binding, whereas if the latter alternative holds true, associations are to be specific to a greater or lesser extent. Hence when binding phenomena are studied one of the first questions to be asked is about specificity. Specific binding can have the following functional consequences:

1. Activation or inactivation (partial inhibition) of individual enzymes, i.e. changes in V_{max} and K_m, through microenvironmental effects [785].
2. Alignment of several enzymes to form a scaffolded enzyme array with all the functional repercussions associated with enzyme clustering, i.e. metabolite channelling, coordinate regulatory effects, catalytic facilitation, etc.
3. Positioning of enzyme(s) for direct coupling to matrix function. This is the case when the associating enzyme gets its substrate from, or discharges its product into, the matrix. In essence, this is a proximity effect with all its corollaries.

As long as the binding forces are weak, small perturbations, e.g. conformational changes induced by ligands, can influence the degree of association. It is easy to conceive the regulatory potentialities of such an arrangement, given the above three types of functional modulations.

Prompted by similar considerations and by the accumulating experimental evidence of loose enzyme–structural element associations, Wilson [1324] proposed the adjective "ambiquitous" for an enzyme that is reversibly distributed between soluble and particulate forms under the influence of ligands whose concentration, in turn, reflects the metabolic state of the cell. Recently Nemat-Gorgani and Wilson [863] gave a brief description of ambiquitous behaviour, along with some specific examples.

Since most enzymes that have been shown to exhibit such behaviour belong to glycolysis, we are going to discuss the relation of the glycolytic system to the contractile apparatus and the cell membrane, as well as to the outer mitochondrial membrane.

GLYCOLYTIC ENZYMES AND THE CONTRACTILE APPARATUS

The first systematic study came from Pette's laboratory showing that muscle proteins were able to bind a variety of glycolytic enzymes [29]. F-Actin had far the highest binding capacity, compared with myosin, actomyosin and stroma-protein, and of the enzymes aldolase had the greatest affinity toward F-actin, closely followed by glyceraldehyde-3-phosphate dehydrogenase (GAPD). The other enzymes that exhibited some propensity for binding were phosphofructokinase, 3-phosphoglycerate kinase, pyruvate kinase and lactate dehydrogenase. Binding was sensitive to ionic strength; 150 mM KCl effected complete desorption. Furthermore, the substrates of aldolase and GAPD also interfered with binding [30]. Association with F-actin altered the catalytic properties of aldolase: its V_{max} increased two-fold and K_m for fructose-1,6-bisphosphate rose about one order of magnitude [30]. Some of the glycolytic enzymes: glycogen phosphorylase, phosphoglucomutase, phosphoglycerate mutase and enolase displayed no affinity to F-actin [27]. Aldolase was found to give with G-actin a 1:1 complex that forms a gel at high concentration; thus aldolase resembles in this respect α-actinin [27]. Histochemical studies revealed that most of the glycolytic enzymes were localized in the I-band [28, 1114]. While the above studies were made with rabbit muscle enzymes, extraction experiments from chicken breast muscle by Hultin and coworkers [274, 334, 811] also revealed the selective retention by particulate matter of glycolytic enzymes, especially GAPD and lactate dehydrogenase. The kinetic parameters of these two enzymes were affected by binding: both had higher V_{max} in the free form, and the K_m (GAP) of GAPD also increased. With lactate dehydrogenase inhibition by pyruvate was warded off by binding. (For an account see Pette [936]).

   The rabbit muscle system has been analysed extensively by Master's group. In early experiments, observing the increase of apparent molecular weight of aldolase, lactate dehydrogenase and pyruvate kinase [221] in sedimentation profiles of a myogen preparation (the type of analysis made by Földi *et al.* [385] by frontal analysis gelchromatography), they suspected a glycolytic multienzyme complex in muscle. However, it was soon realized [222] that some protein of the contractile apparatus was necessary for complex formation. The best adsorbent proved to be the complete thin filament, F-actin-tropomyosin-troponin, for the enzymes phosphofructokinase-aldolase, pyruvate kinase, lactate dehydrogenase, glucose-6-phosphate dehydrogenase and GAPD. In contrast, only weak binding was observed with triosephosphate isomerase, 3-phosphoglycerate kinase, phosphoglycerate mutase, enolase and hexokinase [223]. With several enzymes Ca^{2+} increased the adsorption. Other enzymes present in muscle in large amounts, such as creatine kinase and adenylate kinase, displayed no adsorption to the thin filament. Importantly, at high protein (myogen or "decorated" actin) concentration (30–50 mg/ml) glycolytic enzymes remained bound to a considerable, but varying extent even at 0.15 M salt concentration at 25 °C, as found in sedimentation experiments [223, 706a]. Therefore the objection that these binding phenomena are incompatible with the conditions prevailing in the muscle cell seems largely eliminated. The kinetic parameters of aldolase were markedly changed when bound to F-actin-tropomyosin-troponin, but only slightly when troponin was missing from the filament [1278]. The K_m value increased by two orders of magnitude, whereas V_{max} rose four-fold relative to free aldolase. Ca^{2+} decreased the K_m of bound aldolase to one-fourth, but only slightly diminished its V_{max}. The substrate (FBP) saturation curves of aldolase under these conditions are illustrated in Fig. 6.14. It has been noted that the overall catalytic facilitation effect of Ca^{2+} might be a mechanism synchronized with muscle contraction. More details about the above phenomena can be found in reviews [224, 786].

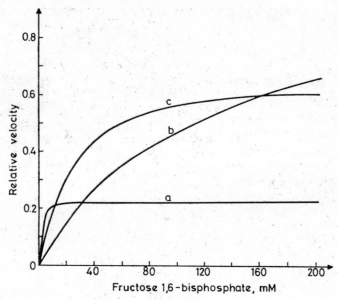

Fig. 6.14. Substrate saturation curves of free and bound rabbit muscle aldolase. (*a*) free enzyme; (*b*) and (*c*) enzyme bound to thin filament (100% binding as checked by sedimentation and analysis of supernatant for activity). In (*b*) Ca^{2+} was removed by 0.2 mM EGTA; in (*c*) 0.1 mM $CaCl_2$ was added. The curves were *calculated* from the K_m and V_{max} values determined experimentally. Activity is expressed in units relative to that of aldolase bound to thin filament in the absence of Ca^{2+}. (From Walsh *et al.* [1278].)

It seemed then that aldolase bound to the thin filament through the T and I components of the troponin complex. This had been indicated by electron microscopic observation of the two-dimensional lattice formation in a mixture of aldolase and thin filament [845], whose neighbouring actin double helices were crosslinked at about 38 nm regular intervals (Fig. 6.15), which corresponds to the spacing of troponin complexes along the thin filament [1017]. Such an ordered lattice could not be formed if either the I or T component of troponin was absent from the filament. Apparently, both are required to form the lattice, whereas troponin C is indifferent. It is pertinent here, that the interaction of aldolase and tropomyosin-troponin had been detected earlier by means of ultracentrifugation [225].

However, more recent studies on binding, using moving-boundary electrophoresis [1279] and electron microscopy combined with computerized image analysis [1182], show that the interaction of aldolase with thin filament is a more complex phenomenon. Tropomyosin also seems to be directly involved in binding. There are multiple binding sites on aldolase (at least three) for the regulatory proteins, which is not surprising in view of the fact that aldolase is an isologous tetramer. The binding affinity for aldolase is the greatest with F-actin alone (dissociation constant, $K_d = 5 \times 10^{-6}$ M) and is less with F-actin-tropomyosin-troponin ($K_d = 6 \times 10^{-5}$ M).

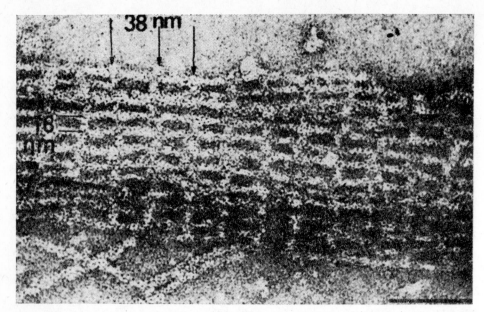

Fig. 6.15. Lattice structure formed in a mixture of F-actin-tropomyosin-troponin and aldolase. The dimensions of the lattice are indicated as centre-to-centre spacings. Bar = 100 nm. (From Morton *et al.* [845].)

On the other hand, binding stoichiometry for the former is one aldolase molecule per fourteen actin monomers, whereas for the latter it is 4 : 14 [1279]. The electron microscopic analysis of aldolase-(complete thin filament) paracrystals shows two transverse bands every 38 nm: the major band is thought to be aldolase binding to troponin, while the minor one is presumably aldolase crosslinking the filaments. The minor band is Ca^{2+} sensitive: it is most pronounced at low Ca^{2+} concentration. Crosslinking by aldolase must be a cooperative phenomenon, as the interaction of the enzyme with the individual proteins is much weaker than with the filament. Interestingly, the paracrystalline arrays could also be produced if instead of aldolase the other glycolytic enzymes phosphofructokinase, GAPD or pyruvate kinase were added to the filament preparation.

The cross-bridges produced by aldolase probably cannot be formed in muscle, since the spacing of actin filaments in muscle is different. In this sense, the paracrystalline arrays are to be regarded as artefacts which, however, reflect the propensity of thin filament for binding glycolytic enzymes and may underlie the early observation that glycolysis is located over the I-band in muscle [28, 1114]. As actin is ubiquitous in cells [676], just like glycolysis, and non-muscle actin is very similar to muscle actin [463], it can be reasonably assumed that similar interactions between glycolytic enzymes and cytoskeletal proteins may also occur.

F-Actin and the thin filament have been found by Lion and Anderson [743] to interact with rabbit muscle phosphofructokinase influencing its catalytic properties.

Thus, there is an activation effect which mainly consists in the reversal of inhibition by high ATP concentration and a decrease in the K_m for fructose-6-P. Importantly, activation by F-actin was also observed at physiological ionic strength. In these studies F-actin had a more pronounced effect than the thin filament. In the muscle cell, of course, only the latter is relevant.

The crucial question about these interactions, some of which at least are likely to exist *in vivo,* is whether they serve any physiological end. As is perhaps apparent from the foregoing, we are not in want of attractive hypotheses ranging from the regulation and supramolecular organization of glycolysis up to the direct involvement in the cell's motile functions [1182]. It must be clear, however, that compelling evidence for these functional roles is still lacking. On the other hand, this is no wonder in view of our rather poor understanding of the subtleties of metabolic regulation at the level of macromolecular interactions.

GLYCOLYSIS AND THE ERYTHROCYTE MEMBRANE

The relationship between the erythrocyte membrane and glycolytic enzymes has puzzled biochemists, cell biologists and others concerned with red cell structure and function for a long time. The easy availability of red blood cells from a variety of species including man has rendered erythrocytes popular objects of study. Apart from this practical aspect, research on erythrocytes has been fostered by its immediate relevance to medicine. The techniques of blood preservation in blood banks for transfusion have been developed on the firm ground of understanding, at least in broad outlines, the metabolic processes of erythrocytes. This challenge still prevails and urges workers to investigate the subtleties of erythrocyte function.

Mature mammalian erythrocytes are much simpler than a normal eukaryotic cell. They are highly specialized "containers" for haemoglobin, the oxygen and carbon dioxide transport protein, possessing only a rudimentary metabolic system just to provide sufficient energy to maintain biconcave cell shape and osmotic integrity during the average 120 days life-span of a circulating erythrocyte. Thus they have no cell nucleus, which became dissolved in the *reticulocyte* stage, and are unable to divide. They also lack the respiratory chain, and depend almost exclusively on glycolytic ATP production, metabolizing blood glucose to lactate, which is then discharged into the blood plasma.

An important difference between glycolysis in erythrocytes and other cells is that in the former 1,3-bisphosphoglycerate can be converted to 2,3-bisphosphoglycerate (2,3-BPG), an allosteric effector, decreasing the oxygen affinity of haemoglobin. This 2,3-BPG "by-pass" discovered by Rapoport and Luebering [982] in 1950 (Fig. 6.16) is under strict control, and supplies the allosteric effector in amounts dictated by the metabolic state of the organism. During O_2 shortage induced, for example, by high altitude, more 2,3-BPG is synthesized, and binds between the four subunits of the haemoglobin tetramer decreasing its O_2 affinity by stabilizing the

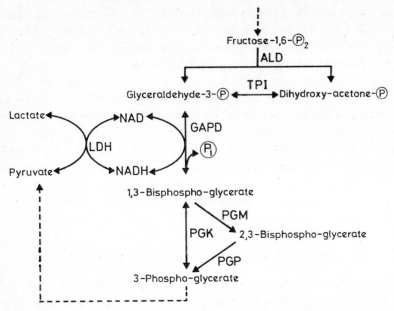

Fig. 6.16. Part of the glycolytic pathway in erythrocytes illustrating NAD-NADH recycling and the 2,3-bisphosphoglycerate by-pass. Abbreviations: ALD, aldolase; PGK, 3-phosphoglycerate kinase; PGM, phosphoglycerate mutase; PGP, phosphoglycerate phosphatase; LDH, lactate dehydrogenase.

deoxy-conformation. Decreased O_2-affinity, in turn, results in greater oxygen release from haemoglobin in the tissues, which eventually improves the oxygen supply of cells.

Thus glycolysis is needed by erythrocytes for two main reasons: ATP production to cover energy demands and 2,3-BPG generation for modulating haemoglobin function. We may add the formation of NADH which is partly used by methaemoglobin reductase (EC 1.6.2.2) keeping haemoglobin's iron in the reduced (ferrous) state. Erythrocytes also have the enzyme system of the *pentose–phosphate shunt*. The main use of this pathway for red cells is the production of NADPH in the enzymatic steps of glucose-6-phosphate dehydrogenase (EC 1.1.1.49) and 6-phosphogluconate dehydrogenase (EC 1.1.1.44), needed mainly by glutathione reductase (EC 1.6.4.2), a vigorous defence mechanism to keep functionally important SH-groups in he reduced form in a highly oxidative milieu.

Considering the relative simplicity of the red cell Rapoport, one of the pioneers in erythrocyte research, has foretold [980] that probably the erythrocyte will be the first cell whose metabolism will be completely understood. Indeed, many of its enzymes have been purified and studied enzymologically. A computer simulation of erythrocyte glycolysis using kinetic parameters of individual enzymes gave results in satisfactory agreement with experimental observations [523]. It might then appear that any supramolecular organization of red cell glycolysis is unnecessary in order to describe the behaviour of the system. However, given the "tolerance" of computer simulation of multi-step reaction sequences and a number of observations to be described in the next section, we may say with confidence that the picture is more complex than it appeared to be.

Evidence for enzyme–membrane interaction: The case history of GAPD

The first line of evidence suggesting some kind of a connection between the cell membrane and glycolysis came from the study of incorporation of $^{32}P_i$ into glycolytic intermediates and ATP, some 30 years ago. It was found by several authors [439, 468, 958] that ATP was labelled faster than intracellular P_i. This finding could only be explained by the preferential utilization of extracellular P_i, rather than intracellular P_i, in the GAPD reaction:

$$GAP + NAD^+ + P_i \rightleftharpoons 1,3\text{-}BPG + NADH + H^+$$

which is the site of entry of P_i into glycolytic intermediates. The authors reasoned that GAPD must be located at or in the membrane, thereby having ready access to penetrating P_i. These incorporation data were later challenged [1020, 1069, 1223], but in the latter experiments the extracellular P_i concentration was set rather high, which produced a strong P_i influx probably outweighing uptake through the GAPD reaction. In more recent studies the preferential labelling of ATP was confirmed for both erythrocytes and HeLa cells [873]. Compartment analysis of $^{32}P_i$ uptake in intact erythrocytes also indicated the presence of a membrane-located ATP-generating system [720].

Another line of evidence came from the early observations that after hypotonic haemolysis the particulate material, called "stroma" or "ghosts" and corresponding to the cell membrane, contained some or the whole set of glycolytic enzyme activities [26, 476, 825, 1067, 1068]. ATP synthesis by ghosts was detected in the presence of appropriate glycolytic intermediates, which indicated that at least GAPD and 3-phosphoglycerate kinase were present on the membrane.

These studies gained a new impetus from investigations of red cell membrane structure in the early seventies, exploiting the then new technique of SDS polyacrylamide gel electrophoresis. One of the proteins of white (haemoglobin-free) ghosts, the band-6 protein according to the designation of Steck and coworkers [363, 609], could be easily eluted from the membrane by salt solutions of moderately high ionic strength. It was shown, first by peptide mapping [1209], then by selective chemical modification [169], that the band-6 polypeptide corresponded to the subunit of GAPD. On this ground GAPD was dubbed to be a *peripheral protein* of the red cell membrane [609].

The characterization of this enzyme–membrane association has been undertaken in several laboratories. It has been shown by testing inside-out and right-side-out vesicles [1169] prepared from erythrocyte membranes that GAPD only bound to the cytoplasmic membrane face [609]. Binding is reversible [609, 801, 1136], but the quantitative parameters of binding reported vary, presumably owing to the dissimilarity of reaction conditions. Thus Kant and Steck [609] and Solti [1134] found a single dissociation constant of $\sim 10^{-7}$ M and a maximum binding capacity of 1.5 to 2.0×10^6 GAPD molecules per ghost. In contrast, McDaniel et al. [801] discerned two types of binding site; the high affinity site could bind 3×10^5 enzyme molecules per ghost with $K_d < 10^{-8}$ M, while the low affinity site bound 6×10^6 enzyme molecules with $K_d > 10^{-7}$ M. A number of physiological substances affect binding; it is weakened by ATP [609, 1103], NADH and, to a lesser extent, NAD [609, 731, 1103] and bilirubin [450]; it is potentiated by GAP in the presence of various adenine nucleotides [609].

Binding is not specific for the erythrocyte enzyme, as witnessed by the practically equal affinity toward rabbit or pig muscle GAPD [801, 1136] (Table 6.3). Altogether

TABLE 6.3.

Comparison of the affinity of human erythrocyte ghosts to the homologous and mammalian muscle GAPDs

| No. | GAPD added to ghosts | | Depletion of radioactivity in the supernatant |
	[14C] Carboxymethylated	Carboxymethylated	
1	Human erythrocyte	Rabbit muscle	65
2	Rabbit muscle	Human erythrocyte	75
3	Human erythrocyte	Pig muscle	60
4	Pig muscle	Human erythrocyte	77

Pure enzymes were treated with [14C] iodoacetate so that about four carboxymethyl groups were attached per tetrameric molecule. An equimolar mixture of two enzymes (0.2 mg/ml each) was incubated with 4 mg/ml of white ghost suspension at 0°C in 7 mM sodium phosphate buffer, pH 7.4, for 1 hour, then centrifuged and the supernatant analysed for radioactivity. Depletion of radioactivity $= 100 - X$, where X is radioactivity of the supernatant after centrifugation in per cent of radioactivity of the suspension before centrifugation. (From Solti and Friedrich [1136].)

this is not surprising as human erythrocyte GAPD is very similar to the various mammalian muscle enzymes in many respects [330, 777, 1280]. On the other hand, yeast GAPD was bound to a much lesser extent [801].

The binding site was identified by Yu and Steck [1345, 1346] as the cytoplasmic part of the so-called Band-3-Protein (B-3-P), the predominant polypeptide of the human erythrocyte membrane. This is a glycoprotein [550] of apparent molecular weight 93,000, and it exists as a dimer ($M_r = 180,000$) both when solubilized and in the membrane, the latter evidenced by ready crosslinking *in situ* to dimer [1346]. It spans the red cell membrane giving rise to the intralipid particles seen on freeze-

Fig. 6.17. Schematic disposition of band-3-protein in the erythrocyte membrane as revealed by proteolytic fragmentation. Three major integral fragments are generated by digestion with chymotrypsin (CH). Trypsin (TR) cleaves the polypeptide into a 52,000-dalton integral segment and three major cytoplasmic-surface fragments not anchored in the lipid bilayer. S-Cyanylation (CN) of a cysteine residue (SH) cleaves a 23,000-dalton piece from the cytoplasmic, amino-terminal region. (From Steck [1168].)

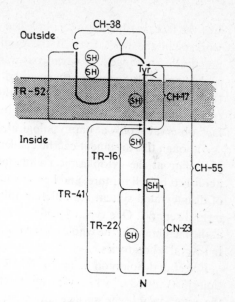

fracture electron micrographs [1294, 1330] and it is thought to be involved in anion transport [550, 1024, 1029, 1330]. B-3-P has been dissected into fragments [998, 1170] and much of its primary structure and intramembrane topography is known. A diagram of its disposition in the membrane is given in Fig. 6.17. The N-terminal part of the polypeptide, corresponding to about 40,000 molecular weight, is sticking out on the cytoplasmic side of the membrane; within this stretch the N-terminal fragment of $M_r = 23,000$ is highly acidic and seems to constitute the binding site for GAPD [1171]. It should be added that B-3-P associates also with other proteins on the cytoplasmic surface: it is linked via band 2.1 (ankyrin [65]) to spectrin [64], forming a submembrane reticulum. For a review on B-3-P see Steck [1168] and about oligomeric membrane proteins and transport consult Klingenberg [664].

The impact of membrane binding on the catalytic properties of GAPD is not clear. Wooster and Wrigglesworth [1335] when studying the binding of the enzyme to phospholipid vesicles found that with negatively charged liposomes the K_m increased and V_{max} decreased, whereas with positively charged liposomes the K_m decreased and V_{max} did not change. These observations are probably explained by the effect of the liposome surface on the concentration of the negatively charged substrates of GAPD. With the erythrocyte membrane itself an increase in both V_{max} and K_m was measured [1336]. Eby and Kirtley [330] found that both the specific activity and stability of the human erythrocyte enzyme were increased by washed red cell membranes. In contrast, Solti and Friedrich [1136] reported a partial reversible activity loss when GAPD was bound to ghosts. Recently it is claimed that binding is accompanied by complete inhibition of the enzyme [1239].

All the above detailed characterization of the binding of GAPD to red cell membrane had, however, been performed while having a skeleton in the cupboard. This was the fact that binding was abolished in isotonic ($\mu = 0.15$) media, as found by all workers in the field. Even if the original intracellular bulk concentrations were practically unchanged, as when lysis was brought about by ultrasonic irradiation of packed red cell suspensions (haematocrit: 100), no GAPD binding could be detected to the membrane fragments [776]. On this basis some groups of workers came to the conclusion that the binding of GAPD to the red cell membrane was an *in vitro* artefact, observable only at unphysiological, low ionic strengths [411, 776].

Although this argument can be challenged by claiming that the microenvironment prevailing at the cytoplasmic membrane face of the intact red cell is irreversibly abolished by cell rupture, and hence isotonicity in the intact cell and that in any kind of disintegrated system are different things, such a rebuttal clearly asks for experimental support. One type of evidence requested is whether GAPD is indeed located at the membrane in the whole erythrocytes. This question has recently been addressed in several laboratories.

Fossel and Solomon using ^{31}P-NMR technique have made a series of investigations [390, 391, 392] which were interpreted as indicative of some connection between the outer membrane surface and a putative multienzyme complex, consisting of 3-phosphoglycerate kinase + phosphoglycerate mutase + GAPD, on the cytoplasmic membrane surface of human erythrocytes. This conclusion was drawn from the shift of phosphorus resonance of intracellular 2,3-BPG induced by the binding of ouabain to the ($Na^+ + K^+$) ATPase on the outside of cells. Resonance shifts of 2,3-BPG were also observed with phosphoglyceromutase + 3-phosphoglycerate kinase in solution [390]. When the extracellular K^+/Na^+ ratio was varied a concomitant change in the phosphorus resonance of GAP was observed, which pointed to conformational changes in GAPD [392]. The authors also claimed to have isolated a glycolytic multienzyme complex of about 10^6 molecular weight by chromatography on a Biogel A-5M column [391], analogously to the work of Mowbray and Moses [851] on *E. coli*. However, in the present case the assignment of a multienzyme *complex* seems at least questionable, as ghosts or membrane debris were also present in the high molecular weight fraction containing the glycolytic enzymes. Therefore binding to, or trapping in, membrane vesicles could also have occurred. In a closely similar experiment where membrane entrapment was excluded, Solti and Friedrich [1137] found negligible GAPD activity in the large molecular weight fraction. More recently, the above ^{31}P nuclear magnetic resonance data have been questioned by Momsen *et al.* [830] arguing that the resonance shifts observed by Fossel and Solomon were due to small variations of pH and were, as shown by error analysis, within the noise level of the measurements.

Szabolcsi, Cseke and coworkers [272, 1203] have utilized the molecular sieving property of the erythrocyte membrane, i.e. the phenomenon that during mild, reversible haemolysis pores are opened on the membrane through which proteins pass in an inverse relation to molecular weight [579, 761]. In these studies the release of several glycolytic enzymes, as well as haemoglobin, was examined during a short

Fig. 6.18. Kinetics of redistribution of glyceralde-hyde-3-phosphate dehydrogenase (GAPD) after lysis of intact erythrocytes. Fresh human red cells were diluted (final haematocrit: 0.25%) in 10 mM Tris-HCl, pH 7.4, containing 0.1% saponin and the concentration of NaCl as indicated at 37°C. Filtrates were assayed for haemoglobin content (not shown) and for GAPD activity. (From Kliman and Steck [665].)

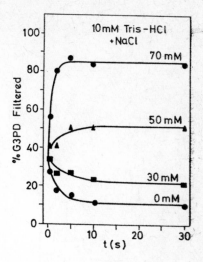

exposure to moderate hypotonicity. Whereas some proteins conformed to a linear plot of molecular radius vs. amount released, GAPD, lactate dehydrogenase and 3-phosphoglycerate kinase were discharged in greater amounts than predicted by this correlation. It was suggested that preferential liberation might be due to the localization of these enzymes near the cell membrane.

A detailed analysis of the kinetics of saponin-induced GAPD release from intact red cells has recently been carried out by Kliman and Steck [665]. A rapid filtration technique made it possible to monitor the efflux and re-equilibration of the enzyme, from 0.5 s onwards, after dilution in buffer. The type of experiment is illustrated in Fig. 6.18. It is seen that the amount of enzyme filtered, i.e. released, reaches a plateau value in 5 to 10 s, indicating the attainment of equilibrium. The equilibrium values strongly depend on the ionic strength of the lysis buffer, but all curves converge to about 35% enzyme released at 0 time. At 0 and 30 mM NaCl the time curves bend downward indicating GAPD uptake, because at low ionic strength the permeable cells bind more enzyme than they did originally. Haemoglobin emerged from the cells in an entirely different manner, being released to 100% right from the beginning. The latter finding suggests that the kinetics of GAPD release was not limited by holes, but rather by the association to the membrane. From the measurements three parameters of GAPD-membrane interaction could be estimated: k_{+1}, the apparent second-order rate constant of association; k_{-1}, the apparent first-order rate constant of dissociation; and y_0, the fraction of enzyme soluble at 0 time. As NaCl concentration was increased from 0 to 70 mM, k_{-1} increased 19-fold, k_{+1} decreased 3.5-fold, and consequently the dissociation constant ($K_d = 6 \times 10^{-9}$) rose 67-fold. The value of y_0 was approximately 35 in all cases, as also seen in Fig. 6.18, which means that in the intact cell two-thirds of the GAPD content is bound to the membrane.

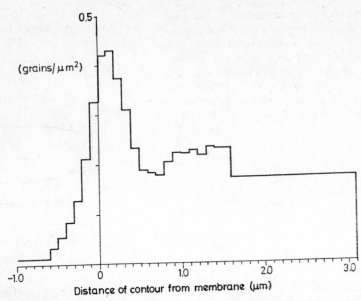

Fig. 6.19. Autoradiographic analysis of the localization of [³H] carboxymethylated glyceraldehyde-3-phosphate dehydrogenase in thin sections of fixed human erythrocytes. The histogram shows grain density as a function of distance from the membrane. Negative abscissa represents distances of contours outside the membrane. The peak at the membrane proved significant by a Kolmogorov–Smirnov non-parametric statistical test on the relative cumulative frequency distribution. (From Solti *et al.* [1135].)

A conclusion compatible with the above picture has been reached by Solti *et al.* [1135] from the autoradiographic localization of GAPD in thin sections of fixed whole erythrocytes. GAPD was labelled with [³H] iodoacetate in intact cells according to an optimized protocol in which the total cellular GAPD content became tagged. This amounted to half of the total label incorporated into the cells, whereas the other half was nearly equally distributed over glutathione, haemoglobin and carbonic anhydrase. Evaluation of a great number of electron microscopic autoradiographs by a computerized program that calculated the density of developed grains as a function of distance from the membrane gave the distribution histogram of Fig. 6.19. It is seen that grains accumulate near the membrane. The diffusion of iodo-acetate within the red cell is fast enough not to maintain an intracellular concentration gradient. The minor labelled species had little or no affinity to the membrane, at least none of them were comparable to GAPD in this respect. Taking these facts together, the grain distribution in Fig. 6.19 most likely indicates that the majority, if not the total amount, of GAPD is located near the membrane in the human erythrocyte.

In conclusion, the idea of membrane-localization of GAPD in human erythrocyte has got considerable experimental support. Although none of the individual findings would perhaps be convincing, they build up a body of evidence that seems now compelling. The great question is, of course, what the physiological significance of this topology is. We shall make an attempt to give a tentative answer after having reviewed the connection of other enzymes with the red blood cell membrane.

Other glycolytic enzymes and haemoglobin

Aldolase has been long recognized to display affinity toward the erythrocyte membrane [476, 1103, 1224]. As in the case of GAPD, rabbit muscle aldolase could bind just as well as the erythrocyte enzyme [1187]. The first indication that aldolase and GAPD bind to the same site on the membrane came from Solti and Friedrich [1136]. They observed a characteristic loss of activity when these enzymes became attached to the membrane. In Fig. 6.20 it is seen that the activity loss of aldolase when incubated with white ghosts could be fully reverted by the addition of an equimolar amount of GAPD, while GAPD activity diminished. This finding is consistent with GAPD displacing aldolase from the membrane. Indeed, analysis of the mixture after sedimenting the ghosts revealed that GAPD, but not aldolase, activity was enriched in the membrane pellet fractions (Fig. 6.21). The above data indicated that (*i*) enzyme binding and consequent activity loss were reversible, (*ii*) GAPD and aldolase competed for the same binding site(s), and (*iii*) the affinity of membrane for GAPD was considerably greater than that for aldolase.

Steck and coworkers [855, 1187, 1188] confirmed the loss of catalytic activity upon binding of aldolase and identified the binding site as the cytoplasmic part of B-3-P. Then the competition between GAPD and aldolase for binding, as described above

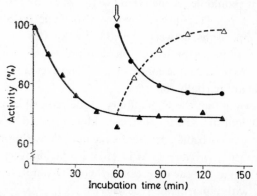

Fig. 6.20. Reversal of aldolase activity loss due to association with membrane by GAPD. Rabbit muscle aldolase (0.5 mg/ml) was incubated with white ghosts (4.2 mg/ml) at 0°C in 7 mM sodium phosphate buffer, pH 7.4. At appropriate times samples were taken for aldolase assay (▲). At 60 min, indicated by the arrow, rabbit muscle GAPD (to 0.5 mg/ml concentration) was added to the suspension and the changes in aldolase (△) and GAPD (●) activities were followed. The total amounts of enzyme activity added were taken as 100%. (From Solti and Friedrich [1136].)

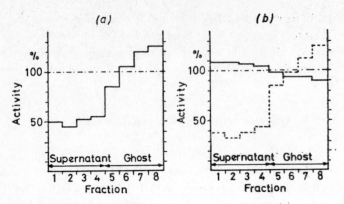

Fig. 6.21. Distribution of aldolase and GAPD activities in sedimented ghost–enzyme mixtures. Th two panels refer to points in Fig. 6.20 at 120 min with aldolase alone (*a*) and with aldolase + GAPD (*b*). The suspensions were centrifuged at $10,000 \times g$ for 1 hour at 4°C in narrow tubes. The liquid column was fractionated by successive withdrawal of 30 μl aliquots that were analysed for enzyme activity. ————, Aldolase activity; – – – –, GAPD activity. 100% = original activity level in the unsedimented mixture calculated from the amount of enzyme activity added to ghosts. Note that in (*b*) the total amount of aldolase activity is recovered, indicating full release from the membrane. (From Solti and Friedrich [1136].)

[1136] and later confirmed by others [538], seems to be plausible. Bound aldolase could be eluted by high ionic strength or specifically by its substrate FBP. As to stoichiometry, one aldolase tetramer was bound per band-3-polypeptide chain. For erythrocyte aldolase there were about 1.2×10^6 binding sites per ghost, characterized by $K_d = 3.7 \times 10^{-8}$ M. This dissociation constant is about the same as that for GAPD, but they cannot be directly compared because of the different experimental conditions. It should be recalled that GAPD was found to bind much stronger than aldolase (Fig. 6.20), [1136], a statement also made by Strapazon and Steck [1188]. Clearly, comparisons between the affinities of this, or any other, pair of enzymes for the membrane are preferably made by direct-competition experiments. Some recent experiments [1325a] seem to suggest, however, that the binding sites of aldolase and GAPD are partially different.

It is worth mentioning that the high affinity of the red cell membrane at low ionic strength for GAPD, and to a lesser extent for aldolase, could be exploited by applying ghosts as affinity adsorbents for these enzymes. Thus the membrane has been used for the preparation of pure GAPD [1039] and aldolase [1188, 1343] from various sources, as well as for the detection of GAPD–aldolase conjugates produced by crosslinking with glutardialdehyde [497].

Yeltman and Harris [1344] have addressed the question of aldolase location in the intact red cell. They resorted to the same approach, crosslinking with glutardialdehyde, applied by others to GAPD [642]. The results were also similar: up to 92% of the total cellular aldolase could be covalently attached to the membrane, which supports the idea that aldolase is located at the cytoplasmic side of the membrane *in vivo*. Importantly, the amount of aldolase immobilized by glutardialdehyde could be decreased by preincubation with cytochalasin B or colchicine, agents that disrupt microfilaments and microtubules. Although microtubulues have not been found in red cells, actin microfilaments are present (about 5×10^5 actin monomers per cell [1167]). Furthermore, a high-affinity cytochalasin B binding site, unrelated to sugar transport, has been tentatively identified in erythrocytes as the spectrin–actin complex [740]. These data pointed to the possible involvement of cytoskeletal actin in the binding of aldolase. Indeed, Yeltman and Harris [1344] observed a tight complex between red cell spectrin–actin (1 : 1 ratio, w/w) and aldolase, with $K_d \approx 10^{-9}$ M, which could be disintegrated by cytochalasin B. Practically the same was found with red cell F-actin and aldolase. This strong and apparently specific complex formation, however, could be prevented by an equimolar amount of GAPD: the latter enzyme completely displaced aldolase from actin. This phenomenon is in full agreement with the experiment shown in Fig. 6.20.

In light of these data actin has to be considered as a potential binding site for aldolase in red cells. Aldolase is less abundant than GAPD; in human erythrocytes there are about 3.5×10^4 aldolase molecules per cell [1343], i.e. about one order of magnitude less than there are actin monomers. A recent review on erythrocyte cytoskeletal proteins has been given by Marchesi [775].

Phosphofructokinase (PFK). Human erythrocyte PFK is a hetero-oligomer composed of erythrocyte-type (E, $M_r = 80,000$) and muscle-type (M, $M_r = 85,000$) subunits [613]. It exists in several aggregational forms ranging from a large aggregate, consisting of a dodecahedral E-subunit core of $M_r \approx 3 \times 10^6$ surrounded by M-subunit dimers, to a tetramer of $M_r = 330,000$, the smallest enzymatically active form. The human erythrocyte membrane can bind the enzyme [612] and the binding has much in common with that of GAPD and aldolase. Thus PFK only binds to the inner side of the membrane, it can be removed by a saline wash and the dissociation constant is about 5×10^{-8} M; there are 4×10^5 binding sites per ghost [538]. The effects of ligands on binding and the influence of binding on the kinetic parameters of PFK are noteworthy: FBP, Mg^{2+} and ATP favour association, ADP and 2,3-BPG do not. Membrane-bound PFK is not inhibited by ATP and 2,3-BPG, and its fructose-6-P saturation is hyperbolic. It has been proposed that the desensitization toward inhibitors may be important for securing the glycolytic flux in red cells [612].

Competition experiments indicate that PFK also binds to B-3-P [538]. One mole of enzyme ($M_r = 330,000$) is bound per mole of B-3-P. Both aldolase and GAPD compete with PFK, and both can effect the release of bound PFK. While aldolase-induced release is stoichiometric (inasmuch as one PFK molecule is released by two aldolase molecules), GAPD is less efficient: ten to sixty dehydrogenase molecules could replace one PFK molecule. This suggests that PFK and aldolase sites are largely overlapping whereas the GAPD site is more remote. It is pertinent here that *haemoglobin* was also shown to interact with the red cell membrane [380, 825, 1043,

1090, 1091, 1205]. The binding forces are mainly electrostatic. There seem to be two types of site: the high affinity sites had a $K_d \approx 10^{-8}$ M and are probably located on B-3-P, as shown by competition with GAPD. On the other hand, haemoglobin did not compete with aldolase [1091] or PFK [538]. The low affinity sites appear non-specific. Spectrin was proposed to participate in binding [181], but this was later questioned [173]. Phospholipids, mainly phosphatidyl serine [1205], may also contribute to haemoglobin binding.

3-Phosphoglycerate kinase. There have been reports about the association of some 3-phosphoglycerate kinase activity with the erythrocyte membrane [476, 1067, 1070, 1224]. When discussing GAPD, we have already referred to the synthesis of ATP from glycolytic intermediates by ghosts, which implied the presence of 3-phosphoglycerate kinase [1068]. Parker and Hoffman [910] suggested a direct connection between this enzyme and $(Na^+ + K^+)$ATPase on the basis that the ouabain-sensitive transport process influenced the ATP-generating system. The possibility was raised that during rapid cation transport the enhanced energy requirement of the pump was directly supplied by 3-phosphoglycerate kinase via ATP channelling. Subsequently, Proverbio and Hoffman [960] provided further support for this coupling, and for the existence of a membrane compartment of ATP, by analysing the ^{32}P-phosphoprotein labelling pattern in ghosts under various conditions. Latzkovits *et al.* [719, 720] also postulated a membrane ATP pool to explain their P_i incorporation data. Feig *et al.* [364] found a sharp decrease in ion transport in glucose-depleted cells, where energy was derived from 2,3-BPG. From these studies the attractive hypothesis emerged that ATP generated at the membrane by phosphoglycerate kinase was utilized in pumps, while pyruvate kinase made "soluble" ATP used to phosphorylate sugars in the early steps of glycolysis.

In contrast to the above ideas, erythrocytes deficient in phosphoglycerate kinase could operate ion pumps normally [1081]. Beutler and coworkers [79, 210] did not find any tight coupling between the enzyme and $(Na^+ + K^+)$ATPase, and found a single functional ATP pool by tracer kinetic methods, with no sign of a membrane-ATP compartment. The experiments of De and Kirtley [294] likewise failed to support this coupling, although the association of phosphoglycerate kinase with the membrane was confirmed. In particular, binding was rather tight ($K_d = 7 \times 10^{-10}$M) but the maximum capacity was only about 500 enzyme molecules per ghost. Binding was also found to be reversible, as tested with the rabbit muscle enzyme, and was unaffected by whether the original GAPD complement of the ghost was present or not. Increasing ionic strength eluted the two enzymes in a parallel manner. As for the connection between ion transport and phosphoglycerate kinase activity, De and Kirtley [294] observed no influence by ouabain on the rate of production of 3-phosphoglycerate by membrane-bound phosphoglycerate kinase, which is at variance with the earlier claim that the ion pump would influence lactate production [910]. It may be added, however, that direct coupling need not involve mutual stimulation (or inhibition) of the two processes, but merely a chanelling of ATP between the two systems. The latter is not ruled out by available evidence. Therefore experiments testing ATP compartmentation are to be performed to answer this question.

Tentative scheme of enzyme organization at the erythrocyte membrane

The data described above strongly suggest that there is a supramolecular structural organization of the central part of glycolysis under the red cell membrane. As pointed out already, this idea has been reiterated in the literature and topological schemes of various complexity have been put forward (e.g. [960, 1044, 1067, 1224]). Admittedly, it is difficult not to draw up a scheme, as we shall also succumb to the lure of this exercise. It must be clear, however, that even with our current background of knowledge, any such scheme can only be regarded as a working hypothesis. It is not without reason that caution is advised in this respect. If the reader forgives a personal digression, in the mid-seventies while seeking experimental approaches to check the membrane localization of GAPD in the intact red cell we were often confronted, mainly on the part of membrane biologists, with a puzzled attitude toward our endeavours. The source of perplexedness was that after GAPD had been dubbed a peripheral membrane protein, to many colleagues it looked like splitting hairs to scrutinize, and question, this "cornerstone" of red cell membrane topology. As fate had it, those who had no scruples now seem to be justified. However, justification would have never come without the efforts of those who did have scruples. Likewise, organizational schemes for other glycolytic enzymes at the erythrocyte membrane, no matter how attractive or even reasonable they may appear, are still to be proved.

The tentative scheme of Fig. 6.22 is presented with these reservations in mind. It is based on the evidence enumerated above and relies on many considerations set forth in the papers cited. It emphasizes the relation of enzymes to B-3-P, as deduced from the competition experiments and the possible involvement of cytoskeletal proteins (actin) in binding GAPD and, to a lesser extent, aldolase. Haemoglobin is also included in the diagram, though its presence is not considered functionally. The idea illustrated in Fig. 6.22 is that the sequential enzymes of the central part of glycolysis are juxtaposed by a bunch of B-3-P "tails" on which there are fairly specific sites for the binding of each enzyme. The microenvironment created by this scaffolded enzyme array would facilitate metabolite flow from fructose-6-P to 1,3-BPG, with a possible coupling to phosphoglycerate kinase and, in turn, to the transport ATPase(s).

The model of Fig. 6.22 contains many arbitrary assumptions and simplifications. The position of phosphoglycerate kinase relative to the other enzymes is entirely speculative, just as is the location of ATPase. The positioning of phosphoglycerate kinase only conforms to the observation that it is apparently unrelated to B-3-P

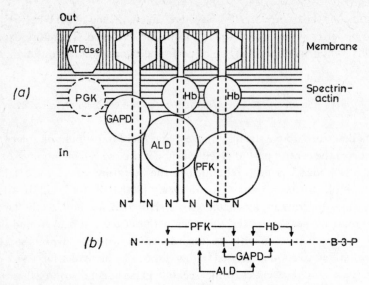

Fig. 6.22. Putative scheme of enzyme organization at the erythrocyte membrane (*a*). The B-3-P
dimers are the membrane-embedded hexagonal structures, with the long N-terminal segment sticking
out toward the cytoplasm and forming a channel for anions through the lipid bilayer. The spectrin-
actin reticulum is not detailed; as spectrin tetramers crosslinked by actin monomers [944, 975] leave
a considerable part of the membrane surface accessible, it is implied that penetration by enzymes,
haemoglobin and B-3-P occurs through the holes of the reticulum. Only the enzymes and haemoglobin are drawn to scale, inasmuch as they are represented by the cross-sections of spheres whose
volumes are proportional to the respective molecular weights. (*b*) Tentative, approximate positioning
of bindings sites for various proteins on the N-terminal segment of B-3-P, as deduced from literature
data. For details see the text.

and the binding site of GAPD. The stoichiometries of the various enzymes, both
relative to each other and to B-3-P, are not illustrated. It is tacitly implied, however,
that two GAPD and aldolase molecules, whereas only one PFK, can bind per B-3-P
dimer. A major objection to this scheme may be that there are great differences
in the molar proportions and binding sites of the various enzymes (Table 6.4). The
most conspicuous is the scarcity of phosphoglycerate kinase binding sites. Two
aspects may be considered here: (*i*) in the real, three-dimensional array one bound
phosphoglycerate kinase might be "serviced" by a number of GAPD and other
enzyme molecules; (*ii*) a membrane-organized, "pump-oriented" enzyme array may
only be one of the modes of occurrence for these enzymes in the cell. Apart from
the pumps, ATP is also needed for the kinase reactions and for shape maintenance.
A subfraction of glycolytic enzymes providing for these needs is not necessarily
organized as proposed in Fig. 6.22, but may be organized in another way or not at
all. In fact, the kinetics of saponin-induced haemolysis suggest that one-third of
total cellular GAPD is free in the intact cell [665]. Probably any single topological
model would be too restrictive to be generally valid for the whole red cell.

TABLE 6.4.

Quantitative characteristics of some membrane-related proteins of human erythrocytes

Protein	Molecular parameters		Cellular parameters**		References
	M_r	Diameter* nm	Number of molecules per cell	Number of binding sites per cell	
Band-3-protein	195,000	–	1.2×10^6	–	1167
Phosphofructokinase	330,000	8.6	2×10^4	4×10^5	538, 613
Aldolase	160,000	6.8	3.5×10^4	1.2×10^6	1343
Glyceraldehyde-3-P dehydrogenase	146,000	6.6	2×10^5	$1.5-2.0 \times 10^6$	609
Phosphoglycerate kinase	48,000	4.6	5×10^3	500	294
Haemoglobin	68,000	5.0	3×10^8	1.2×10^6	1043, 1090, 1091

* Diameters of the spheres proportional to the M_r values.
** Approximate values.

On the other hand, the association of glycolytic enzymes, especially of GAPD, with the anion transport protein (B-3-P) might explain the early $^{32}P_i$ incorporation data according to which, at a moderate rate of P_i influx, GAPD had ready access to entering $^{32}P_i$. It has been calculated that the number of B-3-P molecules in the cell is sufficient to accommodate all the relevant enzymes [538]. The competitions observed (cf. above) refer to experiments where total enzyme/B-3-P ratios were >1. Figure 6.22 would suggest interactions between the enzymes themselves and hence some cooperativity in binding. Although this has not been observed, the available data are not quite adequate, as cooperativity may manifest itself only with the full set of enzymes at concentrations when B-3-P sites are not limiting.

A fraction of the peroxide-scavenging enzyme *catalase* is also bound to the cytoplasmic face of the red cell membrane: there are about 10^6 binding sites per cell. Membrane-bound catalase retains its enzymatic activity. It is of interest from the topological point of view that GAPD, aldolase and haemoglobin interfere with catalase binding [37a].

It is clear that much more knowledge of macromolecular organization at the cytoplasmic face of the erythrocyte membrane is needed before a meaningful enzyme organizational pattern can be established or, alternatively, refuted. Work to this end is underway in a number of laboratories (see review by Gillies [448]). Nevertheless, the problem might prove even more elusive than we see it today, after so many years of effort. We are not yet in full grasp of the means, both technical and conceptual, to tackle the extremely complex interplay of loose macromolecular assemblies. Srere [1157] has drawn attention to the fact, in connection with the mitochondrial matrix, that the functional description of highly concentrated protein mixtures might be more successfully approached from the side of crystalline state than from the behaviour in solution. We may add one more thought here. In the mature erythrocyte, haemoglobin ($\sim$330 mg/ml) is in a quasicrystalline state. If the interaction forces between haemoglobin molecules are greater than those between haemoglobin and glycolytic enzymes, the latter will tend to be extruded from the haemoglobin mass, necessarily to the cell periphery. Such a mechanism would greatly contribute to the building up of a submembrane glycolytic "shell", reinforced by the anchoring of some of the individual enzymes at integral membrane proteins. Obviously, such "extrusion forces", if they are operative at all, are hopelessly lost once the cell is disrupted.

THE "GLYCOSOME"

A unique arrangement of glycolytic enzymes has been found in the African trypanosome, *Trypanosoma brucei,* by Opperdoes, Borst and coworkers [886, 887, 888]. This organism, a pathogen in cattle, is closely related to the human parasite causing African sleeping sickness, *T. gambiense* and *rhodesiense* [1172]. In its energy metabolism the bloodstream form of *T. brucei* is entirely dependent on glycolysis, as the enzymes of the Krebs cycle and respiratory chain are repressed [117]. The terminal step of glycolysis is modified, inasmuch as the NADH produced at the GAPD reaction is reoxidized by O_2 through NAD-linked glycerol-3-phosphate dehydrogenase (I) and glycerol-3-P oxidase (II) [470]:

The oxidase has been located in the single, repressed promitochondrion, whereas the dehydrogenase has been found in microbodies [887, 888]. In the latter, the presence of glycerol kinase and seven glycolytic enzymes catalysing the steps from glucose to 3-phosphoglycerate were also detected [886]. It appears, then, that in *T. brucei* the whole set of glycolytic enzymes is packed into a membrane-bordered organelle for which Opperdoes and Borst [886] suggested the name "glycosome". The closed, intraorganellar localization, as opposed to the external attachment to membrane debris, of this somewhat altered glycolytic system is indicated by the fact that the majority of enzyme activities in the microbodies were cryptic, but could be "activated" by disrupting the microbodies by freeze–thawing or Triton X-100 treatment. Tracer kinetic experiments [1271] have shown two pools of glycolytic intermediates in *Trypanosoma brucei:* the fast-labelling pool amounting to about one-fourth of total cellular metabolites probably belongs to the glycosome, whereas the slow-labelling pool may be in the cytosol. The communication between the two pools across the glycosome membrane is relatively slow. The glycosome has also been found in *T. cruzi* and *Crithidia fasciculata* [1211], two other species of *Kinetoplastida*. It has been suggested that the glycolytic organelle may prove to be unique to this order of flagellated protozoa.

This type of arrangement for glycolytic enzymes need not involve functionally significant enzyme–membrane interactions, nevertheless the membrane has an important role in creating a microenvironment characterized by high local concentration of enzymes and metabolites. Further, one should distinguish the trypanosoman "glycosome" from the glycolytic enzyme assembly found in *E. coli* extracts [851], as the latter had apparently no membranous scaffolding. On the other hand, one cannot help comparing trypanosomes with mammalian erythrocytes, both cell types relying on glycolysis for energy production. Although red blood cells have no well-defined glycolytic organelles, the submembrane glycolytic shell may fulfil the same role as microbodies do in trypanosomes.

HEXOKINASE AND THE MITOCHONDRIAL MEMBRANE

Hexokinase (EC 2.7.1.1) catalyses the first reaction of the metabolism of glucose:

$$\text{D-Glucose} + \text{ATP} \xrightarrow{\text{Mg}^{2+}} \text{D-Glucose-6-P} + \text{ADP}.$$

It is found in all cells, but is of special significance in tissues, such as the brain [1131], that almost entirely depend on glucose as energy source. Several isoenzymes have been described for hexokinases [619, 620]. Type I, the predominant form in general, is found in brain and kidney, type II in fat pad and muscle, and both types I and II are present in heart and intestine. Type III (together with I) is encountered in erythrocytes [565]. Glucokinase, present only in liver [481], is sometimes called hexokinase IV. A brief review on mammalian hexokinases has recently been given by Trager [1230]. Hexokinase is generally regarded as one of the control points of glycolysis. In the context of our discussion it deserves interest by being one of the earliest recognized participants in a reversible enzyme–membrane association. It was Crane and Sols [260] who in 1954 first provided evidence for its loose association with the membrane of brain mitochondria. Since then a considerable body of evidence has accumulated regarding the structural–functional aspects of this connection. In fact, Wilson [1324] coined the term "ambiquitous enzyme" mainly inspired by the example of brain hexokinase. The binding of hexokinase to mitochondria was also shown for liver [368, 1018], heart [386] and small intestine [797]. No significant association was found between hexokinase and the erythrocyte membrane [368], although the latter binds several enzymes of glycolysis (cf. preceding section).

In the following we are going to discuss briefly the relationship of this enzyme to mitochondria in brain, the system known in greatest detail. Then a highly speculative theory of insulin action invoking an interaction between hexokinase and the mitochondrion through the hormone will be described. Both the former [1325] and the latter [77] topics have recently been reviewed at length.

Brain hexokinase

The enzyme from brain mitochondria is a monomer (single polypeptide chain) of $M_r = 98,000$ [216]. It is regarded as distinct from brain cytoplasmic hexokinase, but the evidence is not quite convincing [1325]. Its most important regulatory properties are inhibition by glucose-6-P (and the other hexose-6-Ps) and by ADP (cf. [962]).

The characterization of hexokinase association to brain mitochondria came from Wilson [1323], following the description of the analogous phenomenon in ascites tumour cells by Rose and Warms [1018]. Binding is specific to the outer mitochondrial membrane. Importantly, ATP (as well as other nucleoside triphosphates in the absence of Mg^{2+}) and glucose-6-P promote solubilization, whereas P_i counteracts the glucose-6-P effect. On the other hand, in the bound state hexokinase is more active, because it is less susceptible to the inhibitory effects of glucose-6-P and ATP, and has a lower K_m for Mg-ATP. On the basis of these properties a regulatory scheme was envisaged [1323, 1325] according to which in metabolic states requiring rapid glycolysis the metabolite concentrations change in a manner that shifts the soluble-bound equilibrium in favour of the particulate form, i.e. activates the enzyme.

As for the structural requirements of enzyme–membrane association, binding is thought to be mainly electrostatic, with attractive and repulsive forces acting in a hydrophobic milieu [370]. Increasing ionic strength elutes hexokinase from the mitochondrion, but the effect is biphasic inasmuch as up to about 20 mM ion concentration binding is strengthened by salt. Divalent cations were found to be much more potent than monovalent ones in this "glueing" effect, whereas the opposite holds for elution by more concentrated salt solution. It has been proposed that divalent cations act by neutralizing repulsive forces between negatively charged groups.

Mild chymotryptic treatment of hexokinase completely abolishes binding ability [1018, 1325], though only a small part of the N-terminus is removed [369]. Catalytic activity is not affected by this treatment. It was therefore suggested that the N-terminus played an important role in binding. A specific "hexokinase-binding protein" occurs as an integral protein of the outer mitochondrial membrane, with an $M_r = 31,000$ [368]. The scheme for hexokinase–membrane interaction as suggested by Wilson [1325] is seen in Fig. 6.23. The binding of hexokinase to the mitochondrion seems to cause metabolite channelling, as well as activating the enzyme. Thus, evidence has been presented that intramitochondrially generated ATP has a preferential access to hexokinase relative to exogenously added ATP [466, 467, 1268]. This arrangement would effectively recycle ADP to the mitochondrion, at the same time relieving the inhibition of oxidative phosphorylation by ATP. Thus the reversible association of hexokinase to the mitochondrial membrane apparently serves

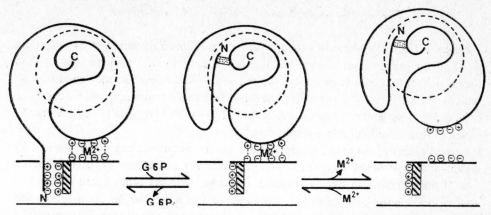

Fig. 6.23. Current view of the binding of hexokinase to the outer mitochondrial membrane. The region encircled by the dashed line comprises the catalytic domain, whereas another domain including the N-terminus is involved in electrostatic interaction with the membrane. The striped rectangle in the membrane represents the hexokinase-binding protein. On the addition of glucose-6-P (G6P, stippled rectangle) a conformational change occurs in the enzyme accompanied by activity loss and release from the membrane. Glucose-6-P as effector binds probably not at the active site. M^{2+}, divalent metal ion. (From Wilson [1325].)

an interesting dual purpose: the regulation of hexokinase activity, and thereby glycolytic flux, and the control of terminal oxidation through the respiratory chain.

The question again is, of course, whether these observations *in vitro* apply to the conditions *in vivo*. There are data suggesting that at least some of these phenomena also occur in the living cells. Thus Knull and coworkers have demonstrated in chicken for a variety of metabolic states including galactosaemia [669], ischaemia [669] and insulin-induced hypoglycaemia [670], conditions all characterized by a fall in glucose-6-P and ATP levels, that the amount of particulate hexokinase increased. Bachelard [38] found analogous changes in ischaemic mouse brain. Furthermore, Kriegelstein and associates [80, 81, 140] have observed that a number of anaesthetics reversibly removed hexokinase from mitochondria both *in vitro* and *in vivo,* when applied at clinical concentrations, which do not cause extensive membrane damage [1007]. A parallelism was found between the proportion of hexokinase solubilized and the depth of anaesthesia. It has not been proved yet whether the lower glycolytic rate during anaesthesia is indeed elicited by the detachment of hexokinase from the mitochondria. At all events, although the primary regulatory significance of the "ambiquitous" behaviour has been questioned [77, 961], the above data strongly support the view that hexokinase reversibly associates with mitochondria in brain cells, to an extent depending on the metabolic state.

A possible role of the hexokinase–mitochondrial system in the metabolism of tumour cells has been raised [154, 155].

The hexokinase–mitochondrial theory of insulin action

Bessman [73, 74] advanced a theory for the mechanism of action of insulin that is based on the phenomenon of hexokinase binding to the mitochondrion. This theory asserts that insulin promotes the binding of hexokinase to mitochondria by directly forming a linkage between the mitochondrial membrane and hexokinase (Fig. 6.24). This would facilitate the recycling of ATP/ADP in the mitochondria, which in turn would stimulate respiration.

Several pieces of evidence, albeit none of them conclusive, can be mentioned in favour of this proposal, as recently done by Bessman and Geiger [77]. Hexokinase type II predominates in tissues sensitive to insulin, whereas types I and III are favoured in insulin-insensitive cells, such as nerve and red cells. As discussed in the previous section, brain hexokinase is largely bound to the mitochondria, possibly through a receptor protein, without insulin. In contrast, in insulin-sensitive tissues there is an inverse relation between insulin effect and the fraction of hexokinase bound to mitochondria *before* hormone administration. Insulin promotes binding of hexokinase to the mitochondrion and some of the insulin effects are abolished by blocking the hexokinase pathway.

As part of this theory Bessman [74, 75] assigned a role to another enzyme as well, to creatine phosphokinase (EC 2.7.3.2). This enzyme catalyses the reaction

$$ATP + Creatine \rightleftharpoons ADP + Creatine\text{-}P$$

which serves to store energy in muscle for times of vigorous exercise. A considerable portion of creatine phosphokinase is bound on the outer surface of the mitochondrial inner membrane [585]. As creatine influences mitochondrial respiration [76, 585] and physical exercise has an insulin-like effect on metabolism, Bessman postulated that creatine phosphokinase may substitute for hexokinase in recycling ATP–ADP on the mitochondrial surface. In the hypothetic "comparticle" (Fig. 6.25) two adenylate kinase molecules are also involved to explain ^{32}P distribution data (cf. [77]).

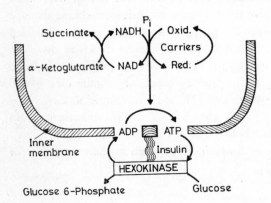

Fig. 6.24. Scheme of the proposed role of insulin in attaching hexokinase to the mitochondrial inner membrane. Anchoring of hexokinase would facilitate recycling of ADP. (From Bessman and Geiger [77].)

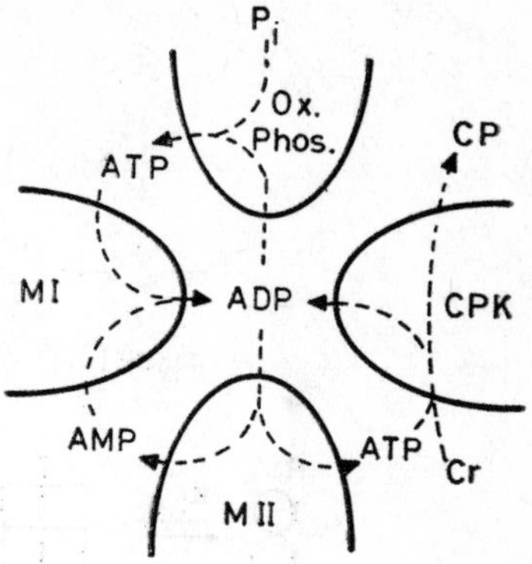

Fig. 6.25. Tentative arrangement of enzymes in a "comparticle" around the site of oxidative phosphorylation site in the mitochondrion. Symbols: Ox. Phos., site of oxidative phosphorylation M_I and M_{II}, two molecules of adenylate kinase (myokinase); CPK, creatine phosphokinase; Cr, creatine; CP, crestine phosphate. According to the hypothesis all enzymes in the "comparticle" would use a common ADP pool. (From Bessman and Geiger [77].)

The creatine phosphokinase system would be operating during exercise (without insulin), whereas muscle would need insulin during rest to restore the hexokinase–mitochondrial system. Furthermore, as creatine phosphokinase is also bound to the M-band of striated muscle, a creatine–creatine-P shuttle was postulated to operate transporting energy from the mitochondrion to the functioning muscle [78].

It is pertinent to mention here that with liver mitochondria added hexokinase can interfere with the functioning of carbamyl-P synthase (EC 2.7.2.5) probably by withdrawing newly generated ATP via the ATP–ADP translocator [973]. It has been postulated that carbamyl-P synthase is loosely bound to the inner mitochondrial membrane. One may visualize that in the mitochondria of different cell types different enzymes make up a "comparticle".

The complex model described above is an attempt to combine enzyme–membrane organization with hormone function and raises a new aspect in the ill-understood mechanism of action of insulin. However, the theory is far from being universally accepted (cf. [273]). As emphasized above, none of the evidences or even their sum total is conclusive. On the other hand, recent investigations showing that insulin becomes internalized by target cells [457, 1062] and interact not only with the cell membrane receptor, as thought earlier, give scope for such speculations. Nevertheless, further experiments are needed to decide whether the interaction of hexokinase and mitochondria has anything to do with insulin action.

THEORETICAL MODELS OF ENZYME REGULATION BY REVERSIBLE ADSORPTION

In the preceding sections several cases have been described where enzymes reversibly associated with cellular surfaces. In many instances the functional consequences of adsorption–desorption are not clear, because the experimental evidence is inconclusive or controversial. It may help to clarify the picture, at least in some cases, if we are aware of the predictable behaviour of such systems.

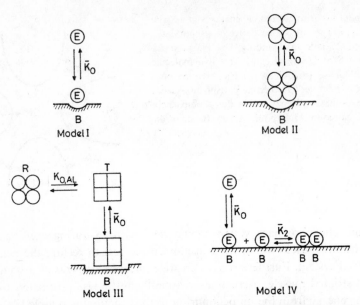

Fig. 6.26. Models of adsorptive enzyme systems. E is the enzyme molecule, B is the binding site on the support and K_0 is the equilibrium constant of the association reaction. For the description of the various models see the text. (From Kurganov and Loboda [704].)

Kurganov and Loboda [704] considered various models and examined what regulatory properties, such as positive or negative cooperativity, are to be expected from them. The scheme given by the above authors is seen in Fig. 6.26. In model I a monomeric enzyme is bound to some support. If the bound enzyme is inactive, e.g. because of masking the active centres, negative kinetic cooperativity will be observed (Hill coefficient <1) in the substrate saturation curve. Effectors that promote or hinder binding act then as inhibitors and activators, respectively. In model II an oligomer is bound whose kinetic behaviour, again assuming that binding has an inactivating effect, will depend on the intraoligomer coupling of protomers. This may result in a positive kinetic cooperativity (sigmoidal saturation curve). In model III the support acts as an allosteric "macroeffector" interacting with one of the conformations (R or T) of the free enzyme. This case is formally analogous with the concerted allostery model for binding [834]. Thus, for example, if the inactive T-state is preferred by the support, then binding will enhance positive kinetic cooperativity. Model IV treats the case when the enzyme is able to self-associate on the support while becoming catalytically inactive. The behaviour of such a system is rather complex: it may range from positive cooperativity to mixed negative and positive cooperativity depending on the actual values of the constants involved and on the enzyme concentration.

The mathematical formulas derived for the above models imply rapid equilibrium in adsorption–desorption. If these processes are slow, hysteretic behaviour (cf.

Chapter 3, p. 71) may be observed. Another feature worth recalling is that reversible binding may be a means to amplify cooperative response in a small oligomer (such as a dimer).

The use of the above and further conceivable models is two-fold. First, they are heuristic when analysing real systems. Nevertheless, the complexity is not to be underestimated when enzyme kinetics are to be measured in the presence of a biological macrostructure. Second, the theoretical model may inspire workers to carry out model experiments to check the theories. Endeavours to this end have been made, for example, with lactate dehydrogenase M_4 binding to artificial supports like carboxymethyl-cellulose [549] and dextran sulphate [703]. The results showed that, indeed, binding to dextran sulphate gave rise to positive kinetic cooperativity, which was not observed with the enzyme in solution.

CHAPTER 7

METABOLITE COMPARTMENTATION

> *"The compartmentation of small molecules in eukaryotic cells is easy to believe, but hard to prove."*
>
> Davis [292]

This chapter on metabolite compartmentation is well introduced by the above statement of R. H. Davis, who described the "surface model" of channelling in 1967, when studying arginine metabolism in *Neurospora crassa* [291]. Metabolite channelling, a synonym for microcompartmentation, had been observed already in the tryptophan pools of *Neurospora* by Matchett and DeMoss [788], and the very term "channelling of metabolites" can be traced back to Vogel and Bonner [1273] who applied it to the sequestration of proline and ornithine biosynthesis in the same organism.

It is characteristic of the field that Davis opened his talk [292] with the above words at a *recent* symposium on metabolic channelling [876]. Thus, no breakthrough occurred during the past 15 years in our experimental strategy for detecting compartmentation. Nevertheless, several potentially useful approaches have been devised that are worth summing up.

In earlier chapters metabolite compartmentation has already been mentioned in general, as a possible corollary of enzyme juxtaposition (Chapter 4), and in connection with the various enzyme assemblies discussed. The aim of this chapter is to place metabolite compartmentation in the proper perspective, to raise ideas about novel mechanisms, and to survey experimental approaches.

THE CELL AS BASIC COMPARTMENT

Compartmentation is one of the fundamental principles in the organization of living matter. The very existence of cells is a manifestation of biological compartmentation. The cell boundary, as a rule, a lipid bilayer interspersed with proteins and protein complexes having specialized functions, endows the cell with the integrity needed for its balanced performance, sharply distinguishing the "within" from the "without". The latter is for unicellular organisms, whether prokaryotic or eukaryotic, a widely varying, harsh ambience against the deleterious effects of which the organism had to develop sophisticated defence mechanisms. For example, Gram-negative bacteria, as *E. coli,* have in addition to an inner cytoplasmic membrane an outer cell envelope, separated from the former by a periplasmic space, which is a complex lipoprotein–matrix protein–peptidoglycan sheet acting as a sieve against harmful substances such as antibodies [313]. In contrast, the cells of higher plants and animals organize into tissues, where the ambience is fairly constant and mild, so that cells do not need special means of protection. Moreover, they interact with each other up to the extent where the cell contents directly communicate. This is exemplified by the gap junctions between liver cells [747], which are made up of hexameric (cyclic) protein oligomers embedded into, and traversing, the cell membrane; two such hexamers of contacting liver cells form an aqueous channel of 20 Å diameter [768]. Material flow through these channels is cut off by high Ca^{2+} concentration. It follows that one may regard liver tissue as a compartment distinct from other organs. Indeed, there is a supracellular level of compartmentation as well, involved in the development and morphogenesis of multicellular organisms. Our concern, however, will not be extended to these phenomena but will be confined to the intracellular space.

INTRACELLULAR COMPARTMENTS
AND METABOLITE POOLS

The division of space does not stop at the boundaries of cells. On the contrary, spatial sequestration is highly developed within all eukaryotic cells. The means of separation is partly well known: the interior of higher cells is subdivided by bio-membranes into a variety of intracellular spaces such as the nucleus, mitochondria, endoplasmic reticulum, etc. These are more or less clearly defined morphological entities, recognizable under the microscope or electron microscope and are charac-terized by their specific content of macromolecules and small molecules. De Duve's [296] "postulate of single location" held that each enzyme in the cell had its unique site of occurrence. By "site" these, mainly membrane-bounded bodies called cell organelles are meant which consequently represent a definite compartment. (There are two exceptions to the membrane-boundedness of organelles or cytosomes: the ribosome and the nucleosome.) Since organelles are large relative to the size of an average metabolite, we shall refer to them as *macrocompartments.*

The permeability of membranes is selective to various degrees. In general, they decide what to let into, or out of, the compartment and at what rate and stoichio-metry. The selectivity of the different membranes has been thoroughly characterized, but the mechanisms underlying this specificity are only now being unravelled. What-ever the mechanism, the resultant of the membrane transport processes will determine the composition and size of the various *pools,* i.e. the content of the compartment. The terms "compartment" and "pool" are closely related, sometimes even used synonymously. It is more appropriate, however, to define a metabolic compartment as *a subcellular region of biochemical reactions kinetically isolated from the rest of the cellular processes* [633] and the pool as *the amount of compound(s) subject to this kinetic isolation.* In other words, a pool is accommodated in a compartment, even if the compartment is virtual rather than a morphological entity (cf. below).

Beside membrane-bounded macrocompartments, there are further subcellular regions that meet the criteria of a compartment (Table 7.1). Such is a "channel" in

TABLE 7.1.

Intracellular metabolic compartments

Macrocompartments (organellar)		Microcompartments (molecular)	
Name	Family type of boundary membrane[a]	Name	Nature of boundary
Nucleus	1	Enzyme–enzyme channel	Protein surface, closed[b]
Endoplasmic reticulum	1		
Golgi apparatus	1	Debye–Hückel layer	Membrane surface, open
Cytosomes (peroxisomes, glyoxysomes, endosomes, primary and secondary lysosomes, akanthosomes (coated vesicles), synaptic vesicles, melanosomes, aminosomes, cuprosomes)	1	Metabolite binding site	Protein surface, open
Mitochondrion			
Inner compartment (matrix)	2		
Outer compartment (intermembrane space)	1		
Plastids			
Inner compartment (thylakoids)	3		
Outer compartment	1		

[a] The family types 1, 2 and 3 are characterized by the presence of sterols, cardiolipin and trans-3-hexadecanoic acid *plus* galactolipids, respectively (see the text).
[b] Closedness is, of course, not complete, as input and output metabolites have to be able to enter and leave the channel, respectively.

multienzyme clusters, whose physical equivalent is an irregular space formed by the clefts (impressions, crevices and the like) on the clustered enzymes or a Debye–Hückel layer [1231] on a membrane surface. Furthermore, the definition of compartment fits a population of binding sites that can reversibly bind metabolite(s), inasmuch as the bound fraction will behave differently from the free species, from the kinetic point of view. Since all these compartments are in the order of the size of the metabolite, we call them *microcompartments*. It is to be noted that apart from the enzyme–enzyme channel, the other microcompartments are no longer enclosures surrounded by walls, but are open: the physical boundary of the compartment is virtual. A distinction of a similar kind has recently been canvassed by Hess [533].

In the following sections, after a brief introduction to macrocompartments we shall focus on the properties of microcompartments. Consequently, many aspects of cell compartmentation will not be dealt with. Our guideline is to review compartmentation phenomena related to supramolecular enzyme organization. However, since this relationship is not always clear, our selection will be somewhat arbitrary. The reader interested in compartmentation in the fields of cell morphogenesis, development, gene expression and a number of inter-organellar processes should consult the proceedings of recent symposia [876, 1109, 1158].

MACROCOMPARTMENTS: PLASMATIC AND NON-PLASMATIC

Communication between macrocompartments proceeds at two levels. For small molecules there are special transport facilities, i.e. channels, translocators, pumps, etc. made up of proteins, often glycoproteins, that allow permeation through the boundary membrane. Interestingly, apart from these specific means there are also non-specific routes, for example, for water to cross the red cell membrane. Sha'afi [1089] has determined that while about 90% water flux passes through channels formed by oligomeric proteins, about 10% diffuses via statistical defects in the lipid bilayer. This occurs through thermally induced kink-formations in the phospholipid alkyl chains, which produce mobile packets of free volume, filled with H_2O molecules (Fig. 7.1).

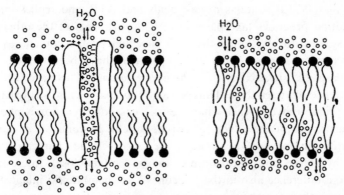

Fig. 7.1. Two ways of water transport across lipid membranes. On the left, a pore (channel) made up of integral membrane proteins provides for a hydrophilic route; this is the main path of water across membranes. On the right, water molecules travel across the membrane in small packets that occupy free volumes ("kinks") generated in the lipid layer by thermal fluctuations. (From Sha'afi [1089].)

In contrast, macromolecules, as a rule, are unable to cross membranes along routes available to small compounds. Nevertheless, they can be transported from one macrocompartment to the other by means of *membrane flow,* i.e. fusion and separation of membrane-bounded compartments. These phenomena are collectively named *infracytosis.* Such is, for example, the mechanism of formation of any cytosome of Table 7.1 or of their fusion, as the generation of a secondary lysosome (digestive vacuole) from an endosome and a primary lysosome. The processes by which macrocompartments communicate with the extracellular space are *endocytosis* (uptake) and *exocytosis* (discharge), the latter involving fusion with, the former detachment from, the cell membrane. A clear and brief description of these phenomena is found in Sitte's review [1121].

It has been observed, however, that not all kinds of membranes can fuse with each other. In spite of their diversity, a kinship exists between certain membranes on the basis of which a system of "membrane families" could be established [798, 1119]. This term denotes classes of membranes, the members of which can be derived from each other or fuse with each other directly or indirectly [1121]. There are two membrane families in animal cells and three in plant cells. Membranes belonging to the same family can be very different in appearance, functions and composition, yet each has a characteristic lipid constituent: *family 1* sterols, *family 2* cardiolipin, and *family 3* trans-3-hexadecanoic acid and galactolipids. In Table 7.1 the family type of each macrocompartment (organelle) is indicated. It is noteworthy that the only representatives of *families 2* and *3* are the mitochondrial and plastid inner membranes, respectively, which contrasts the abundance of cytomembranes in *family 1.* As pointed out by Sitte [1120] this is a strong argument for the external (symbiontic) evolutionary origin of these organelles.

Thorough examination of the bewildering variety of cellular macrocompartments revealed that they can be divided into just two classes. One has the characteristics of active "plasma": it contains nucleic acids and ATP, and replication–transcription–translation processes take place in it. Compartments of the other type never contain these constituents, though they may have high protein content, as, for example, plant vacuoles. The two types of compartment have been termed "plasmatic" and "non-plasmatic", respectively, for the above reasons. There are further characteristic differences between them. Thus plasmatic compartments usually have enzymes with pH optima above 7, their characteristic catabolic reaction is phosphorolysis and the glucans are of the α-type (glycogen, starch). In contrast, in non-plasmatic compartments the enzymes' pH optima are below 7, the characteristic catabolic reaction is hydrolysis and glucans are of the β-type (cellulose). Interestingly, there seems to be a difference also in the "sidedness" of membranes: integral membrane proteins expose their C-termini to the interior of plasmatic compartments, but their N-termini to the interior of non-plasmatic compartments [1121].

It was originally postulated by Schnepf [1066] in 1964 that every biomembrane separates a plasmatic phase from a non-plasmatic one. This postulate has been

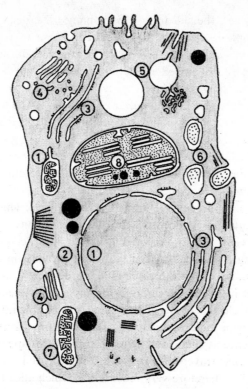

Fig. 7.2. Macrocompartments of an idealized eukaryotic cell. Plasmatic phases are grey, non-plasmatic phase is blank. (1) Nucleoplasm, continuous with the cytoplasm (2). In the cytoplasm, without limiting membranes, are microtubules, centrioles and free polysomes (below the nucleus), tonofilaments (shown at the half-desmosome, left) and oleosomes (black). Non-plasmatic compartments, bounded by different cytomembranes: (3) rough endoplasmic reticulum and nuclear envelope; (4) Golgi dictyosomes; (5) vacuoles (on the right a pulsating or contractile vacuole is engaged in elimination of excess of water); (6) lysosomes, and other cytosomes. Mitochondria (7) and a plastid (chloroplast) (8) are surrounded by non-plasmatic spaces. Their plasmatic bodies, the mitoplasm and plastoplasm, respectively, will never fuse with true cytoplasm (2), only with their like. (From Sitte [1121].)

confirmed by accumulating evidence. Indeed, the well-established asymmetry of biomembranes as regards both proteins and phospholipids [66, 1028] calls for different environments on the two sides. It follows, then, that two plasmatic compartments cannot be separated by a *single* biomembrane (only by two, with a non-plasmatic compartment intercalated). The macrocompartmental system of an idealized eukaryotic cell as envisaged by Sitte [1121] is seen in Fig. 7.2.

Two important generalizations can be made from the foregoing, as is partly apparent also in Fig. 7.2. First, the plasmatic compartments will only fuse with plasmatic ones, and non-plasmatic compartments only with the like. Thus, in addition to the family types, the sidedness of membranes is crucial for fusion. Second, the concept of cytoplasm has to be reappraised. There is a nucleocytoplasmatic continuum, which behaves as a single entity, for example, during mitosis, while the mitochondria and plastids, embedded in the cytoplasm but separated from it by narrow non-plasmatic spaces, function quite independently with their own DNA and ribosome machinery [112]. The nuclear envelope, with its holes of about 100 Å diameter, has a limited role in compartmentation and certainly cannot sequester average-sized metabolites [1107].

The volumes of cellular macrocompartments cover a range of about ten orders of magnitude, but the specific surface, i.e. the surface/volume ratio, of plasmatic compartments falls into a conspicuously narrow range centring around 60 μm^{-1}, which corresponds to a sphere of 100 μm in diameter [1121]. This empirical fact gave rise to various speculations concerning the possible reasons for this convergence. From our point of view, the argument that membranes act as "protein (enzyme) collectors" producing high local concentrations of proteins in an ordered array is of particular interest. Although the value (100 μm diameter) cannot be deduced by any meaningful consideration, it certainly means a high membrane density and, consequently, a very close packing of integrated and peripheral proteins (enzymes). Thus about forty enzyme molecules of average size (~ 8 nm diameter) could be placed along the circumference at the widest cross-section of this sphere and about twelve enzyme molecules could traverse its interior. The potential role of membranes for enzyme "scaffolding" is therefore very remarkable, which adds further weight to the concept of enzyme organization through binding to cytomembranes.

In this bird's-eye view of cellular macrocompartments we have again bumped into the question of enzyme–membrane interactions. Indeed, supramolecular enzyme organization is an extremely complex group of phenomena, so that wherever we pull a thread, the whole web is tugged. To add one more aspect to the organizing role of biomembranes, we must recall that biomembranes are, in a sense, immortal. They apparently cannot arise *de novo,* but need a primer. Disassembled membranes when allowed to reassemble *in vitro,* do not recover their original asymmetry [746]. According to Blobel and Dobberstein's *signal hypothesis* [94, 95] proteins that have to be transferred through and/or built into membranes are synthesized from mRNAs having a special sequence whose translation product ensures the attachment of ribosomes to specific receptor protein complexes pre-existing in the membrane. This direct injecting mechanism would determine how the nascent polypeptide chain is placed in position. (For a recent review cf. [685].) In other words, membrane biogenesis presupposes the existence of membranes. What all this boils down to is that membranes, these three-dimensional entities, form a continuum also *in time,* which is equivalent to their carrying information indispensable for life.

ARGININE METABOLISM IN *NEUROSPORA*: A CASE OF REGULATION
VIA MACROCOMPARTMENTS

A clear-cut example of metabolite compartmentation has been provided by Davis and coworkers in the arginine–ornithine metabolism of *Neurospora crassa*. (For a recent review see [292]). The scheme of the relevant part of amino acid metabolism in the *Neurospora* cell is seen in Fig. 7.3. Ornithine can be found in three cellular compartments. It is generated in the mitochondrion, where it is used for citrulline synthesis by ornithine carbamoyl transferase. It occurs in the cytosol, where it can be metabolized by ornithine decarboxylase or utilized for proline synthesis via ornithine aminotransferase. Cytosolic ornithine may stem from outside, when the organism is fed ornithine, or from the mitochondrion or from a membrane-bounded vesicle, apparently a storage site.

Early genetic studies (briefly summarized in [292]) revealed that mutant strains requiring low levels of ornithine for normal growth contained large amounts of endogenous ornithine. This finding immediately suggested the compartmentation of endogenous ornithine. The ornithine compartment was slightly leaky and this leakiness was much enhanced in a double mutant, so that this organism became prototrophic.

Biochemical work identified this compartment with a vesicle (cf. Fig. 7.3) which could be isolated from broken cells and also contained most of the cellular arginine, lysine and histidine [1296]. Tracer kinetic experiments with [^{14}C] ornithine added to growing cells yielded a remarkable distribution according to which 1% of cellular ornithine was in the mitochondria, another 1% was in the cytosol, whereas 98% was accumulated in the vesicle [118, 614]. Importantly, the vesicle also contained large amounts of polyphosphate.

The latter substance, polyphosphate, was found in large amounts also in the vacuole of the yeast, *Saccharomyces cerevisiae* [1250]. Wiemken and coworkers have

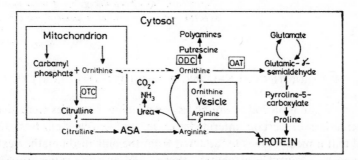

Fig. 7.3. Compartmentation of arginine and proline metabolism in *Neurospora crassa*. The location of the proline enzymes is not known, though they are shown in the cytosol. The conversion of glutamic semialdehyde to glutamate is a catabolic step, independent of the biosynthetic steps in the reverse direction. Abbreviations: ASA, argininosuccinate; OAT, ornithine aminotransferase; ODC, ornithine decarboxylase; OTC, ornithine transcarbamylase. (From Davis [292].)

convincingly demonstrated that the polyphosphate and arginine content of vacuoles changed in a parallel manner [1316] and concluded that polyphosphates bind arginine much like an ion-exchange resin [328]. In *Saccharomyces* the vacuole also has a lysosomal function as it contains most hydrolytic enzymes. The secluded storage of arginine is advantageous for the cell as in this way it does not interfere with arginine biosynthesis, which is repressible by low concentrations of the amino acid (cf. the review by Wiemken [1315]). Clearly, yeast sees to it that an essential nitrogen source be accumulated while nutrients are abundant, in order to survive lean years.

The analogy between the *Saccharomyces* and *Neurospora* systems is compelling. The vesicle of the latter apparently piles up all kinds of basic amino acids on the polyphosphate, but was shown to release arginine at an appropriate rate to support growth in case of need [1298]. On the other hand, added arginine is partly split by arginase to ornithine and is taken up into vesicles, inducing the release of ornithine from the vesicles (cf. Fig. 7.3). The ornithine thus generated and released into the cytosol is then consumed by the catabolic pathway [119, 1297].

Evidently, this neat and economic compartmentational scheme of *Neurospora* requires the active participation of the vesicle membrane. Little is known as yet about the control of vesicle function. Work is in progress with mutant strains defective in the accumulation of basic amino acids [257]. It remains to be seen whether the uptake of these amino acids is secondary to the accumulation of polyphosphate. Obviously, if the accretion of arginine is due to simple ionic trapping, the question shifts to how the polyphosphate content of vesicles is regulated. In all events, the problem is that of membrane transport between a plasmatic and a non-plasmatic compartment.

MICROCOMPARTMENTS

> *"The theories serve to satisfy the mind, prepare it for an 'accident', and keep one going. I must admit that most of the new observations I made were based on wrong theories."*
>
> Albert Szent-Györgyi [1204

METABOLITE CHANNELLING IN MULTIENZYME CLUSTERS

In Chapters 5 and 6 several, more or less convincing, examples were given for the compartmentation of certain intermediates in multienzyme clusters. Once it is established that a metabolite is sequestered from the bulk medium, the next question is how this comes about. Unfortunately, this second question is unanswered in most cases, owing to the lack of adequate structural information. The only instances where we have a fairly concrete idea about the mechanism of channelling is when covalently bound intermediates are involved, as with the pyruvate dehydrogenase complex or fatty acid synthase, which is no "channelling" in the strict sense. Evidently, to describe the subtleties of direct metabolite transfer between two enzymes, first of all the detailed molecular architecture of all parties, as well as of their assembly, ought to be known. Such information can only be expected from X-ray crystallography, which will certainly provide some in the not too distant future, at least for the simpler complexes like tryptophan synthase. However, unclouded optimism seems ill-founded. For even if the X-ray structure is available, the mechanism of facilitated transfer may not be obvious, just as it is a long way to establish the catalytic mechanism for a single enzyme from X-ray and other data. Furthermore, the prospect that one day the models of the ten-odd glycolytic enzymes might be put together as a three-dimensional jigsaw puzzle, provided that the glycolytic enzyme assembly (cf. Chapter 5) holds true, is at present but an attractive dream. To solve this putative, irregular "magic cube" would be much more tantalizing than unscrambling the regular one [562], though already the latter is tricky enough. Canvassing these difficulties is not at all meant to deter X-ray crystallographers from endeavours to this end, which some of them certainly cherish as a desired goal. Biochemists, however, apparently have to content themselves with more modest objectives. What should they be?

A meaningful way to proceed is to attempt to establish the physiological significance of a given channelling effect. As stated in the motto of this chapter, compartmentation is disarmingly appealing and once proved, the physiological implications seem, as a rule, to be only too much at hand. In reality, however, the assignment of a definite role may be a very hard task indeed. Some pertinent reservations will be discussed at the end of this chapter.

DYNAMIC COMPARTMENTATION

Metabolite sequestration in macrocompartments and in channels of multienzyme clusters has the common feature that the physical structure underlying the compartment is fairly stable, at least its changes (dissociation, turnover, etc.) are much slower than the sequence of catalytic steps involved. Channelling would severely deteriorate if the enzymes of a complex dissociated or a membrane-bound enzyme array were in a rapid equilibrium between adsorbed and free species. It seems trivial that a channel should last, otherwise it is no channel but a pipe-break.

Notwithstanding this apparently irrefutable simple idea, it may be useful to consider the opposite situation, i.e. when compartmentation is entailed without a rigid structure. One reason to do so stems from the considerable body of evidence according to which there are multiple pools of various metabolites within macrocompartments but the mechanism (structure) that would be responsible for this pool multiplicity is at best hypothetic. It would make a rather long list if all such claims made over the past 20 years were enumerated. Let it suffice to refer to reviews on metabolic compartmentation [876, 1158, 1161] and, rather arbitrarily, to a bunch of original papers [41, 207, 648, 690, 987, 1134, 1223, 1226]. It is beyond our scope to scrutinize the reliability of these reports. We only want to draw the conclusion that there are probably cases when metabolite compartmentation in a "soluble" system occurs. If one recalls the description of glycolytic enzyme organization in the cytoplasm or Krebs-cycle enzyme organization in the mitochondrial matrix, then one cannot help thinking that the elusiveness of the problem is due to our looking for something that does not really exist in the given system: a stable multienzyme complex. That a glycolytic particle *may* occur in *E. coli* is no guarantee for its presence in muscle, liver or red cells. Nevertheless, compartmentation effects seem to prevail, yet well-documented complexes fail to emerge. In the following we advance a hypothesis for compartmentation without rigid clustering, which may be heuristic, at least in the genesis of more sophisticated hypotheses.

Transient enzyme–enzyme complexes: A hypothetic model for dynamic compartmentation

The salient feature of the hypothesis put forward by Friedrich [404] is that consecutive enzymes may accomplish direct metabolite transfer, i.e. channelling, even if their complex is very loose (K_d large) and the rate constants for both association (k_{+1}) and dissociation (k_{-1}) are high:

$$E_1 + E_2 \underset{k_{-1}}{\overset{k_{+1}}{\rightleftharpoons}} E_1E_2 \qquad (7.1)$$

$$K_d = \frac{[E_1][E_2]}{[E_1E_2]} . \qquad (7.2)$$

This implies that any individual complex is short-lived (residence time small). Alternatively, we may characterize the system as two enzymes engaged in random, *almost elastic* collisions; the slight "stickiness" of the two enzymes gives rise to a finite residence time. Based on this tenet, models of different complexity can be built.

Direct metabolite transfer in the "complementary cage"

The following postulates are made:

(a) The contact areas of functionally adjacent enzymes, designated as *foreign recognition sites,* are located in the vicinity, or even around, the active site. Hence in the complex the two active sites become juxtaposed, a real or virtual *"complementary cage"* is formed. In spite of the high K_d, the affinity towards adjacent enzyme(s) is significantly greater than toward functionally irrelevant enzymes in the same macrocompartment.

(b) The frequency of productive, i.e. complex-forming, collisions between E_1 and E_2 is at least as high as the frequency of product release from the first enzyme, viz. the rate of catalytic reaction.

The combination of the above two conditions may result in metabolite compartmentation as follows. If the formation of complementary cages is properly timed, the release of product from E_1 would occur into the cage where its diffusional freedom is severely restricted and therefore the metabolite will bind to the active site of E_2 with great probability. As the system is dynamic, there will always be free active sites on E_1 to pick up its substrate. As collision frequencies are statistical, not all products from E_1 are expected to be channelled this way. However, much depends here on the stoichiometry of the two enzymes. If E_2 is in excess, there may be little chance for any E_1-product to go astray.

One can make a rough calculation about the reality of the above model. In yeast, about 65% of soluble protein is constituted by glycolytic enzymes [535] and the computed time between two collisions of any glycolytic enzyme is 30 nsec [534]. Hence the collision frequency is about $3 \times 10^7 \text{ sec}^{-1}$, and for two functionally adjacent enzymes, out of the ten-odd enzymes of glycolysis (taking them now as an equimolar mixture for the sake of simplicity), the collision frequency is approximately 2×10^6 sec^{-1}. As the turnover number of, for example, yeast GAPD is 90 sec^{-1} [656], about 2×10^4 collisions take place with an adjacent enzyme during one catalytic cycle. However, in the living cell most enzymes work far below their maximum capacity, with red cell GAPD the utilized catalytic potential was estimated as 1% [821]. Even if this estimate is an extreme, we may accept that in the cell the number of collisions per catalytic cycle can be in the range 10^5–10^6. In other words, to ensure compartmentation of any one metabolite of glycolysis only about every 10^5th collision between the two relevant enzymes need be productive.

It is of interest to compare this value with that one can derive from observed interprotein reactions. If two proteins associate reversibly, $A + B \rightleftharpoons AB$, with a second order rate constant of $10^5 \text{ M}^{-1} \text{ sec}^{-1}$ [486], the rate of reaction at steady-state concentrations $[A] = [B] = 10^{-4}$ M will be 10^{-3} msec^{-1}. If the collision frequency of these two proteins is taken to be $2 \times 10^6 \text{ sec}^{-1}$ (cf. above), then we obtain for the number of collisions per litre per sec the value 1.2×10^{26}, whereas the number of AB complexes formed per litre per sec is 6×10^{20}. Thus, one complex is formed in every 2×10^5 collisions. The requirement that about every 10^5th collision be productive for compartmentation to occur does not therefore seem unrealistic.

Alternating complementarity in an enzyme sequence

In a metabolic sequence consisting of three steps catalysed by enzymes E_1, E_2 and E_3, respectively, both the substrate and product of E_2 can be compartmented if E_2 displays complementarity, i.e. possesses foreign recognition sites, for E_1 as well as E_3. This could occur also if the enzyme structure is rigid, but it seems more feasible if E_2 undergoes conformational changes during catalysis. Three further postulates are made:

(a) The intermediate enzyme, E_2, can exist in three different conformations:

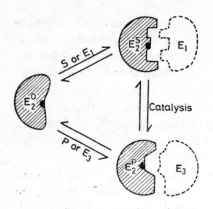

Fig. 7.4. Conformational transitions and alternating complementarity of enzyme E_2. S and P are the substrate and product of E_2, respectively, whereas E_1 and E_3 are the functionally adjacent enzymes of E_2. For further details see the text. (From Friedrich [404].)

E^0, the state of the unliganded enzyme, and E^s and E^p, the states characteristic of the ES and EP complexes, respectively.

(b) The foreign recognition sites on E_2 for E_1 and E_3 are created when the enzyme assumes the E^s and E^p conformations, respectively.

(c) Transition into one or the other conformation is promoted by both the appropriate metabolite and enzyme partner.

Figure 7.4 illustrates the above postulates. During catalysis the enzyme oscillates between conformations E_2^s and E_2^p and thereby exhibits *alternating complementarity* toward E_1 and E_3. The mechanistic scheme of compartmenting the substrate (M_2) and product (M_3) of E_2 is visualized in Fig. 7.5. The model does not exclude that conformations E^s and E^p be available to the unliganded enzyme, i.e. that there is a pre-equilibrium between the three states in the absence of metabolites. It only asserts that the appropriate ligand (metabolite or enzyme) *promotes* conformational adjustment in an induced-fit manner.

An important consequence may follow from the adjustment of conformation by the partner enzyme. Namely, if state E_2^p is induced by E_3, then E_3 may energetically contribute to the completion of a catalytic cycle, which in turn means that complex formation between the two enzymes will trigger the release of P. Such an arrangement may greatly increase the efficiency of metabolite channelling.

Examining the above postulates against experimental evidence, we can reassuringly accept that conformational changes take place in enzymes during catalysis. Already early models of enzyme action by Lumry [754] and Hammes [507] were based on structural changes in the protein during catalysis. Furthermore, for the tenet that substrates and products induce (or stabilize) certain, often different, conformational states in enzymes Citri [220] was able to compile an enormous body of data already in 1973. Postulate (a) is therefore amply corroborated.

It is another question, whether the other two assumptions ever combine with the first one. Admittedly, there is no rigorous evidence for the validity of this model in any real system. Davis's [292] statement at the head of this chapter is particularly relevant here, perhaps with the modification that dynamic metabolite compartmen-

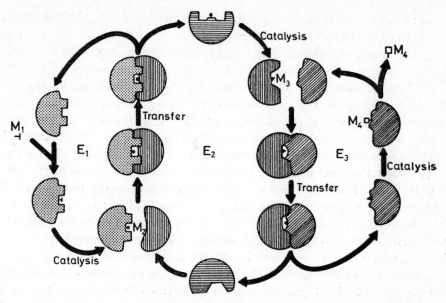

Fig. 7.5. Dynamic compartmentation of substrate and product of an enzyme through the "cage" effect ensured by alternating complementarity. Of the three consecutive enzymes (E_1, E_2 and E_3) of a pathway one, E_2, is assumed to undergo the conformational transitions required for alternating complementarity, whereas E_1 and E_3 are rigid. M_1, M_2, M_3 and M_4 are the metabolites transformed by the three enzymes. The substrate and product of E_2, i.e. M_2 and M_3, are compartmented. (From Friedrich [404].)

tation via alternating complementarity is not all that easy to believe. To my knowledge there is one experimental observation which could be explained by such a mechanism. We have already discussed the finding of Patthy and Vas [916] in Chapter 5 that syncatalytic inactivation of aldolase and GAPD by nascent hydroxy-pyruvaldehyde-P derived from the aldolase reaction is about one order of magnitude more efficient with GAPD than with the parent enzyme, aldolase. Now, if collision with GAPD triggers the release of product from aldolase, such a phenomenon may come about. It belongs to truth, however, that the above authors did not raise this possibility, rather contented themselves with the explanation that in the aldolase–GAPD complex an essential arginine in GAPD is, for steric reasons, more readily attacked by the nascent modifier than the one in aldolase.

Non-random distribution of soluble enzymes in macrocompartments: Development of enzyme gradients

The third limb of our hypothesis [404] offers a mechanism for the non-random distribution of enzymes in macrocompartments. This would favour collisions between functionally adjacent enzymes and hence augment the compartmenting effect. Two postulates are made:

(a) E_1 and E_3 bind weakly but selectively to different structural elements inside a cell, the A-wall and Z-wall, respectively.

(b) E_2 tends to self-associate, i.e. it bears *self-recognition sites* on its surface.

In such a system the distribution of enzymes will be non-random, inasmuch as E_1 and E_3 accumulate at the A and Z walls, respectively, while E_2 is concentrated in the middle. Thus an enzyme gradient develops making "useless" collisions, such as between E_1 and E_3, scarce. The distance over which such a gradient may exist cannot be long, and the gradient's shape will depend on the actual parameters (association constants, concentrations, etc.) determining binding. Considering the high surface/volume ratio found in cellular macrocompartments (on the average 60 μm^{-1}, cf. p. 186), there is really no need to span long distances between two "walls". The great variety of enzyme–membrane and enzyme–structural protein (muscle) interactions discussed in previous chapters may well be symptoms of such a "sandwich-like" enzyme organization. If the various recognition sites are formed as a result of liganding with substrates, one may visualize the development of an enzyme gradient to be regulated by metabolic flux. This would mean that the metabolic state is reflected in the organizational pattern of enzymes. It is worth mentioning that in the description of small systems composed of only a few enzyme molecules the stochastic, rather than the deterministic, approach must be followed [25].

It goes without saying that if any of these distributional phenomena occurs *in vivo*, disruption of the cell is very likely to destroy it beyond recognition. Therefore special approaches are required to tackle the problems; some of them will be dealt with later, in this and the next chapter.

Biochemical oscillations

A special kind of dynamic compartmentation is generated by oscillatory biochemical reactions. They create heterogeneous spatial distribution of compounds fluctuating in time. Nicolis and Prigogine [872] have expounded the conditions and manifestations of self-organization in dissipative, non-equilibrium systems. In chemical systems far from equilibrium diffusion and chemical reaction are coupled insofar as diffusion tends to homogenize the spatial distribution of the reactants while the chemical reaction works against this equilibration. The competition between the two processes can be characterized in terms of the critical length, λ_c, as a coupling function between them. For oscillatory biochemical reactions λ_c is between 10^{-4} and 10^{-2} cm. Oscillatory phenomena are common in biological systems at the infracellular, cellular and supracellular levels [72, 536].

Glycolytic oscillations have been studied in considerable detail by Hess and coworkers [103, 104, 105, 533, 536]. By the aid of a UV-sensitive television camera they could continuously monitor changes in NADH absorption in a cell-free yeast extract [105]. Phosphofructokinase, a complex allosteric enzyme, seems to be the pacemaker of glycolytic oscillations. Computer simulations gave satisfactory agreement with experiments.

Oscillatory reactions are thought to play important roles in a variety of periodic phenomena from nerve impulses to clock functions and morphogenesis. In respect of metabolite compartmentation and supramolecular enzyme organization, sustained oscillations may create alternating reactant gradients, which have been invoked in the suppression of *futile cycles* in glycolysis, i.e. enzyme couples whose simultaneous functioning would lead to wasteful ATP loss. Such enzyme couples are hexokinase-glucose-6-phosphatase and phosphofructokinase-fructose-bis-phosphatase [106]. For further details the reader is referred to the literature cited.

HETEROGENEITY OF METABOLITE POOLS DUE TO BINDING SITES

It has been pointed out in earlier chapters that the concentration of enzymes in various macrocompartments is high. As a rough estimate, enzyme concentrations are close to their substrates' intracellular concentration and both approximate the respective K_m values (10^{-4}–10^{-5} M). Sols and Marco [1133] were among the first to realize that in such circumstances the usual way of calculating intracellular concentration of metabolites is rather questionable. Namely, if the amount of extractable material is divided by the volume of the space (cell or macrocompartment) it is accommodated in, we imply that the metabolite is homogeneously distributed. However, since a metabolite is, in the majority of cases, acted upon by at least two enzymes, the latter will have specific binding sites for the compound. It follows, then, that under the conditions of the living cell a major portion of metabolites will be bound and does not contribute to the reactions of the free fraction. For instance, in rat skeletal muscle the sum of D-glyceraldehyde-3-phosphate (GAP) binding sites was calculated to be one or two orders of magnitude higher than the total concentration of GAP found in resting muscle [1133]. Under such conditions little free GAP is expected to be present. Later it was proposed that GAP existed in muscle not as the free aldehyde (or its hydrated form), but rather was covalently bound to GAPD as a 3-phosphoglyceroyl moiety [96]. This more recent finding, however, does not affect the qualitative picture that a large pool of triose phosphate is sequestered by binding to enzyme(s).

It is pertinent here that the bound state in the living cell may not only hold for metabolites, but also for macromolecules. Kempner and Miller [634, 635] subjected the protozoan *Euglena gracilis* to high centrifugal fields (up to $10^5 \times g$). Most of the cells were crushed by the high hydrostatic pressure, but the few whose cell walls were strong enough displayed an interesting intracellular stratification: within the cell envelope the different organelles were layered according to their densities. Importantly, all macromolecules were in the sedimented particulate fraction, the top "soluble" phase contained no protein as witnessed by fluorescent microscopy and tests for various enzyme activities. The authors concluded that no "free" protein existed *in vivo*. It ought to be added that the stratified cells did not perish, but recovered and grew normally after their adventurous spinning.

Turning back to small molecules, our present-day understanding of intracellular concentrations is likely to be very simplistic. The classical approach is characterized by the puzzle of Sols and Marco [1133]: What is the highest possible pH in the mitochondrial matrix? The answer is pH 8. For taking into account the volume of the matrix, a single H^+ would give a $[H^+] = 0.01$ μM. Beyond this only $pH = \infty$ follows. Clearly, this does not make physicochemical sense, but it does make sense critically to revise the application of statistical parameters developed for large systems to the molecular dimensions.

Demonstration of heterogeneity in the NAD pool of a human erythrocyte system

As an example, the study of coenzyme pool in sonicated human erythrocytes will be described. When discussing the evidence concerning enzyme organization in the red cell (cf. p. 157), it was pointed out that the mature cells rely for energy supply almost entirely on glycolysis, which involves the recycling of $NAD \rightleftharpoons NADH$. Under normal conditions NAD is greatly predominant over NADH [981]. Apart from the cell membrane, there are no other membraneous structures (organelles) in the red cell.

On the outer surface of the erythrocyte membrane is anchored the enzyme NAD-glycohydrolase (EC 3.2.2.5). This enzyme hydrolyses NAD to nicotinamide and adenosinediphosphate-ribose irreversibly and has been characterized in some detail by Hofmann and coworkers [559, 560, 561]. The strategy adopted by Solti and Friedrich [1137] was the following. They subjected packed red cells to mild sonication so that full haemolysis was achieved. By this procedure the NAD-glycohydrolase originally sequestered from intracellular NAD gained access to its substrate. The time course of NAD+NADH decay was followed (Fig. 7.6a). The complex decay curve could be graphically resolved into three first-order reactions, corresponding to three NAD-pools (fractions) discernible on the basis of reactivity towards glycohydrolase (Table 7.2). The three pools could be more straightforwardly distinguished in another way, as pool I was fairly selectively depleted at 0°C, followed by the depletion of pool II at 37°C, whereas the very sluggishly reacting pool III vanished rapidly after the addition of detergent (Fig. 7.6c).

Pool I showed reactivity similar to that of free NAD, consequently this NAD is likely to be an unbound fraction. Pool II was identified as NAD bound to GAPD. This conclusion was reached by monitoring the sensitivity of GAPD to inactivation by iodoacetate as a function of NAD-decrease due to glycohydrolase action. Here the same property of GAPD has been exploited as in the selective labelling with the reagent for the electron microscopic localization, by means of autoradiography, of

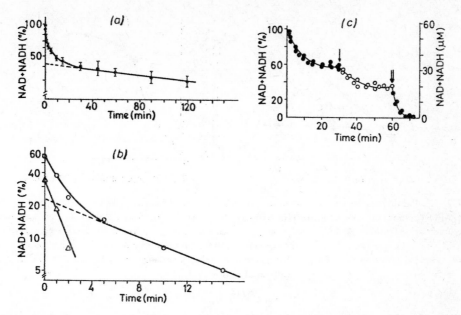

Fig. 7.6. Demonstration of heterogeneous reactivity of the NAD pool toward NAD-glycohydrolase in human erythrocyte sonicate. Packed red cells were mildly sonicated to achieve complete haemolysis. After sonication aliquots were taken from the slurry over a period of 2 hours and the NAD+NADH content (which was practically equal to NAD) was measured by a sensitive recycling technique [874]. (*a*) Semilogarithmic plot of coenzyme decay (100% corresponded to about 50 μM) at 37°C. The overall decay curve (I, ●) could be resolved graphically into three first order reactions (cf. panels *a* and *b*), which indicates the existence of three NAD pools with parameters seen in Table 7.2. Curve II (○) was derived by subtracting the extrapolated slow phase from curve I; curve III (△) was obtained in a similar manner from curve II. (*c*) Alternative way of distinguishing the three NAD pools. ●, incubation at 0°C; ○, incubation at 37°C; ●, at 37°C after addition of 0.01% Triton X-100. ↓, transfer to 37°C; ⇓, addition of detergent. (From Solti and Friedrich [1137].)

TABLE 7.2.

Kinetically distinguishable NAD pools in human erythrocyte sonicate

Pool No.	Identity	Per cent of total NAD	Half-life[a] at 37°C (min)
I	Free NAD	35	~1
II	NAD bound to GAPD	23	7
III	Unknown	42	240

[a] Half-life of decomposition under the effect of NAD-glycohydrolase. The data were derived from the experiments shown, in part, in Fig. 7.6. For the details see the text and [1137].

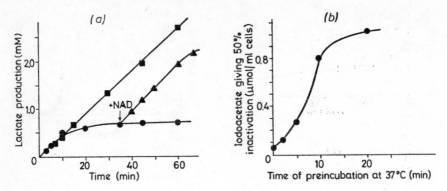

Fig. 7.7. Parallel decline of lactate production (*a*) and development of iodoacetate resistance of GAPD (*b*) in erythrocyte sonicate. (Haematocrit: 90%) (*a*) lactate production at 37°C from 5 mM glucose by intact cells (■) and sonicate (●); in sonicate after the addition of 100 μM exogenous NAD at 30 min (▲). (*b*) The increase in the amount of iodoacetate required for 50% inactivation reflects the decrease of reactivity of Cys-149 SH-group due to the "stripping" of NAD from GAPD. (From Solti and Friedrich [1137].)

the enzyme in thin sections of fixed erythrocytes (cf. Chapter 6, p. 164). This feature is that the reactivity of the Cys-149 group of the enzyme [514] to iodoacetate is markedly increased by bound NAD [271, 966], owing to the orienting effect of the positively charged pyridinium moiety on the negatively charged reagent. Thus the degree of saturation with NAD could be estimated by measuring the rate of alkylation, manifested in inactivation, of the enzyme. The type of evidence obtained is demonstrated in Fig. 7.7. The iodoacetate-resistance, expressed as amount of reagent giving 50% inactivation of GAPD in the sonicate where an about 10^5-fold molar excess of other SH-groups (belonging mainly to glutathione and haemoglobin) is present, rapidly increased as a function of preincubation time at 37°C and reached a final level in about 15 min (Fig. 7.7*b*), corresponding to the time of depletion of NAD pool II (Fig. 7.6*c*). At the same time, lactate production from glucose also came to a standstill (Fig. 7.7*a*), but could be restarted at any time by adding exogenous NAD. The experiments suggest that pool II is the NAD bound to GAPD and, furthermore, that for normal lactate production only this pool is needed. It should be noted that the depletion of pool I did not interfere with lactate production, whereas pool III, which is hardly affected after 15 min, cannot maintain glycolysis. The nature of pool III is not known as yet; it is not coenzyme entrapped in membrane vesicles, as shown by gel chromatography (cf. [1137].) The retarding effect of muscle GAPD on NAD decomposition by glycohydrolase from *Neurospora* and pig brain has earlier been observed by Astrachan *et al.* [35] and Bernofsky and Pankow [70, 71], respectively. It may be fortuitous but in muscle homogenate about 20% of total NAD was bound to GAPD [71], a value close to that found in the red cell sonicate [1137].

The above experiments illustrate that multiple binding sites can cause heterogeneous reactivity within a macrocompartment. The degree of heterogeneity will depend on the number of types, their relative amounts and the binding parameters (dissociation constants, association and dissociation rate constants) of the sites.

Pool heterogeneity, compartmentation and metabolic regulation

The immediate question that may arise about the above phenomena is what role, if any, they play in the living cell; in other words, have they any significant contribution to metabolic regulation. For it may happen that a certain pool heterogeneity or other kind of micro- or macrocompartmentation is indifferent and only we endow it with physiological import.

The answer is as difficult as the question is obvious. The difficulty resides partly in our insufficient knowledge of metabolism *in vivo,* and partly in defining what we mean by metabolic regulation. The first point needs little explanation. It is often hard to tell the significant from the incidental in the highly ordered "mess" that makes up a cell. More precisely, the significance of a particular, known interaction or compartmentation may lie in a metabolic consideration of which we are still ignorant. The second point, what constitutes regulation, is fraught with conceptual problems, as Stebbing [1166] has recently pointed out. He cautions against the indiscriminate assignment of "control properties" to observed interactions. Indeed, bioregulation enjoys such high priority among biology projects that almost any interaction between biomolecules gets dubbed "regulatory". While many examples probably have minimal significance, lacking appropriate criteria we are at a loss when it comes to selection.

Computer simulation of metabolic pathways, in spite of its inevitable simplification, neglections and assumptions—which make it an obscure exercise in the eyes of many a pure-bred enzymologist—can be of considerable help in assessing relative roles. (For mathematical modelling of pathways cf. also Chapter 8.) Although simulations may be less meaningful in describing large systems in their full complexity, nevertheless they can definitely provide useful answers in simpler systems with a limited number of variables and constants. For example, it has been suggested by Sols and Marco [1133] and later invoked by Masters [785] that sequestration of FBP by aldolase may regulate metabolic flux rates. Ottaway [806, 892, 893, 894] has made a computer simulation study of the aldolase, triose-P isomerase and GAPD system which yielded the partition of substrates between binding sites and free solution at different, selected aldolase and FBP concentrations. The unexpected result was that under all conditions aldolase bound hardly any FBP, but it was nearly saturated with dihydroxyacetone-P when aldolase and FBP concentrations were high. Thus sequestration of FBP by binding to aldolase seems not to occur. Furthermore, flux control practically could not be exercised by bound dihydroxyacetone-P either, as no matter how much of this triose-P was bound to aldolase, the free dihy-droxyacetone-P pool was replenished from the infinite substrate reservoir. There were conditions, however, (high aldolase and low FBP concentration) when the control strength [604] was shared between aldolase and GAPD and free dihydroxyacetone-P decreased significantly, which might affect metabolite flow toward glycerol-3-P dehydrogenase. On the whole, however, in an open system with infinite inflow and outflow, metabolite sequestration by binding could only possess major control power through the regulated addition and removal of the binding species. This is not likely to occur in a living cell over a short time scale.

The case is different for a system including conserved species like NAD/NADH or acetyl-CoA/CoA [895], where the total amount of the compound is fixed [893]. Here sequestration may have a marked impact on flux control provided that the bound species is not the minor component of a couple. For example, at very high NAD/ /NADH ratio sequestration of NADH has no effect since the free NADH pool is replenished from the abundant NAD. However, at a ratio of unity sequestration will have detectable effects, which become more pronounced as the NAD/NADH ratio further decreases. Taking the example of the red blood cells, the binding (sequestra-tion) of NAD at the high NAD/NADH ratio may be expected to affect flux rates, as shown above experimentally for lactate production. Other NAD-requiring func-tions, such as methaemoglobin reductase and the Touster-cycle enzymes [1229] are likely to be affected. Studies with intact or quasi-intact cells in which pathway fluxes are measured as a function of various perturbations of pool distribution are needed.

Another, quite general comment about the possible role of microcompartmentation in control is pertinent here. Kacser and Burns [604, 605] have pointed out that living matter is a continuum and control may be shared among several parties. In a sense, *every* enzyme is regulatory inasmuch as it fulfils a given role at a given point in a metabolite network and its alteration may have far-reaching consequences. In re-spect of control strength enzymes differ from one another quantitatively, rather than qualitatively, and even this distinction is not absolute but depends on the parameters of the system. (The approach of Kacser and Burns is dealth with in more detail in Chapter 8.) By analogy, all binding phenomena in a living cell are potentially sig-nificant as elements of a "microcompartmentational spectrum", the perturbation of which may alter the relative rates of different metabolic pathways.

SURVEY OF METHODS FOR THE DETECTION OF METABOLITE
COMPARTMENTATION

Isotope tracer kinetics

One of the most reliable ways to detect metabolite compartmentation is through the application of radioactive tracers. In fact, *compartment analysis* is a well-developed mathematical tool, which by monitoring the kinetics of appearance and disappearance of an exogenously added label can provide information about the number and sizes of pools, rates of conversions and transport between compartments. This type of analysis refers mainly to cellular macrocompartments (or even supracellular entities, e.g. the extracellular fluid in higher organisms), but may also be useful in studying microcompartmentation phenomena like pool heterogeneity. The mathematical formalism has been worked out for a variety of cases from one up to many compartments and can be found in specialized handbooks (e.g. [1105]). Obviously, the more compartments are invoked, the greater is the uncertainty of the conclusions drawn. It is a similar case to any kinetic model with a number of assumed constants. Earlier in this book (Chapter 6) we have already referred to the compartment analysis of P_i uptake into human red cells, where the experimental data could be reconciled by assuming three different intracellular phosphate compartments [720]. Recently, a general purpose programme called TFLUX for tracer experiments in compartment analysis has been developed, intended for the naive computer user [1098].

In this section, however, we are rather concerned with the application of isotope tracers to test metabolite channelling in an isolated system, let it be a multienzyme complex, cluster or membrane-bound enzyme array. The question of pool size is irrelevant here as it is, or is assumed to be, of the dimension of a single metabolite molecule. This makes the mathematics much simpler, at the expense of inflating experimental difficulties. For the structural entity has to be isolated, it should withstand *in vitro* manipulations, a prerequisite of performing with it any tracer kinetic study. Many loosely associated macromolecular systems do not lend themselves to these practices, whereas cells or even whole animals are unlikely to fall apart during compartment analysis.

The simplest possible system to be tested for metabolite channelling is a two-enzyme complex and can be visualized as follows:

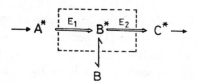

The broken-line box represents the microcompartment or channel, E_1 and E_2 are the two consecutive enzymes in a sequence that carries out the transformation of metabolites A to C. The common strategy is to add an isotopically labelled form of the first metabolite, A*, and non-labelled intermediate B, to the enzyme cluster and follow in time: (*i*) the concentration changes in A, B, and C, and (*ii*) the specific radioactivity changes of A, B and C. The types of result that can be obtained are illustrated qualitatively in Fig. 7.8.

The following main alternatives are possible:

(1) *Tight channelling:* Added B cannot enter the reaction sequence, its concentration remains unchanged. A is consumed and C is produced stoichiometrically.

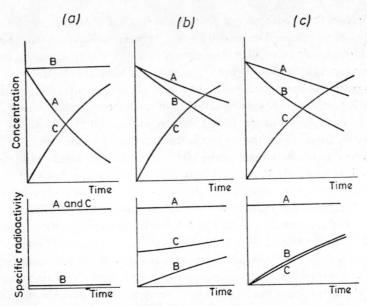

Fig. 7.8. Time course of concentration and specific radioactivity changes in a two-step enzyme cluster (E_1–E_2) catalysing the reaction $A \rightarrow B \rightarrow C$. Radioactively labelled A is added to test metabolite (B) channelling. (*a*) Tight channelling; (*b*) loose channelling; (*c*) no channelling. The concentration of exogenously added B was chosen to be equal with the initial concentration of A. The ratio of concentration changes of A and B in panels (*b*) and (*c*) is arbitrary; it may be the opposite depending on the values of rate constants, sensitivity of E_1 to product inhibition by B, etc. In case (*c*) fairly strong product inhibition by B is assumed. The main diagnostic feature is the qualitative ratio of specific radioactivities. For further details see the text.

The specific radioactivities of A and C are the same; B does not become labelled (Fig. 7.8*a*).

(2) *Loose channelling:* B can enter the channel and is converted to C; the consumption of A is slower than in case (1), because exogenous B interferes with the transfer of endogenous B from E_1 to E_2. The existence of channelling, i.e. preferential acceptance of endogenous B by E_2, is indicated by the fact that the specific radioactivity of C is higher than that of B. Much depends on the concentration of exogenous B applied. The lower B concentration is needed to depress the specific activity of C, the looser is the channelling (Fig. 7.8*b*).

(3) *No channelling:* B produced by E_1 freely equilibrates with exogenous B. The concentration of B is simply the function of kinetic parameters of the enz.ymes The specific radioactivities of B and C start from 0 and run parallel (Fig. 7.8*c*).

Two-enzyme systems are unlikely to produce tight channelling, as witnessed by the case of tryptophan synthase from *E. coli* and *Neurospora crassa* (cf. Cihapter 5) It should be pointed out that "loose channelling" is a designation derved from. experiments *in vitro* and does not mean that the compartmentation is leaky *in vivo*. Rather, it is probable that it ensures complete sequestration of the intermediate under normal metabolic conditions, where flooding with intermediate B does not occur. The experimental strategy outlined above may be varied a good deal, adapted to the system under study. One may apply non-radioactive A and labelled B. Sometimes it is informative to use double labelling (e.g. ^{14}C and ^{32}P, one in A, the other in B). We only refer to the studies already mentioned in earlier chapters, where compartmentation was detected by radioactive tracer technique [141, 790].

Transient-time measurements

Metabolite channelling decreases the transient time, as described in detail in Chapter 4. As a diagnostic test it can be applied to systems (complexes) that are amenable to study in both associated and dissociated states. Thus it is a routine approach to follow when looking for interactions between "soluble" enzymes, as done by Hess and Boiteux [534] for yeast glycolytic enzymes, almost exclusively with negative results and by Bryce et al. [141] who, in contrast, detected no lag time in the coupled reaction of mitochondrial aspartate aminotransferase and malate dehydrogenase, which is indicative of tight channelling of the intermediate, oxaloacetate (cf. Chapter 5).

Transient-time measurements have been quite illuminating in model studies with artificially juxtaposed multienzyme systems. Mosbach and coworkers [794, 846, 847, 848] pioneered these endeavours, the essence of which is to immobilize two or three (or possibly more) consecutive enzymes by covalent coupling to some matrix (Sephadex, Sepharose, etc.) or entrapment in polyacrylamide gel, and to study the kinetic behaviour of the system in comparison to the equivalent mixture of free enzymes. Two parameters are usually measured: the lag period in attaining the steady-state (i.e. the transient time) and the rate in the steady-state. The latter is related to *transit time* as distinct from transient time (cf. Chapter 4). The decrease of transit time may result in an increase of steady-state velocity and belongs to the family of phenomena referred to as *catalytic facilitation*. Importantly, immobilization as a rule decreased transient time [794] or increased the steady-state rate [1160] or did both [673, 846], or enhanced the feed-back inhibition by the end-product [723]. The main interest of these model experiments is that the above phenomena were produced in spite of an obviously random orientation of enzymes on the support, which is to be distinguished from the specific orientation of parties in natural multienzyme clusters. Srere and Mosbach [1161] attributed these effects to the proximity of enzymes on the support, which creates a microenvironment that favours the running of the reaction sequence (cf. also [785]). This microenvironment would raise diffusion barriers to the intermediate (Fig. 7.9), which, together with the exclusion effect of the macromolecular matrix, may give rise to a higher local concentration of the intermediate.

More recently, Koch-Schmidt et al. [673] addressed the question of the relative roles of proximity and microenvironment. They prepared soluble enzyme conjugates

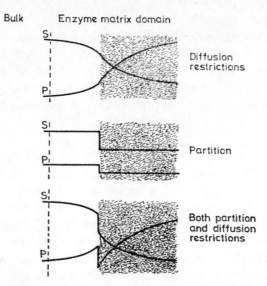

Fig. 7.9. Concentration profiles of substrate, S, and product, P, in a porous medium containing an immobilized enzyme and in the surrounding bulk solution. Top, diffusion restrictions are present; middle, the species are distributed between the two phases as a result of partition effects, but the reaction is kinetically controlled; bottom, both diffusion restrictions and partition occur. If more than one enzyme is in the matrix catalysing the sequence S→P→P′ and diffusion restrictions are operative, then the intermediate P has higher concentration in the matrix than in the bulk solution. (From Mosbach and Mattiasson [848].)

of malate dehydrogenase and citrate synthase by crosslinking with glutardialdehyde and compared their behaviour to that of the same species, or just of free enzymes, immobilized on Sepharose beads. The answer was clear: proximity by itself, as ensured by crosslinking, did not entail any catalytic facilitation, whereas entrapment both reduced transient time and increased steady-state velocity. Apparently, the microenvironmental factors are operative in co-immobilized enzyme systems.

Although the above results about transient time and, in effect, transit-time measurements refer to artificial systems, they may have an important bearing on enzyme organization *in vivo*. Namely, these model studies indicate that even fairly unspecific juxtaposition on the surface of some cellular structure may confer metabolite channelling and perhaps there need not be a precisely tailored multienzyme cluster in all cases where catalytic facilitation is to be achieved.

Trapping of intermediates

A long-used approach for testing metabolite channelling is to apply some agent that can trap the metabolite in question. The rationale is simple: if the intermediate enters the bulk medium, it is bound to react with the appropriate trapping agent.

The reaction need not be a chemical one, it may also be physical, for example, solvation. This approach was adopted in the classical studies of Yanofsky and Rachmeler [1341] testing indole channelling in *Neurospora* tryptophan synthase (cf. Chapter 6). The trapping agent was toluene, in which indole is readily soluble but the substrates of the enzyme complex are not. These authors found that only trace amounts of indole appeared in the toluene layer while running the reaction indole-glycerol-P→tryptophan, which indicated that indole did not mix with the bulk medium. It is to be noted that intermediate trapping can only establish, or refute, the fact of metabolite channelling, but cannot distinguish between tight and loose channelling.

A ubiquitous trapping agent is water for intermediates that have a propensity for hydration. This is the case with various oxo-compounds whose hydrated forms are, as a rule, not transformed by the corresponding enzymes. An example of exploiting the trapping ability of water was given in Chapter 7. Ovádi and Keleti [899] studying the aldolase-GAPD coupled reaction could reconcile their kinetic data only by assuming that glyceraldehyde-3-P was channelled between the two enzymes, as it apparently was protected from the rapid hydration reaction.

When studying complex systems, such as reconstituted metabolic pathways or crude extracts at high enzyme concentrations, it is mandatory to employ a specific and potent trapping agent. Specificity is required to be sure of the point of attack, whereas high reactivity is needed to compete successfully, in the case of *no* channelling, with the "distal" enzyme for the given metabolite. These criteria are best met by an enzyme. Indeed, enzymes have been used as trapping agents in testing metabolite compartmentation [70, 1137, 1232].

A strategy aimed at detecting metabolite channelling in loose, associating–dissociating enzyme systems, by the aid of a "probe-enzyme", has been devised by Friedrich *et al.* [406] in an attempt to test for metabolite channelling around the enzyme GAPD in human red cell glycolysis. In Chapter 6 we have discussed in detail the question of supramolecular enzyme organization of erythrocyte glycolysis, with special reference to the cell membrane. In the present study [406] disrupted packed cells were used, just as when detecting the heterogeneity of the NAD pool (cf. above [1137]). In this "90% haematocrit" sonicate the dilution of cell content is negligible so that one may expect loose macromolecular associations to prevail. With the help of some exogenous NAD this system produces lactate from glucose, i.e. operates glycolysis, linearly in time for over 1 hour (Fig. 7.10). As trapping enzyme the authors employed rabbit muscle glycerol-3-P dehydrogenase, which is absent from mature red cells. However, its substrates dihydroxyacetone-P and NADH are being constantly produced during glycolysis (cf. Fig. 6.16). Thus if an appropriate amount of glycerol-3-P dehydrogenase is added to the sonicate, it may divert metabolite flow from lactate to glycerol-3-P, as the equilibria of both dehydrogenase reactions are much in favour of these products, provided that the enzyme gains access to its substrates. As seen in Fig. 7.10, in the concentrated sonicate glycerol-3-P dehydrogenase

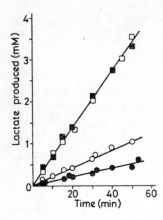

Fig. 7.10. Effect of exogenous glycerol-3-phosphate dehydro-genase as "probe-enzyme" on the lactate production of concentrated and diluted human erythrocyte sonicate. Lactate production was measured in concentrated (haematocrit = 90%) sonicate without (□) and with (■) 0.5 mg/ml glycerol-3-phosphate dehydrogenase or in diluted (haematocrit = 30%) sonicate without (○) and with (●) 0.17 mg/ml "probe-enzyme". (From Friedrich *et al.* [406].)

had no effect on lactate production. In contrast, if the whole sonicate was diluted threefold, glycerol-3-P dehydrogenase depressed lactate production to about one-half. The authors interpret these data in terms of a dilution-sensitive microcompartment that effectively shields metabolites from the probe-enzyme in the concentrated sonicate, but which becomes altered, and leaky, when the proportion of free water is substantially increased.

Other methods

There are a variety of other ways to detect metabolite channelling; in fact, the method chosen depends upon the ingenuity of workers to devise approaches specially suited to a particular problem. With microbial multienzyme clusters there are the possibilities offered by mutants. Thus if one takes the cluster from a strain defective in the first step and another defective in the second step of a two-step enzyme reaction sequence, their combination will ensure normal reaction rate if there is no intermediate channelling, but only a very slow rate if compartmentation is inherent in the system [1341]. Of course, in such studies the formation of wild-type hybrids has to be excluded. Another way applicable when the two components can be isolated, as with tryptophan synthase [1341], is to run the reaction with the two enzymes separated by a semipermeable membrane. If channelling is normally not involved, the diffusion of the intermediate through the membrane will not effect a severe decrease in overall reaction rate. However, this method cannot distinguish between channelling of the intermediate and mutual activation induced in the enzymes by association.

The syncatalytic inactivation found with the aldolase-GAPD system in the presence of oxidizing agents [916] (cf. Chapter 5, p. 119) is an elegant way of demonstrating between two enzymes, but does not prove metabolite channelling. Anyhow, its scope of application is limited by the chemistry of such reactions, as pointed out earlier in this book.

CHAPTER 8

CURRENT TRENDS IN THE STUDY OF ENZYME SYSTEMS

"The specialist concentrates on detail and disregards the wider structure which gives it context. The new scientist, however, concentrates on structure on all levels of magnitude and complexity, and fits detail into its general framework."

E. Laszlo [718]

MATHEMATICAL MODELLING OF ENZYME SYSTEMS

The endeavour to construct mathematical models of enzyme action, for a single enzyme or a set of enzymes catalysing consecutive reactions, has a long past and an expanding future. Perhaps the most fruitful model proposed so far is that of Michaelis and Menten [816], which dates back to 1913. Although a literature survey of steady-state enzyme kinetics showed "anomalous" behaviour in most cases and led the authors [540] to query whether any enzyme really followed the Michaelis–Menten equation, the heuristic value of the model is beyond doubt. After all, the very formulation that many enzymes behave "anomalously", i.e. exhibit deviations from linearity in double reciprocal plots of initial velocity vs. substrate concentration, implies that for many enzymes the basic idea is valid at least over a fairly wide range of substrate concentrations. For the intricacies of the Michaelian behaviour of enzymes see Keleti [628].

It must be clear, however, that few if any enzymatic mechanisms are as simple as portrayed by the Michaelis equation. Many enzymes have more than one substrate, discrete steps along the reaction coordinate may be discerned, various effectors may influence the kinetic properties (usually expressed in terms of parameters V_{max} and K_m toward a given substrate), possibly by shifting the enzyme molecule between different conformational states. The ensuing complexity attains such a degree where the adequate model can no longer be handled by pencil and paper, but calls for computer technique.

As recently claimed by Garfinkel [424], computerized modelling of enzymes is expected to rapidly expand in the near future, partly because of the foreseeable drop in the cost of hardware and in part due to the development of specific "friendly" software. A "friendly" program spares the user from bothering with routine mathematics and helps in several other ways, e.g. to recover from mistakes. The program PENNZYME fits steady-state kinetic data to enzymatic equations [674]. As an example of the application of computer simulation to the study of a single enzyme, the work on pyruvate kinase by Hess and coworkers [107, 780] can be mentioned. This tetrameric enzyme has a number of effectors and seemed to fit two types of concerted allosteric models equally well. By the aid of simulation Markus *et al.* [780]

determined the conditions under which the two, closely similar predicted behaviours would differ to the greatest extent and then did the crucial experiment. In this way they could distinguish between the two models. This use of computer simulation is, in a sense, akin to the guidance given by X-ray crystallography to enzyme structure–function studies. The common feature is that in both cases many alternative possibilities can be ruled out, which may dramatically shorten the path to the incisive experiments.

It is only too obvious that if one enzyme may be complex enough to be preferably tackled by the computer, a system of enzymes is much more so. It is for this reason that there have been many attempts to simulate whole metabolic pathways. There were, among others, early attempts to model glutamate metabolism [422] and the Krebs cycle [423] in rat brain or to describe glycolysis in ascites cells [425] relying on a compilation of enzyme kinetic data. More recently, red cell glycolysis has been simulated with the simplifications that all near-equilibrium enzymes were regarded as fast enough not to play a role in flux control and only the kinases and the enzymes connected with 2,3-BPG were considered [983]. The best detailed review of the field was given by the same authors, Heinrich, Rapoport and Rapoport [523], based upon their above work.

Attempts to simulate even more complex systems have also been made. Thus the photosynthetic Calvin cycle was represented by seventeen reactions [822], the carbohydrate metabolism in Dictyostelium was constructed from nineteen reactions [1337]. The most heroic undertaking so far is the modelling of energy metabolism in the heart by Achs and Garfinkel [1, 2] including sixty-five enzymatic and membrane transport steps, which has recently been upgraded to some eighty reactions [424]. The latter model was made for medical purposes, in particular to evaluate the relative significance of the various reactions in the generation of heart ischaemia (shortage in O_2 supply), a primary factor in heart failure.

It must be added, however, that many enzymologists look askance at these elaborate exercises. Their scepticism, shared to some extent by the author, stems partly from a lack of familiarity with computer technique and, more fundamentally, from the uncertainty necessarily generated when handling a great number of variables along with ambiguous constants. While the second of these reservations seems, regrettably, valid, one must get accustomed to the idea that the intuitive faculties of the human brain, even if assisted by the tools of classical enzyme kinetics, are inadequate to the extreme complexity prevailing inside a living cell. Therefore if we ever want to handle this complex system cognitively, there seems no way out but by recourse to the computer. Nevertheless, reality must be approached on two limbs, in a highly interrelated manner: computer modelling of a system (preferably of moderate complexity) followed by experimental testing, which in turn would lead to a refined model whose predictions can again be tested experimentally, and so forth. As already mentioned in Chapter 7, simulation of a small system of enzymes may provide valuable information about expectable responses to variations in flux, metabolite pools, etc., and may be instrumental in weeding out intuitively appealing but physicochemically unsound hypotheses.

In the simulation of multienzyme systems the phenomena of metabolite compartmentation and enzyme–enzyme interactions should be, and in some instances already have been, taken into account. Undoubtedly, the multifarious repercussions of supramolecular enzyme organization do increase the complexity and consequent uncertainty. To remain on relatively safe ground, these interactions are to be tackled, at least at first, in simple systems consisting of the minimum number of components. There are a number of theoretical treatises dealing with enzyme–enzyme interactions. The statistical mechanical approach was adopted by Hill to examine the effect of enzyme–enzyme interactions on steady-state velocity in a lattice made up of a single type of enzyme in various conformational states that are in quasi-equilibrium [544, 545] or forming aggregates and displaying hysteresis [546] or being far from quasi-equilibrium [547]. Hill's analysis also covered interactions within heterologous and isologous dimers [548], as well as multienzyme complexes [542, 543].

More practically oriented are the attempts to describe the steady-state kinetics of two interacting enzymes. Thus Nichol *et al.* [871] used Maclaurian polynomials [689] to solve rate equations for consecutive enzyme reactions and so provided explicit expressions for detecting interactions (i.e. activation or inhibition of either enzyme) between two consecutive enzymes by measuring their steady-state velocities separately and in a coupled assay.

Bartha and Keleti [48] have considered the following system:

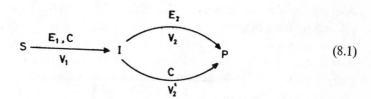

$$(8.1)$$

where S and P are the substrate of the first enzyme (E_1) and the product of the second enzyme (E_2), respectively, I is the intermediate metabolite and C is the equimolar complex of E_1 and E_2. If complex formation between the two enzymes affects the activity of the second enzyme but does not influence the first, then the expression derived by the authors allows one to determine the dissociation constant of the enzyme–enzyme complex from the plot transient time vs. enzyme concentration. If the second enzyme, whether free or in complex, follows Michaelian kinetics, then by measuring the steady-state velocity with and without added intermediate I, one can tell under certain conditions whether the interaction only entails kinetic (i.e. V_{max} and/or K_m) changes in the second enzyme or also involves metabolite (I) channelling between the two enzymes. The above examples contain several restrictions, which severely confine their applicability. It seems that the development of experimentally useful diagnostic tests for interacting multienzyme systems is lagging. Endeavours to this end are needed, combined with the discipline that models should remain practicable and authors are not to be carried away by the lure of the almost infinite potentials of computer technique.

STUDY OF ENZYMES *IN VIVO*

The ever-increasing appreciation of the dichotomy between the conditions of enzymological experiments *in vitro* and those prevailing in the living cell, led to endeavours to study enzymes *in vivo* or at least under circumstances resembling their natural ambience. In usual test-tube experiments enzyme concentrations are low and the substrate/enzyme ratio high, whereas in the cell it is more or less the other way round. The various compartments and microenvironmental effects that occur in the cell are absent in the homogeneous solution of a purified enzyme. It is clear that full understanding of enzymatic pathways will only be attained if all these factors are taken into account.

We have been scrutinizing these factors in the major part of this book. Nevertheless, most approaches outlined were of the type when one selects a limited number of possible interacting partners, mixes them and sees what happens. In other words, we screen for interactions in a basically analytic setup. A more direct and holistic strategy is to study enzymes in the intact or quasi-intact cell, tissue or organ, when necessarily all interactions are simultaneously at work and what we observe is undoubtedly related to the physiological milieu. The snag is, of course, that it may prove difficult to make head or tail of such observations. When studying isolated enzymes we ask questions and seek answers at the molecular or even atomic level and the system studied is congruent with this level. On the other hand, the description of enzyme behaviour *in vivo* should likewise be molecular, but the system is cellular or higher in complexity. In spite of the ensuing interpretational difficulties, the science of enzymes is shooting off a new branch, sometimes referred to as *cellular enzymology* or enzymology *in vivo*, which has its specific methodology and even conceptual framework. In this section we are going to survey the experimental part. A recent exposition of the field was given by Sies [1108].

The methods developed for enzymology *in vivo* can be classified depending on whether they are *invasive* or *non-invasive*. The former always involve some extent of disruption of the original structure, at the expense of which a fairly well-defined system can be studied. The latter mainly applies spectroscopic techniques to intact cells or tissues with practically no destruction but, at least as seen at present, with less resolving power at the molecular level.

INVASIVE METHODS

The technique of *in situ* experiments was developed by Sols and coworkers [994, 1085, 1132]. This consists in the permeabilization of cell or organelle membrane by treatment with an organic solvent (usually toluene) or detergent, so that the small molecular weight substrates and products can freely enter or leave the cell or organelle, while the enzymes and other macromolecules are kept within at unchanged concentration. In fact, permeabilized cells had earlier been used by others to monitor various intracellular processes, e.g. replicative and repair DNA synthesis in *E. coli* [146, 849, 850] and *Chlamydomonas* [572] or simply to assay enzyme activities, for example β-galactosidase [833] in bacteria, but the explicit utilization for enzymological purposes stems from Sols, and almost simultaneously from Weitzman [1300, 1301] probably inspired by the earlier studies on toluenized bacteria by Bridgeland and Jones [131].

By this technique the kinetic parameters of a number of enzymes have been determined in bacteria [994, 1300], yeast [1085, 1301], rat liver mitochondria [789] and rat erythrocytes [24, 1132], and compared with those obtained with the isolated enzymes or cell-free extracts. Some of these data are compiled in Table 8.1. It is seen that in several instances no differences could be detected between the free and *in situ* enzymes in respect of the parameter examined. Such results may lend support to the *in vitro* measurements. On the other hand, there were cases when the enzyme behaved differently *in situ*. It is of interest to compare results obtained with citrate synthase in various environments. While the bacterial enzymes were apparently unaffected by the cellular milieu, this was not the case with eukaryotic systems. The data are too scanty to make any generalization apart from the statement that it might be worthwhile to study most enzymes *in situ* as well.

Solubilized cells or organelles are evidently only approximations of the real conditions *in vivo*. The removal of membrane lipids is likely to alter the microenvironment near the membrane, though it does not necessarily affect protein–protein interactions. The chemical crosslinking of membrane proteins needed for the solubilization of animal cells [1132] may also modify the system. On the other hand, this technique preserves all enzymes at their physiological concentrations inside the small "bags", irrespective of the overall enzyme concentration in the actual reaction mixture. This circumstance makes such systems relatively easy to study, at least it dispenses with cumbersome rapid reaction techniques otherwise needed when working at high enzyme concentrations. At all events, the *in situ* approach is perhaps the best among attempts to study *reconstituted metabolic pathways* (e.g. [1077]), (although, or perhaps because, it differs methodologically in that the pathway is never taken apart.)

Subfractionation of cells and tissues by the aid of various solvents and detergents [5] provides information about the distribution of metabolites among various macrocompartments and hence it has been considered by Sies [1108] as an invasive method

TABLE 8.1.

Comparison of enzyme behaviour in situ relative to the free state

Organism/Organelle	Enzyme	Observation *in situ*	References
Bacteria:			
Escherichia coli	Phosphofructokinase EC 2.7.1.11	Decreased cooperativity (Hill coefficient) for fructose-6-P; K_m for ATP increased ten-fold; activation by ADP, inhibition by high ATP and by PEP	994
Aerobacter aerogenes	D-Lactate dehydrogenase EC 1.1.1.28	Sigmoidal pyruvate saturation curve unchanged	1300
Pseudomonas aeruginosa	Citrate synthase EC 4.1.3.7	Inhibition by NADH unchanged	1300
Acetobacter xylinum	Citrate synthase	Kinetic and regulatory properties unchanged (inhibition by ATP, no effect by NADH)	1201
	Glycerokinase	Kinetic and regulatory properties unchanged (inhibition by FBP, substrate specificity)	1201
Yeast:			
Saccharomyces cerevisiae	Hexokinase EC 2.7.1.1	Kinetic parameters unchanged	1085
	Pyruvate kinase EC 2.7.1.40	Kinetic and allosteric properties unchanged (sigmoidal saturation with PEP, activation by FBP)	1085
	Citrate synthase	Marked differences in kinetic parameters (>100-fold increase in K_m for acetyl-CoA, no inhibition by ATP)	1301
	Succinate thiokinase EC 6.2.1.4	K_m for succinyl-CoA hardly changed, K_m for ATP markedly decreased	1301
Mitochondria, rat liver	Citrate synthase	K_m and V_{max} for oxaloacetate unchanged K_m for acetyl-CoA increased about ten-fold Inhibitory effect of ATP, NADPH and tricarballylate diminished	789
Erythrocyte, rat	Glyceraldehyde-3-P dehydrogenase EC 1.2.1.13	K_m for glyceraldehyde-3-P and P_i unchanged, K_m for NAD increased about four-fold	24

of the *in vivo* approach. Since it is essentially a specialized version of classical cell fractionation, we deem it too invasive to be regarded as pertinent here.

By measuring the *steady-state relaxation* [508] of an isolated enzyme, the kinetic parameters determining the flux through that enzyme at reactant concentrations close to equilibrium can be calculated. Then, if the enzyme in question works *in vivo* near its equilibrium, the *in vivo* flux can be calculated, and compared with measured data, in the knowledge of the total enzyme activity in the cell and the extent of deviation of reactant concentrations from equilibrium. The latter quantities can be determined by rapid sampling from the living tissue in various metabolic states. Bücher and Sies [143] examined glycolysis at the enolase step by this approach and found good agreement both in white (fast, tetanic) and red (slow, tonic) rabbit muscles between measured and predicted lactate productions. The steady-state relaxation technique thus provides information about *in vivo* conditions indirectly, by comparing predictions derived from enzymological experiments *in vitro* with flux data determined *in vivo*. It is perhaps due to this circumstantial character that this approach has hardly been used over the past 10 years.

NON-INVASIVE METHODS

These techniques are capable of measuring metabolite concentrations in intact cells, tissues and organs, and thereby report about metabolic flux rates. The optical methods already available are *organ spectrophotometry,* both in the absorbance and reflectance modes, and *organ fluorometry* including two-dimensional flying-spot fluorometry [201], which can detect *intercellular* metabolite heterogeneity. The instrumental and methodological aspects of these techniques, with the scope of application illustrated by several examples, have recently been reviewed in detail by Sies and Brauser [1110]. The potential usefulness of non-invasive methods is great and will certainly increase along with the progress in instrument sophistication. The major problems are *selectivity,* i.e. the identification of signal source at the high degree of complexity, and *absolute quantitation.* As to the latter, the difficulty arises from the uncertainty of optical pathlength, as a consequence of which only relative concentrations, compared to a reference compound, can be determined. Substances having characteristic and intense optical signals, such as pyridine nucleotides [144, 1112], flavoproteins [142], cytochromes [1299] and other haem-proteins, like catalase [891, 1111], particularly lend themselves to such investigations.

Nuclear magnetic resonance spectroscopy is another tool that seems promising in monitoring chemical reactions in intact cells and tissues (for reviews cf. [149, 969]). Busby *et al.* [150] applied phosphorus NMR to study metabolite distribution in rat muscle. They found multiple pools, probably macrocompartments, of P_i and glucose-6-P, but no pool heterogeneity in ATP or creatine-P. This technique will certainly yield meaningful results in the future in respect of enzyme organization *in vivo,* especially if its sensitivity can be increased. At present, the lower limit of detectability is 10^{-3} to 10^{-4} M.

SYSTEMIC APPROACH TO METABOLIC CONTROL

The final goal of enzymology is two-fold: first, to describe in full detail the mechanism of each enzymatic reaction and, second, to understand how these catalytic activities are governed to yield a balanced output. The latter field, i.e. enzyme regulation, is particularly in the foreground of interest as we hope to have a much better grasp of all life processes in man, animals and plants once we know how metabolic pathways are controlled.

The great wealth of data accumulated in this area and the concepts that emerged from them shed light on many aspects of enzyme regulation. We have touched upon them in earlier chapters of this book. However, while our knowledge about the structural details, ligand interactions and subtle local conformational transitions affecting catalytic activity of several enzymes has advanced enormously, our basic concept of metabolic regulation has changed little over the past two decades. This picture, mainly derived from the regulation of biosynthetic pathways in microorganisms, envisages a "key" enzyme, generally the one catalysing the first committed step in a pathway, whose activity is modulated by feedback inhibition. To avoid misunderstandings, several other "regulatory circuits" have been recognized, such as various complex feedback inhibitions, precursor substrate activation, etc. [1163[, but emphasis was usually placed on particular enzyme(s) in a pathway. This way of thinking dubbed certain enzymes as being "regulatory" or having a "key" role in controlling metabolic flux through the pathway, while the others became somewhat second-class members of the enzyme society.

The borderline between "key" and "non-key" enzymes is, however, not at all clear. Perusal of the literature about enzymes may convince the reader that there is hardly any intracellular enzyme that would not have been invoked as a means of regulation in one way or another. While this fact may reflect the authors' bias toward their object of study, or simply their endeavour to put the stamp of "bioregulation" on their article to make it sell better, the arguments advocating the regulatory importance of practically any enzyme are nevertheless thought-provoking. Clearly, certain enzymes are very likely to be more important in flux control than others, e.g. phosphofructokinase and pyruvate kinase in glycolysis. But, after all, are not all enzymes equal in a sense, with the qualification applied to democratic societies, that some members are more equal?

This question boils down to the way one attempts to understand metabolic control. In the *analytic approach* one takes each enzyme in a sequence separately, determines its kinetic properties in the greatest possible detail and accuracy and then tries to find out how it works when put together with the other enzymes. For this the computer should be used. However, the uncertainty is great, since one ought to have known the absolute concentrations of all pertinent molecular species in the cell. We will hardly ever cope with this requirement to any satisfactory extent.

On the other hand, one may follow the *systemic approach*, which takes the *whole* and tries to deduce from it inferences about the *parts*. The roots of this approach, which is different both conceptually and methodologically from the analytic one, can be recognized with several authors [148, 539, 603, 1050, 1051, 1052], yet the clear-cut formulation came from Kacser and Burns [604, 605]. Subsequently, partly similar ideas were put forward by Heinrich and Rapoport [524] and Newsholme and Crabtree [256, 864, 865, 866, 867, 868]. We shall follow up briefly the reasoning of Kacser and Burns [605].

Let us consider a pathway consisting of n enzymes and starting from "external" substrate X_0 and ending with "external" product X_n, with S_{n-1} intermediate metabolites:

$$X_0 \xrightarrow{\;E_1\;} S_1 \xrightarrow{\;E_2\;} S_2 \xrightarrow{\;E_3\;} S_3 \longrightarrow S_{n-1} \xrightarrow{\;E_n\;} X_n \qquad (8.2)$$

Let the system be in steady-state, i.e. the flux F through the pathway, dX_n/dt, is constant. It is clear that under such conditions the rates at all enzymic steps are equal. Now if the amount of activity A_i, of the ith enzyme is changed and this is expressed as a fractional change $\Delta A_i/A_i$, a new steady-state will be reached with a different flux, and this change is similarly expressed as $\Delta F/F$. The ratio of these fractional changes, at the limit that ΔA_i is infinitesimal, is the *sensitivity coefficient, Z*:

$$\frac{\Delta F}{F} \bigg/ \frac{\Delta A_i}{A_i} \longrightarrow \frac{dF}{F} \bigg/ \frac{dA_i}{A_i} = \frac{d \ln F}{d \ln A_i} \equiv Z_i. \qquad (8.3)$$

The coefficient Z_i of any enzyme, E_i, in the sequence is a measure of the influence of that enzyme on flux. Z_i can assume values from 0 to 1. An enzyme with a higher Z value will have a greater impact on flux than another enzyme with lower Z. If for an enzyme $Z \approx 1$, this enzyme will exclusively determine the flux. However, it should be clear that there is a continuum of Z values along the sequence and control may be shared by several enzymes.

Experimentally, the Z values can be estimated by modulating the amount of activity of a given enzyme, in the intact pathway (organism) by genetic means [443, 604] or, potentially, by using specific inhibitors. Alternatively, computer simulation may be helpful in deducing Z values and checking the *summation property* (cf. below), as done in the case of glutathione synthesis and the citric acid cycle [806].

Kacser and Burns [605] have shown that the sum of sensitivity coefficients within a pathway is unity:

$$\sum_n Z_i = 1. \qquad (8.4)$$

Since all Z_i values are positive (an exception to this would be an enzyme that diverted metabolite flux from the pathway in question), the average Z value expected in a pathway is $1/n$. Deviation from this value is a measure of control power in the pathway. An important consequence of this *summation property* is that if any enzyme activity, A_i (and consequently Z_i), in the sequence is changed, all other sensitivities should change for the new sum to be again 1. In other words, the coefficient is a systemic property that cannot be deduced from the characteristics (allostery, positive or negative cooperativity, etc.) of the single enzyme. This implies that the role of a particular enzyme in control cannot be derived from its individual properties. The most sophisticated allosteric mechanism is just a potential, but no evidence, for regulatory significance. Furthermore, the summation property gives new perspective to the frequently heard opinion that most enzymes are "in excess" in the cell. This assertion is based on the measured or calculated large catalytic potential, expressed in amount of (maximum) activity at substrate saturation, of enzymes, which is of course fallacious, as under natural conditions hardly any enzyme exploits its V_{max}. Another source of the above assertion is that marked reduction in the amount (activity) of some enzymes had negligible effect on flux. According to the present formulation this means a low Z value. Obviously, since pathways usually consist of many enzymes with $\bar{Z} = 1/n$, if one or two have a high Z value the rest will necessarily have very low sensitivities.

In addition to flux, the other systemic variables are metabolite concentrations (pools). Their sensitivity to enzyme activities can also be defined as:

$$\frac{\delta S_j}{S_j} \bigg/ \frac{\delta A_i}{A_i} \rightarrow \frac{\partial \ln S_j}{\partial \ln A_i} = S_{A_i}^{S_j}. \tag{8.5}$$

It can be readily conceived that the enzyme *before* S_j in the pathway will have a positive S value, whereas enzymes *after* S_j will have a negative one. There is no upper limit to S values, as witnessed by the large accumulation of metabolites sometimes observed after only moderate changes in enzyme activities (e.g. [603]).

The substrates (metabolites) of a pathway can be regarded as internal effectors whose concentration changes influence the rates of the respective enzymatic step, v_i. The response of the enzymes can again be expressed as follows:

$$\frac{\delta v_i}{v_i} \bigg/ \frac{\delta S_i}{S_i} \rightarrow \frac{\partial \ln v_i}{\partial \ln S_i} = \varepsilon_{S_i}^{v_i}. \tag{8.6}$$

where ε is called *elasticity coefficient* by Kacser and Burns [604]. (The same coefficient has been named "intrinsic sensitivity" by Crabtree and Newsholme [256].) In fact, there are as many elasticity coefficients for an enzymatic step as there are substrates, products, effectors, etc. acting upon that reaction. While elasticities are related to *individual* enzymes, rather than to the system, their actual value will depend on the constellation of all factors, i.e. on the physiological milieu, which is determined by the *system*.

The flux-sensitivity coefficients Z_i and elasticity coefficients ε_i of adjacent enzymes are quantitatively related through their common linking metabolite, S_c. It can be shown that

$$Z_i \varepsilon_{S_c}^i + Z_j \varepsilon_{S_c}^j = 0 \qquad (8.7)$$

or

$$\frac{Z_i}{Z_j} = -\frac{\varepsilon_{S_c}^j}{\varepsilon_{S_c}^i} \qquad (8.8)$$

i.e. the ratio of two adjacent system flux sensitivities equals the ratio of the elasticities of the two enzymes with respect to the common metabolite pool [604]. This is called the *connectivity principle*. The elasticities describe quantitatively the transmission of metabolite effects through the enzyme and thereby, in a sense, elasticities generate flux sensitivities.

The connectivity principle holds even if more than two enzymes act on a metabolite. In general:

$$\sum_n Z_i \varepsilon_S^i = 0. \qquad (8.9)$$

This relationship gives us the basis for assessing the role of *parts* in the control of the *system*. The experimental strategy to follow is to perturb the system by changing the concentrations of initial (X_0) and end (X_n) substrates and/or the activity of some of the enzymes in a manner that causes small but measurable changes in flux and in the steady-state concentration of metabolites. By the aid of a simple set of equations the elasticities can then be calculated (cf. [605]) from which, according to the connectivity principle, the sensitivities Z_i can be obtained. In this way the control analysis of the system is essentially completed.

In the above analysis no assumptions were made about enzyme mechanisms, concentrations, compartments, etc. These are to be determined by traditional enzyme structure–function studies, by the methods aimed at metabolite channelling, and so forth, i.e. through molecular approaches. The latter complement and interpret, but do not substitute for, the systemic analysis.

The type of information one may get from flux-sensitivity measurements is illustrated by the computer simulation of the glycolytic segment comprising the enzymes aldolase-triose-P isomerase-GAPD, made by Ottaway [893]. The sensitivity coefficients Z_i were determined for the three enzymes at various FBP (initial substrate) and aldolase concentrations (Table 8.2). It is seen that at low aldolase concentrations control is exercised practically by this enzyme alone. If both aldolase and FBP concentrations are high, control shifts to GAPD, whereas at high aldolase and low FBP concentration control is shared by aldolase and GAPD. Note that the summation property is valid for the system with great accuracy.

TABLE 8.2.

Flux sensitivity coefficients Z_i of enzymes in a glycolytic segment

[Aldolase] μM	1	1	100	100
[FBP] μM	1	50	1	50
	Sensitivity Z_i			
Aldolase	0.99	0.99	0.54	0.07
Triose-P isomerase	0.003	0.004	0.13	0.02
GAPD	0.005	0.007	0.33	0.91
Sum	0.998	1.001	1.00	1.00

Computer simulation of the aldolase-triosephosphate isomerase-GAPD system. Aldolase and fructose-1,6-bisphosphate (FBP) concentrations were as indicated. The concentrations of triose-phosphate isomerase (34 μM) and GAPD (100 μM) were held constant. Z_i values are as defined by Kacser and Burns [605]; 0.01 was taken as the perturbing fractional activity change $\Delta A/A$, where A is the original amount of enzyme activity (here expressed as enzyme concentration). For the specific activities of enzymes and details of computation see [892, 893, 894]. From Ottaway [893].

FUNCTIONAL COMPLEMENTARITY BETWEEN ENZYMES

It follows from the systemic view of metabolic pathways that the individual properties of enzymes are subordinated to serve the system as a whole. This subordination has many levels building up a hierarchy on top of which "sits" the organism. Evolution worked at the level of organisms and favoured species whose subordinative system responded best to selective pressures. Although Stebbing [1166] emphasized the pitfalls in considering the biochemical aspects of evolution, it may not be far-fetched to claim that in present-day organisms the relative amounts and kinetic-regulatory properties of enzymes are carefully balanced to give the appropriate output. This idea is supported, for example, by the existence of "constant proportion groups" of enzymes in various muscle types, as pointed out by Pette and coworkers [52, 936, 937, 938]. The enzyme spectrum of muscle is highly characteristic of the metabolic type, i.e. whether short-term high power ("fast" muscles) or long-term low power ("slow" muscles) is required. Fast and slow muscles can be interconverted by changing their innervation pattern, and then the whole enzyme spectrum changes [938].

The proper fit of an enzyme into the metabolic network, whether it involves macromolecular interactions or not, may be called *functional complementarity,* in analogy to structural complementarity in interprotein bonding domains. Functional complementarity implies that the properties of enzymes are tailored to mesh with those of their immediate and farther neighbours. (Thus functional complementarity, without macromolecular interactions, is the manifestation of *functional* enzyme *organization* referred to in earlier chapters.) When interpreting the molecular behaviour of any single enzyme, this ought to be done in view of the features of its fellow enzymes. To take a simplistic metaphor, if one examines a cog-wheel from a disassembled clock, its properties (geometry, number of cogs on the circumference, etc.) only make sense if one knows the place in the clock where that cog-wheel fits. Too often the "physiological interpretations" attached to painstakingly determined characteristics of an enzyme do not apply this test, but rather indulge in readily available *ad hoc* assumptions about regulatory significance through allostery, cooperativity and other catchwords of the craft.

Admittedly, interpretations involving functional complementarity between enzymes are not easy to make and will probably necessitate the use of computers. By intuition alone, one can only *suspect* complementarities, which may therefore eventually prove fallacious. Nevertheless, we present below an intuitive example put forward by Friedrich [405], partly to illustrate the point and partly with a provocative aim in mind.

In Chapter 3 on quaternary structure, we have briefly discussed positive and negative cooperativity in ligand binding, i.e. the phenomenon that in an oligomeric protein (enzyme) the affinity for the ligand increases and decreases, respectively, as the subunits of the oligomer become successively saturated. The physiological significance of positive cooperativity in the oxygen transport of haemoglobin is quite obvious: by making the saturation curve steep in the physiologically relevant oxygen concentration range it reduces the amount of protein needed for oxygen transport. By analogy, it has been rationalized that in the case of enzymes positive cooperativity renders enzymes more susceptible to substrate (effector) concentration changes above a threshold level, which may contribute to maintaining a constant level of intermediates [205, 1162]. Negative cooperativity, on the other hand, has been postulated to be beneficial by making reaction rates *less* sensitive to substrate changes at high substrate concentration [245, 276] and more sensitive at low concentration [734], although the latter argument has been challenged [251]. At any rate, these hypotheses are typically non-systemic, they attempt to assess the impact of an enzyme on metabolic flux without knowing the system.

Glyceraldehyde-3-phosphate dehydrogenases (GAPD) from mammalian muscle and yeast behave in an opposite manner in respect of NAD binding: the muscle enzyme exhibits negative cooperativity [63, 102, 245, 277, 308, 814], whereas the yeast enzyme shows positive cooperativity [337, 432, 655]. This difference is immediately somewhat striking, since if these allosteric properties are important in flux control, one would be inclined to expect the same *type* of behaviour from the same enzyme even in different organisms.

The apparent contrast between the two kinds of GAPDs can, however, be rationalized at least qualitatively, if we examine them in the light of the properties of their "counterenzymes" in glycolysis, i.e. the enzymes that recycle NADH to NAD. This enzyme is lactate dehydrogenase (LDH) in muscle and alcohol dehydrogenase (ADH) in yeast. Tables 8.3 and 8.4 list the dissociation constants of NAD and NADH

TABLE 8.3.

Dissociation constants (μM) of mammalian muscle GAPD and LDH complexes with coenzymes

Enzyme	Coenzyme	Site				Reaction conditions		Reference
		1	2	3	4	pH	°C	
GAPD	NAD	$<10^{-5}$	$<10^{-3}$	0.3	26	7–8.5	3	245
		≤ 0.05	≤ 0.05	4	35	8.2	20	102
		0.01	0.09	4	36	7.6	25	63
	NADH	≤ 0.05	≤ 0.05	25	50	8.2	20	102
		0.01	0.06	4	36	7.6	25	63
LDH	NAD	500				7.2	20	1183
	NADH	3.3				7.2	20	1183

TABLE 8.4.

Dissociation constants (μM) of yeast GAPD and alcohol dehydrogenase complexes with coenzymes

Enzyme	Coenzyme	Form		Reaction conditions		Reference
		T	R	pH	°C	
GAPD	NAD	180	14	8.5	20	337
	NADH	240	240	8.5	20	337
ADH	NAD	350		7.0	2–5	310
	NADH	12		7.0	2–5	311

complexes of the above enzymes. It is seen that the values reported for GAPD are rather controversial, the difference between the first and fourth sites originally found by Conway and Koshland [245] to be six orders of magnitude now being claimed to be only about one order of magnitude [1059]. However, negative and positive cooperativities are still characteristic of the muscle and yeast enzymes, respectively.

Now if we consider glycolysis running fast under anaerobic conditions, in both muscle and yeast, the recycling of coenzyme can be visualized as follows:

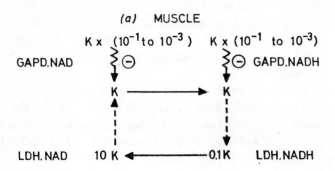

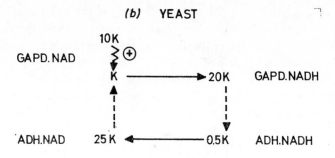

where K is a normalized form of enzyme–coenzyme dissociation constant ($K \approx$ $\approx 4 \times 10^{-5}$ M for muscle and $K \approx 10^{-5}$ M for yeast GAPD) and the rounded multiplying factors are to express the relation between the various species as seen in Tables 8.3 and 8.4. Continuous and broken arrows symbolize enzymatic conversion and transport from one enzyme to the other, respectively; the zigzag arrows with $\oplus$ or $\ominus$ sign mean positive and negative cooperativity in coenzyme binding, respectively. The other substrates of the enzyme reactions are not shown.

It is apparent from the above schemes that there is at least a ten-fold *decrease* in K along the broken arrows. This means that the enzyme which is to take over the coenzyme, let it be the oxidized or reduced form, from the "counter-enzyme" binds it with at least ten-fold greater affinity. Such an arrangement will favour the *vectorial (unidirectional) flow of metabolites*. These relations were established by the intervention of positive and negative cooperativity. It seems reasonable to assume that the two kinds of cooperativities evolved in GAPDs to serve this end. Thus in muscle the inherent affinity of GAPD subunit to NAD–NADH is very great, but too strong a binding may hinder rapid turnover. Negative cooperativity saw to it that this should not occur. Conversely, in yeast the affinity for NAD was possibly lower than desirable and here positive cooperativity helped to adjust the proper relation. It should be mentioned that near pH 7, yeast GAPD binds NAD at its four sites independently, without cooperativity, and the dissociation constant is about 4 μM [828, 1126, 1264]. It appears that at this pH only the high affinity form exists. If so, there might be no need for positive cooperativity. This may well be the case around neutral pH. However, it is an old observation that during glucose fermentation the interior of yeast cells becomes more alkaline [246]. The positive cooperativity in NAD binding observed *in vitro* may be a safety device for such circumstances.

As emphasized at the outset, the above example of functional complementarity is intuitive, and has not much more to its credit than some heuristic character. Obviously, meaningful comparisons would require kinetic constants, as opposed to the dissociation constants considered above, determined under identical experimental conditions. Unfortunately, our knowledge of the kinetic constants of the system is insufficient to check, by the aid of the computer, whether the above cooperativities, some of which also seem to be manifested kinetically [124, 814], indeed contribute to facilitated vectorial metabolite flow or are indifferent. Nevertheless, we feel that attempts to assign physiological significance to molecular kinetic features of enzymes are to proceed along these lines, i.e. by fitting enzymes into the structural–functional milieu they are parts of.

The thermodynamic–kinetic adaptation between functionally coupled enzymes fits the broader framework of the "cytosociological view" of enzyme evolution put forward by Welch and Keleti [1309].

ENZYME ORGANIZATION: SUMMARY AND PERSPECTIVES

THE DIALECTICS OF ENZYMOLOGY: AWAY FROM AND BACK TO THE LIVING CELL

The past half-century has witnessed the emergence of enzymology from biological chemistry, the theoretical and methodological consolidation of the new discipline, and the harvest of a rich crop in this field. During this process enzymology has built up its esoteric world hardly pervadable by the uninitiated and detached itself from most other branches of biochemistry.

Independence, however, was not easily won. The vitalist concept of the past century that enzymatic catalysis is somehow associated with living cells has lingered on, and only the solid evidences provided by the crystallization of enzymes overcame the reluctance to accept the chemical character of enzyme action. (For a historical review see Florkin [383].) As the protein nature of enzymes became acknowledged, the study of enzymes and protein chemistry (and later physics) walked hand in hand, but—adapting Rudyard Kipling's words for the cat—by themselves, and all places were alike to them. Less metaphorically speaking, enzymologists were busy in the purification of ever more proteins that had enzymatic activity and which could be analysed for the genuine interest that a proteinaceous catalyst offers. To understand catalytic action, the compositional and structural properties of enzymes had to be described, for which homogeneous, uncontaminated enzyme preparations were needed. Hence, one of the main ambitions of enzymologists has been to get enzymes in the pure state. According to the current, advanced stage of the craft this is preferably done by a minimum number of purification steps including affinity chromatography and the like. This *trend of separation* yielded many pure enzymes from which our knowledge of enzyme action was derived. Indeed, very much was gained from the study of isolated, pure, and often crystalline, enzymes. Modern enzymology simply would not exist without these endeavours.

On the other hand, much was lost in the pursuit of this research. Most importantly, enzymes were taken out from their natural milieu and placed under very unnatural conditions as regards pH, ionic strength, enzyme and substrate concentrations, the presence of other enzymes, proteins and further macromolecules, etc. We learned a good deal about what enzymes do under a great variety of circumstances *in vitro*, but failed to establish how they work *in vivo*. The cost of knowledge of enzyme action at the molecular and atomic levels was our ignorance about enzyme function at the cellular level. Though in the latter are we interested most profoundly.

It follows that enzymology, after having marched away from the biological toward the chemical and physical, has now to change direction and return to where it came from. In a manner of speaking, the spectrophotometric cell must be replaced by the living cell as the ambience of the enzymes studied. This is by no means a retreat by modern enzymology. On the contrary, enzymology would reach a new level of understanding after completing this circular, or rather spiral, progress, just as envisaged for the development of ideas by Hegelian dialectics. In the latter, the law of "negation of negation" holds that previous ideas are negated at each higher level of understanding: in two consecutive steps the negation made in the first step is negated in the second. This, however, does not mean the complete refusal of earlier tenets but rather their "preservation by cancellation" (Aufheben), i.e. their reappraisal in the new light. In the history of enzymology, the fallacy of a vitalist enzymatic "force" was superseded by the identification of discrete enzymes as protein catalysts. Now the fallacy of discrete enzymes is being superseded by the recognition of supramolecular order in enzyme systems. The basic properties of individual enzymes remain valid in the new context, but their actual behaviour will be such as allowed, or dictated, by the environment *in vivo*.

THE THIRD DIMENSION OF METABOLIC MAPS

On the wall charts displaying metabolic networks, each enzymatic step is represented by an arrow with a list of cofactors, activators, inhibitors, etc., that are needed for, or affect, the given enzyme reaction. Without questioning the usefulness of these charts, one cannot help feeling that they give only the skeleton of the real thing. They are the curious, two-dimensional "fossils" of contemporary living matter, or even worse, since they do not render the original spatial order. Their outlay mainly reflects the ingenuity of the designer in cramming maximum information into a minimum of surface area.

All of us know, of course, that enzymes are not mere "activities", but well-defined three-dimensional objects, the products of the unique folding of their constituent polypeptide chain(s). In this book we followed up their structural–functional organization from the quaternary structure onwards. Most intracellular enzymes are oligomers with definite subunit arrangements that enable them to operate a characteristic system of subunit interactions. The reversible conformational changes mediated through intersubunit contacts underlie the mechanisms of modulating enzyme activity both in true oligomers consisting of identical protomers and in complex enzymes where separate catalytic and regulatory subunits have evolved.

The structural–functional organization of enzymes does not end at the quaternary structure. In many instances two or more enzymes are clustered in order to perform certain catalytic steps more efficiently. In the multienzyme complexes each enzyme is represented by separate polypeptide chain(s), whereas in multienzyme conjugates the polypeptide chains are continuous. There is no major difference between these two ways of clustering in respect of the functional consequences. It seems likely that several multienzyme clusters now regarded as complexes will turn out to be conjugates, because the peptide bonds connecting the different catalytic domains are particularly sensitive to proteolysis and therefore may be split during purification. An example of this is the *arom* multienzyme conjugate. Multienzyme complexes and conjugates may coexist in the same cluster, as we see now in fatty acid synthase from yeast, and a cluster that is a complex in one species may be a conjugate in another species, as is the case for tryptophan synthases. These facts suggest that if enzyme clustering was to be favoured for some reason, Nature provided it, but the protein chemistry of solving this task may have been fortuitous.

Enzyme–enzyme interactions occur not only in fairly tight complexes and conjugates, but also among so-called soluble enzymes of various intracellular compartments such as the cytoplasm and mitochondrial matrix. There is a growing body of evidence for this type of organization and probably much more will be found if systematic search is made in this direction. Furthermore, soluble enzymes interact with intracellular structures such as membranes and the elements of contractile systems, and these surfaces may serve as support for the build-up of enzyme clusters. These loose associations should be distinguished from the case when enzymes are incorporated into structures, as with integral membrane enzymes. Nevertheless, in both varieties the scaffolding role of the support is essential and the enzyme array thus formed is often related to transmembrane processes.

The functional consequences, and thus the possible physiological significances, of enzyme juxtaposition are multifarious. The spatial vicinity of active sites may enable a covalently bound intermediate to visit each active site where successive chemical transformations take place. If the intermediate is diffusible, it may be channelled from one active site to the other thereby preventing mixing in the bulk medium, which would be wasteful if the intermediate is degraded by other enzymes or deleterious to the cell if the intermediate is a highly reactive, and consequently toxic, compound. Metabolite channelling may accelerate the response of a pathway to changes in substrate or effector concentration by decreasing transient time or may increase the overall velocity of a pathway by decreasing transit time. In addition, enzyme clustering may activate the physiologically important reactions of the respective enzymes and suppress unfavourable reactions. Last but not least, coordinate regulatory effects may take place in enzyme clusters, activation or inhibition, via conformational changes induced by a ligand in one enzyme and transmitted through the heterologous contacts to the other enzyme(s).

The structural organization of enzymes to different degrees of complexity, superimposed on the abundance of membranes and other structural elements in the cell, results in the compartmentation of metabolites. The macrocompartments correspond to the various subcellular membrane-bounded organelles, whereas the microcompartments are formed by enzyme–enzyme channels, by water layers at structure surfaces different from bulk water, and by binding sites for the metabolites (when the latter are not in a large excess over the former.) All these compartments subdivide the intracellular space and restrict the motion of metabolites to lesser or greater extents. The picture that emerges is that of an organized, thick maze, rather than an open plain, where the metabolites cannot freely roam.

It follows that we have to discard the fallacy of a wire-diagrammed metabolic plan and put the "flesh" on the skeleton. We should add a third dimension to the metabolic maps, in which the bulky elements are no longer the substrate formulas but rather the space-filling models of enzymes and their complexes. This new representation needs to be a topographical rendering of cellular macromolecular assemblies in the lacunae of which small molecules move subject to probabilistic limitations. We certainly lose much in clarity and simplicity if we try to adopt this new image, but gain enormously much in depth of understanding. The complexity seems bewildering, yet it is the only base on which the questions that have proved so far tantalizingly elusive can be asked with the hope of getting an answer.

THE DIVIDENDS OF COMPLEXITY

On the following pages we shall briefly examine a few areas on which our novel view of metabolism may shed some light. Such examples were chosen to show that the greater complexity of the approach may bear fruits in fields immediately relevant to mankind.

ENZYME ORGANIZATION IN DISEASE AND AGEING

Among the questions that concern us most are those related to disease and ageing. Longevity in good health, that is what most people strive for. Although extreme achievements toward these goals would create social problems, there is still room for safe progress without threatening our social framework.

The involvement of enzymes in many *diseases* is well known. Enzymopathies are pathological states characterized by the lack of or decrease in a certain enzyme activity. They are due to defective genes, hence are hereditary. We know of a great number of such *inborn errors of metabolism*, some of them related to enzymes or enzyme systems dealt with earlier in this book. For example, adult lactose intolerance is a condition that afflicts about one-third of the world's population [1117, 1118]. It is caused by the gradual disappearance of lactase (β-galactosidase) activity from the intestinal brush-border membrane around adolescence. Altogether fifteen enzyme and transport defects have so far been described in connection with the brush-border membrane [258, 259]. The pyruvate dehydrogenase multienzyme complex is affected in Friedreich's ataxia, a neurological disorder [91] and the branched-chain α-ketoacid dehydrogenase multienzyme complexes of skeletal muscle are involved in maple syrup urine disease, a disturbance of leucine, isoleucine and valine metabolism [204, 453]. Orotic aciduria is the defect of the bienzyme conjugate, complex U [1129]. (For reviews on enzyme defects see [567, 974]).

The recognition of enzyme defects of the above kind is equivalent to finding the etiologic factor of certain diseases, and sometimes even suggests therapeutic measures, therefore it is of primary importance. Nevertheless, we believe that there are many pathological states associated with abnormal enzyme function in a more subtle way. Namely, enzymopathies may encompass not only genetic and post-translational (age-related) events affecting catalytic activity, but also the occurrence of enzymes with unchanged catalytic properties *in vitro* but impaired ability to assemble *in vivo*. In practical terms, a metabolic pathway can be handicapped even if the activities of all constituent enzymes, as determined in a cell-free extract, are normal, but the enzymes fail to build up the correct physiological superstructure. There may be "assembly or association mutants" where the amino acid substitution occurs in a heterologous contact surface, rather than an active site. Examples for this are found among microorganisms. The same effect may be achieved by erroneous or spurious post-translational modifications caused by an ageing-related anomaly.

It may not be easy to recognize such assembly defects in man. Our traditional approach of screening for enzyme activity *in vitro* is bound to let most such anomalies pass unnoticed. The concepts and methods outlined in this book, and others yet to be developed, are needed to reveal abnormal enzyme organization.

Ageing occurs, and can be studied, at different levels of complexity from the molecular through the cellular to the organismal. The question of *ageing of protein molecules,* manifested in the deamidation of asparagine and glutamine residues [784, 817] or conformational changes up to denaturation [1093, 1094, 1095], during their life span is controversial. With some enzymes age-related changes could be detected [812, 997, 1031, 1032], whereas no such changes were found with others [484, 968, 1175, 1289, 1300]. (For reviews cf. [440, 1030]). A number of theories have been proposed for *cellular (and organismal) ageing.* These invoke genetic determination [782], "error catastrophe" through the accumulation of damages due to wear-and-tear [521, 564, 889] membrane deterioration [589, 1100, 1350, 1351] possibly through the peroxidation of lipids [331], the racemization of L-amino acids to the D-forms in proteins [300] and the shift of balance between free radicals and antioxidants in favour of the former [324]. (For reviews on the biochemistry of ageing see [555, 556, 557]). This abundance of theories suggests that cellular ageing cannot be traced back to a single causative factor. Indeed, it seems to be a typically *systemic process,* the consequence of upset balance in a number of interactions. The failure to come to grips with ageing is probably due to the simplifying character of the approaches adopted so far. In order to arrive at meaningful conclusions it appears necessary to consider the systemic nature of ageing, the alterations in the various macromolecular and enzyme organizational patterns.

Any complex functioning structure has certain points where it is most likely to go wrong. In the instruction manual of sophisticated instruments there is usually a section on "trouble shooting", reflecting an awareness by the manufacturers of the weaknesses in their product. These Achilles heels are usually the most complex parts, e.g. delicate electronic gadgets, which—by a not so surprising analogy—are often involved in the regulation of the instrument's function. Living organisms are very complex machines whose instruction manual, alas, is not enclosed on delivery, and the pages on trouble shooting have to be written painstakingly by ourselves. It seems probable that many of the blank pages will be filled with future knowledge about enzyme organization. Weak macromolecular interactions are necessarily vulnerable because small energies can perturb them. Consequently, noxious influences may damage them first. Then, if these interactions are of functional significance, their disturbance will impair the life processes.

ENZYME THERAPY

In diseases that are due to a defective enzyme, the causal therapy would be to replace the bad enzyme by the good one. This is easy to conceive but difficult to achieve. Even in the apparently simple case when digestive enzymes are supplemented to patients suffering from pancreatic insufficiency, the practical results are often unsatisfactory. The reason is that the enzyme pills, coated to prevent destruction by acid in the stomach, are not certain to get in contact with the ingested food then and there, when and where they are wanted. However, it is immensely more difficult to replace an intracellular or an integral membrane enzyme. Attempts have been made to entrap enzymes by reversible haemolysis in red blood cells [579]. It was suggested that, for example, the loading of erythrocytes with β-glucocerebrosidase would be beneficial in Gaucher's disease, characterized by the accumulation of β-glucocerebrosides in the cells of the reticuloendothelial system, because the enzyme is then released in the spleen. Although strategies of this type, also applying liposomes, are promising, there remains much to be learned about the fate of the enzymes delivered and about how to influence the proper fit of enzymes *in situ*. These problems are intimately related to supramolecular enzyme organization. Progress in the latter field will certainly contribute to our skill in tinkering with the enzyme composition of cells and subcellular organelles.

ENZYME TECHNOLOGY

Chemical industry, including pharmaceutical, food- and waste-processing industries are being revolutionized by enzyme technologies. The use of immobilized enzymes (matrix-bound, encapsulated, etc.) reduces expenses by the repeated applicability of enzyme preparations. One can readily visualize that artificial or semi-artificial multienzyme systems are constructed for the concerted production of one or more compounds. For the economic running of such systems their behaviour must be thoroughly known. This is the practical reason of current efforts to characterize immobilized enzymes and enzyme systems [442, 1326]. These endeavours cannot be really successful without the adequate conceptual development.

ENZYMOLOGY OF THE 21ST CENTURY

At the end, we may venture a look into the future of enzymology. But really, has it got any future? Many biologists hold the view that enzymology surpassed itself, the hot spots in biochemistry are no longer enzymological in character and therefore enzymologists will gradually be left behind by general progress, toying with their K_m's and V_{max}'s, reactive SH and OH groups in active sites, etc., driven along mainly by their deep-rooted scientific inertia.

I do not share this gloomy view. No doubt, enzymology has to change, too, as times pass, but this change is unlikely to be a decay. It is our luck that, thank goodness, enzymes are involved in practically all biological processes. Hence whether recombinant DNA manipulations, the cybrids, or what have you will be in fashion, those who know how to handle enzymes competently will not remain unemployed.

One way for enzymology to proceed is toward the ever smaller, the ever finer detail. The thorough interpretation of enzyme action in terms of theoretical organic chemistry and possibly quantum mechanics. The other way leads toward the ever greater, i.e. to supramolecular organization. The enzymologist of the next century will probably be more of an interdisciplinarian than his present counterpart. He will work on the enzymological aspects of complex biological phenomena with the main goal to understand, and to be able to influence, that phenomenon rather than for the description of still another enzyme. He will use very sophisticated instrumentation, but will refrain from self-contained exercises under the urge of his team-mates of other expertises. The enzymology and the enzymologists shall survive, if they surrender to the other biological disciplines. Is this "preservation by cancellation"? Good old Hegel again, who never saw an enzymologist, yet raised hope for him. For *him*? Not exactly. Survival will only be furthered if half of the enzymologists are female.

REFERENCES

1. ACHS, M. J. and GARFINKEL, D. (1977) Computer simulation of energy metabolism in anoxic perfused rat heart. *Am. J. Physiol.* **232,** R 164–174.

2. ACHS, M. J. and GARFINKEL, D. (1977) Computer simulation of rat heart metabolism after adding glucose to the perfusate. *Am. J. Physiol.* **232,** R 175–184.

3. ADACHI, O. and MILES, E. W. (1974) A rapid method for preparing crystalline β_2 subunit of tryptophan synthetase of *Escherichia coli* in high yield. *J. Biol. Chem.* **249,** 5430–5434.

4. ADAIR, G. S. (1925) The hemoglobin system. VI. The oxygen dissociation curve of hemoglobin. *J. Biol. Chem.* **63,** 529–545.

5. AKERBOOM, T. P. M., VAN DER MEER, R. and TAGER, J. M. (1979) Techniques for the investigation of intracellular compartmentation. *Techn. Life Sci B2/I: Techn. Metabol. Res.* **B 205,** 1–33.

6. AKIYAMA, S. K. and HAMMES, G. G. (1980) Elementary steps in the reaction mechanism of the pyruvate dehydrogenase multienzyme complex from *Escherichia coli:* Kinetics of acetylation and deacetylation. *Biochemistry* **19,** 4208–4213.

7. AKIYAMA, S. K. and HAMMES, G. G. (1981) Elementary steps in the reaction mechanism of pyruvate dehydrogenase multienzyme complex from *Escherichia coli:* Kinetics of flavin reduction. *Biochemistry* **20,** 1491–1497.

8. ALLEMAND, P., BON, S., MASSOULIÉ, J. and VIGNY, M. (1981) Quaternary structure of chicken acethylcholinesterase and butyrylcholinesterase; Effect of collagenase and trypsin. *J. Neurochem.* **36,** 860–867.

9. ALLEN, G., TRINNAMAN, B. J. and GREEN, N. M. (1980) The primary structure of the calcium ion-transporting adenosine triphosphatase protein of rat skeletal sarcoplasmic reticulum. Peptides derived from digestion with cyanogene bromide, and the sequences of three long extramembranous segments. *Biochem. J.* **187,** 591–616.

10. ALVARADO, F., MAHMOOD, A., TELLIER, C. and VASSEUR, M. (1980) Quantitative analysis of the mixed activating effects of the alkali metal ions on intestinal brush-border sucrase at pH 5.2. *Biochim. Biophys. Acta* **613,** 140–152.

11. ALZIARI, S., TOURAILLE, S., BRIAND, Y. and DURAND, R. (1979) Phosphate transport and protiens with SH groups in rat liver mitochondria. *Biochimie* **61,** 891–903.

12. ANDERSEN, J. P., FELLMANN, P., MØLLER, J. V. and DEVAUX, P. F. (1981) Immobilization of a spin-labelled fatty acid chain covalently attached to Ca^{2+}-ATPase from sarcoplasmic reticulum uggests an oligomeric structure. *Biochemistry* **20,** 4928–4936.

13. ANDERSON, P. J. (1976) The specific activity of aldolase in the livers of old and young rats. *Can. J. Biochem.* **54,** 194–196.

14. ANDREWS, P. R. and JEFFREY, P. D. (1976) The use of sedimentation coefficients to distinguish between models for protein oligomers. *Biophys. Chem.* **4,** 93–102.

15. ANDREWS, P. R. and JEFFREY, P. D. (1980) Calculated sedimentation ratios for assemblies of two, three, four and five spatially equivalent protomers. *Biophys. Chem.* **11**, 49–59.

16. ANDRIA, G., MARZI, A. and AURICCHIO, S. (1976) α-Glutamyl-β-naphthylamide hydrolase of rabbit small intestine. Localization in the brush border and separation from other brush border peptidases. *Biochim. Biophys. Acta* **419**, 42–50.

17. ANGELIDES, K. J., AKIYAMA, S. K. and HAMMES, G. G. (1979) Subunit stoichiometry and molecular weight of the pyruvate dehydrogenase multienzyme complex from *Escherichia coli. Proc. Natl. Acad. Sci. U.S.A.* **76**, 3279–3283.

18. ANGELIDES, K. J. and HAMMES, G. G. (1978) Mechanism of action of the pyruvate dehydrogenase multienzyme complex from *Escherichia coli. Proc. Natl. Acad. Sci. U.S.A.* **75**, 4877–4880.

19. ANGELIDES, K. J. and HAMMES, G. G. (1979) Fluorescence studies of the pyruvate dehydrogenase multienzyme complex from *Escherichia coli. Biochemistry* **18**, 1223–1229.

20. ANGELIDES, K. J. and HAMMES, G. G. (1979) Structural and mechanistic studies of the α-ketoglutarate dehydrogenase multienzyme complex from *Escherichia coli. Biochemistry* **18**, 5531–5537.

21. ANGLISTER, L. and SILMAN, I. (1978) Molecular structure of elongated forms of electric eel acetylcholinesterase. *J. Mol. Biol.* **125**, 293–311.

22. ANTONINI, E., CARREA, G., CREMONI, P., PASTA, P., ROSSI FANELLI, M. R. and CHANCONE, E. (1979) "Subunit-exchange" chromatography of proteins: Behavior of α-chymotrypsin. *Anal. Biochem.* **95**, 89–96.

23. AOKI, T., OYA, H., MORI, M. and TATIBANA, M. (1975) Glutamine-dependent carbamoyl-phosphate synthetase in Ascaris ovary and its regulatory properties. *Proc. Japan. Acad.* **51**, 733–736.

24. ARAGÓN, J. J. and SOLS, A. (1978) Glyceraldehyde-3-phosphate dehydrogenase activity studied under physiological conditions with a linear assay. *Biochem. Biophys. Res. Commun.* **82**, 1098–1103.

25. ARÁNYI, P. and TÓTH, J. (1977) A full stochastic description of the Michaelis-Menten reaction for small systems. *Acta Biochim. Biophys. Acad. Sci. Hung.* **12**, 375–388.

26. ARESE, P., BOSIA, A., PESCARMONA, G. P. and TILL, U. (1974) 2,3-Diphosphoglycerate synthesis by human white ghosts. *FEBS Lett.* **49**, 33–36.

27. ARNOLD, H., HENNING, R. and PETTE, D. (1971) Quantitative comparison of the binding of various glycolytic enzymes to F-actin and the interaction of aldolase with G-actin. *Eur. J. Biochem.* **22**, 121–126.

28. ARNOLD, H., NOLTE, J. and PETTE, D. (1969) Quantitative and histochemical studies on the desorption and readsorption of aldolase in cross-striated muscle. *J. Histochem. Cytochem.* **17**, 314–320.

29. ARNOLD, H. and PETTE, D. (1968) Binding of glycolytic enzymes to structure proteins of the muscle. *Eur. J. Biochem.* **6**, 163–171.

30. ARNOLD, H. and PETTE, D. (1970) Binding of aldolase and triosephosphate dehydrogenase to F-actin and modification of catalytic properties of aldolase. *Eur. J. Biochem.* **15**, 360–366.

31. ASCHAFFENBURG, R., FENNA, R. E., PHILLIPS, D. C., SMITH, S. G., BUSS, D. H., JENNESS, R. and THOMPSON, M. P. (1979) Crystallography of α-lactalbumin. III. Crystals of baboon milk α-lactalbumin. *J. Mol. Biol.* **127**, 135–137.

32. ASENSIO, C. (1976) Molecular ecology. In *Reflections of Biochemistry.* p. 235. (KORNBERG, A., HORECKER, B. L., CORNUDELLA, L. and ORO, J., eds) Pergamon Press, Oxford.

33. ASHMARINA, L. I., MURONETZ, V. I. and NAGRADOVA, N. K. (1980) *Biochem. Int.* **1**, 47–54.

34. ASHMARINA, L. I., MURONETZ, V. I. and NAGRADOVA, N. K. (1981) Immobilized D-glyceraldehyde-3-phosphate dehydrogenase can exist as a trimer. *FEBS Lett.* **128**, 22–26.

35. ASTRACHAN, L., COLOWICK, S. P. and KAPLAN, N. O. (1957) The reactivity of the bound DPN of muscle triose phosphate dehydrogenase. *Biochim. Biophys. Acta* **24**, 141–154.

36. ATKINSON, D. E. (1970) Enzymes as control elements in metabolic regulation. In *The Enzymes*· Vol. 1., pp. 461–489. (BOYER, P. D., ed.), Academic Press, New York–London.

37. AUNE, K. C. and TIMASHEFF, S. N. (1971) Dimerization of α-chymotrypsyn. I. pH-dependence in the acid region. *Biochemistry* **10,** 1609–1617.

37a. AVIRAM, I. and SHAKLAI, N. (1981) The association of human erythrocyte catalase with the cell membrane. *Arch. Biochem. Biophys.* **212,** 329–337.

38. BACHELARD, H. S. (1976) Biochemistry of coma. In *Biochemistry and Neurological Disease.* pp. 228–277. (DAVISON, A. N., ed.), Blackwell, London.

39. BACKMAN, L. and JOHANSSON, G. (1976) Enzyme–enzyme complexes between aspartate amino-transferase and malate dehydrogenase from pig heart muscle. *FEBS Lett.* **65,** 39–43.

40. BAKER, T. S., SUH, S. W. and EISENBERG, D. (1977) Structure of ribulose-1,5-bisphosphate carboxylase-oxygenase: Form III crystals. *Proc. Natl. Acad. Sci. U.S.A.* **74,** 1037–1041.

41. BALASUBRAMANIAM, S., MITROPOULOS, K. A. and MYANT, N. B. (1973) Evidence for the compartmentation of cholesterol in rat-liver microsomes. *Eur. J. Biochem.* **34,** 77–83.

42. BALDWIN, J. M. (1976) A model of co-operative oxygen binding to haemoglobin. *Br. Med. Bull.* **32,** 213–219.

43. BARBOUR, R. L. and CHAN, S. H. P. (1981) Characterization of the kinetics and mechanism of the mitochondrial ADP-ATP carrier. *J. Biol. Chem.* **256,** 1940–1948.

44. BAREA, J. L. and GILES, N. H. (1978) Purification and characterization of quinate (shikimate) dehydrogenase, an enzyme in the inducible quinic acid catabolic pathway of *Neurospora crassa*. *Biochim. Biophys. Acta* **524,** 1–14.

45. BARRACK, E. R., HAWKINS, E. F., ALLEN, S. L., HICKS, L. L. and COFFEY, D. S. (1977) Concepts related to salt resistant estradiol receptors in rat uterine nuclei: Nuclear matrix. *Biochem. Biophys. Res. Commun.* **79,** 829–836.

46. BARRERA, C. R., NAMIHIRA, G., HAMILTON, L., MUNK, P., ELEY, M. H., LINN, T. C. and REED, L. J. (1972) α-Keto acid dehydrogenase complexes. XVI. Studies on the subunit structure of the pyruvate dehydrogenase complexes from bovine kidney and heart. *Arch. Biochem. Biophys.* **148,** 343–358.

47. BARTELS, K. and COLMAN, P. M. (1976) Subunit symmetry of tetrameric phosphorylase a. *Biophys. Struct. Mechan.* **2,** 43–59.

48. BARTHA, F. and KELETI, T. (1979) Kinetic analysis of interaction between two enzymes catalyzing consecutive reactions. *Oxid. Commun.* **1,** 75–84.

49. BARTHOLMES, P., KIRSCHNER, K. and GSCHWIND, H. P. (1976) Cooperative and noncooperative binding of pyridoxal 5′-phosphate to tryptophan synthase from *Escherichia coli*. *Biochemistry* **15,** 4712–4717.

50. BARTHOLMES, P. and TEUSCHER, B. (1979) Cooperative binding of α subunits to the apo-β_2 subunit of tryptophan synthase from *Escherichia coli*. *Eur. J. Biochem.* **95,** 323–326.

51. BARWELL, C. J. and HESS, B. (1970) The transient time of the hexokinase (pyruvate kinase) lactate dehydrogenase system *in vitro*. *Z. Physiol. Chem.* **351,** 1531–1536.

52. BASS, A., BRDICZKA, D., EYER, P., HOFER, S. and PETTE, D. (1969) Metabolic differentiation of distinct muscle types at the level of enzymatic organization. *Eur. J. Biochem.* **10,** 198–206.

53. BATES, D. L., DANSON, M. J., HALE, G., HOOPER, E. A. and PERHAM, R. N. (1977) Self-assembly and catalytic activity of the pyruvate dehydrogenase multienzyme complex of *Escherichia coli*. *Nature* **268,** 313–316.

54. BATES, D. L., HARRISON, R. A. and PERHAM, R. N. (1975) The stoichiometry of polypeptide chains in the pyruvate dehydrogenase multienzyme complex of *E. coli* determined by a simple novel method. *FEBS Lett.* **60,** 427–430.

55. BATKE, J. (1968) Effect of the concentration of D-glyceraldehyde-3-phosphate dehydrogenase on the reactivity of its firmly bound NAD. *FEBS Lett.* **2,** 81–82.

56. BATKE, J., ASBÓTH, G., LAKATOS, S., SCHMITT, B. and COHEN, R. (1980) Substrate-induced dissociation of glycerol-3-phosphate dehydrogenase and its complex formation with fructose-bisphosphate aldolase. *Eur. J. Biochem.* **107,** 389–394.

57. BATTERSBY, M. K. and RADDA, G. K. (1979) Intersubunit transmission of ligand effects in the glycogen phosphorylase b dimer. *Biochemistry* **18**, 3774–3780.

58. BEADLE, F. R., GALLEN, C. C., CONWAY, R. S. and WATERSON, R. M. (1979) Enoyl coenzyme A hydratase activity in *Escherichia coli*. *J. Biol. Chem.* **254**, 4387–4395.

59. BEAM, K. G., ALPER, S. L., PALADE, G. E. and GREENGARD, P. (1979) Hormonally regulated phosphorylation of turkey erythrocytes. Localization to plasma membrane. *J. Cell Biol.* **83**, 1–15.

60. BEECHEY, R. B., HUBBARD, S. A., LINNETT, I. E., MITCHELL, A. D. and MUNN, E. A. (1975) A simple and rapid method for the preparation of adenosine triphosphatase from submitochondrial particles. *Biochem. J.* **148**, 533–537.

61. BEECHEY, R. B., ROBERTSON, A. M., HOLLOWAY, C. T. and KNIGHT, I. G. (1967) The properties of dicyclohexylcarbodiimide as an inhibitor of oxidative phosphorylation. *Biochemistry* **6**, 3867–3879.

62. BEECKMANS, S. and KANAREK, L. (1981) Demonstration of physical interactions between consecutive enzymes of the citric acid cycle and of the aspartate-malate shuttle. A study involving fumarase, malate dehydrogenase, citrate synthase and aspartate aminotransferase. *Eur. J. Biochem.* **117**, 527–535.

63. BELL, J. E. and DALZIEL, K. (1975) Studies of coenzyme binding to rabbit muscle glyceraldehyde-3-phosphate dehydrogenase. *Biochim. Biophys. Acta* **391**, 249–258.

64. BENNETT, V. and STENBUCK, P. J. (1979) Identification and partial purification of ankyrin, the high affinity membrane attachment site for human erythrocyte spectrin. *J. Biol. Chem.* **254**, 2533–2541.

65. BENNETT, V. and STENBUCK, P. J. (1979) The membrane attachment protein for spectrin is associated with band 3 in human erythrocyte membranes. *Naure* **280**, 468–473.

66. BERGELSON, L. D. and BARSUKOV, L. I. (1977) Topological asymmetry of phospholipids in membranes. The distribution of phospholipids in biological membranes is related to that in bilayer membranes of small vesicles. *Science* **197**, 224–230.

67. BERLYN, M. B., AHMED, S. I. and GILES, N. H. (1970) Organization of polyaromatic biosynthetic enzymes in a variety of photosynthetic organisms. *J. Bactertiol.* **104**, 768–774.

68. BERLYN, M. B. and GILES, N. H. (1973) Organization of interrelated aromatic metabolic pathway enzymes in *Acinetobacter calcoaceticus*. *J. Gen. Microbiol.* **74**, 337–341.

69. BERMAN, J. N., CHEN, G.-X., HALE, G. and PERHAM, R. N. (1981) Lipoic acid residues in a takeover mechanism for pyruvate dehydrogenase multienzyme complex of *Escherichia coli*. *Biochem. J.* **199**, 513–520.

70. BERNOFSKY, C. and PANKOW, M. (1971) Role of dehydrogenase binding in protection of nicotinamide adenine dinucleotide from enzymatic hydrolysis. *Biochim. Biophys. Acta* **242**, 437–440.

71. BERNOFSKY, C. and PANKOW, M. (1973) Protein binding of nicotinamide adenine dinucleotide and regulation of nicotinamide adenine dinucleotide glycohydrolase activity in homogenates of rabbit skeletal muscle. *Arch. Biochem. Biophys.* **156**, 143–153.

72. BERRIDGE, M. J., RAPP, P. E. and TREHERNE, J. E. (eds) (1979) Cellular oscillators. *J. Exp. Biol.* **81**, 1–306.

73. BESSMAN, S. P. (1954) A contribution to the mechanism of diabetes mellitus. In *Fat Metabolism*. pp. 133–137 (NAJJAR, V. A., ed.) Johns Hopkins, Baltimore.

74. BESSMAN, S. P. (1960) Diabetes mellitus observations: Theoretical and practical. *J. Pediatr.* **56**, 191–203.

75. BESSMAN, S. P. (1966) A molecular basis for the mechanism of insulin action. *Am. J. Med.* **40**, 740–749.

76. BESSMAN, S. P. and FONYÓ, A. (1966) The possible role of the mitochondrial bound creatine kinase in regulation of mitochondrial respiration. *Biochem. Biophys. Res. Commun.* **22**, 597–602.

77. BESSMAN, S. P. and GEIGER, P. J. (1980) Compartmentation of hexokinase and creatine phosphokinase, cellular regulation, and insulin action. *Curr. Top. Cell. Regul.* **16**, 55–86.

78. BESSMAN, S. P. and GEIGER, P. J. (1981) Transport of energy in muscle: The phosphorylcreatine shuttle. *Science* **211**, 448–452.

79. BEUTLER, E., GUINTO, E., KUHL, W. and MATSUMOTO, F. (1978) Existence of only a single functional pool of adenosine triphosphate in human erythrocytes. *Proc. Natl. Acad. Sci. U.S.A.* **75**, 2825–2828.

80. BIELICKI, L. and KRIEGLSTEIN, J. (1977) Solubilization of brain mitochondrial hexokinase by thiopental. *Naunyn-Schmiedeberg's Arch. Pharmacol.* **298**, 61–65.

81. BIELICKI, L. and KRIEGLSTEIN, J. (1977) The effect of anesthesia on brain mitochondrial hexokinase. *Naunyn-Schmiedeberg's Arch. Pharmacol.* **298**, 229–233.

82. BIESECKER, G., HARRIS, J. I., THIERRY, J. C., WALKER, J. E. and WONACOTT, A. J. (1977) Sequenc and structure of D-glyceraldehyde 3-phosphate dehydrogenase from *Bacillus stearothermophilus Nature* **266**, 328–333.

83. BINSTOCK, J. F., PRAMANIK, A. and SCHULZ, H. (1977) Isolation of a multi-enzyme complex of fatty acid oxidation from *Escherichia coli. Proc. Natl. Acad. Sci. U.S.A.* **74**, 492–495.

84. BIRKENMEIER, E. and ALPERS, D. H. (1974) Enzymatic properties of rat lactase-phlorizin hydrolase. *Biochim. Biophys. Acta* **350**, 100–112.

85. BIRKTOFT, J. J. and BLOW, D. M. (1972) Structure of crsytalline α-chymotrypsin. V. The atomic structure of tosyl-α-chymotrypsin at 2 Å resolution. *J. Mol. Biol.* **68**, 187–240.

86. BISSON, R., AZZI, A., GUTWENIGER, H., COLONNA, R., MONTECUCCO, C. and ZANOTTI, A. (1978) Interaction of cytochrome c with cytochrome c oxidase. Photoaffinity labeling of beef heart cytochrome c oxidase with arylazidocytochrome c. *J. Biol. Chem.* **253**, 1874–1880.

87. BISSWANGER, H., KIRSCHNER, K., COHN, W., HAGER, V. and HANSSON, E. (1979) N-(5-Phosphoribosyl)anthranilate isomerase-indoleglycerol-phosphate synthase. 1. A substrate analogue binds to two different binding sites on the bifunctional enzyme from *Escherichia coli. Biochemistry* **18**, 5946–5953.

88. BISSWANGER, H. and SCHMINCKEOTT, E. (eds) (1980) *Multifunctional Proteins,* John Wiley & Sons, New York.

89. BITONTI, A. J. and COURI, D. (1981) Drug-induced migration of cytoplasmic ornithine decarboxylase into rat liver nucleus is not related to increased RNA polymerase activity. *Biochem Biophys. Res. Commun.* **99**, 1040–1044.

90. BLAKE, C. C. F. (1976) X-ray enzymology. *FEBS Lett.* **62**, Suppl. E30–E36.

91. BLASS, J. P., KARK, R. A. P. and MENON, N. K. (1976) Dehydrogenase deficiency in Friedreich's ataxia. *N. Engl. J. Med.* **295**, 62–67.

92. BLEILE, D. M., HACKERT, M. L., PETTIT, F. H. and REED, L. J. (1981) Subunit structure of dihydrolipoyl transacetylase component of pyruvate dehydrogenase complex from bovine heart. *J. Biol. Chem.* **256**, 514–519.

93. BLEILE, D. M., MUNK, P., OLIVER, R. M. and REED, L. J. (1979) Subunit structure of dihydrolipoyl transacetylase component of pyruvate dehydrogenase complex from *Escherichia coli. Proc. Natl. Acad. Sci. U.S.A.* **76**, 4385–4389.

94. BLOBEL, G. and DOBBERSTEIN, B. (1975) Transfer of proteins across membranes I. Presence of proteolytically processed and unprocessed nascent immunoglobulin light chains on membrane-bound ribosomes of murine myeloma. *J. Cell Biol.* **67**, 835–851.

95. BLOBEL, G. and DOBBERSTEIN, B. (1975) Transfer of proteins across membranes II. Reconstitution of functional rough microsomes from heterologous components. *J. Cell Biol.* **67**, 852–862.

96. BLOCH, W., MACQUARRIE, R. A. and BERNHARD, S. A. (1971) The nucleotide and acyl group content of native rabbit muscle glyceraldehyde 3-phosphate dehydrogenase. *J. Biol. Chem.* **246**, 780–790.

97. BLOCH, K. and VANCE, D. (1977) Control mechanisms in the synthesis of saturated fatty acids. *Annu. Rev. Biochem.* **46**, 263–298.

98. BLOOMFIELD, V., DALTON, W. O. and VAN HOLDE, K. E. (1967) Frictional coefficients of multi-subunit structures. I. Theory. *Biopolymers* **5**, 135–148.

99. BLUNDELL, T., DODSON, G., HODGKIN, D. and MERCOLA, D. (1972) Insulin: the structure in the crystal and its reflection in chemistry and biology. *Adv. Protein Chem.* **26**, 279–402.

100. BOEKER, E. A. (1978) Arginine decarboxylase from *Escherichia coli* B: Mechanism of dissociation from the decamer to the dimer. *Biochemistry* **17**, 258–263.

101. BOEKER, E. A. and SNELL, E. E. (1968) Arginine decarboxylase from *Escherichia coli*. II. Dissociation and reassociation of subunits. *J. Biol. Chem.* **243**, 1678–1684.

102. BOERS, W. and SLATER, E. C. (1973) The effect of substrates, effectors, acetyl phosphate and iodoacetate on the binding of NAD$^+$ and NADH to rabbit-muscle glyceraldehydephosphate dehydrogenase. *Biochim. Biophys. Acta* **315**, 272–284.

103. BOITEUX, A., GOLDBETER, A. and HESS, B. (1975) Control of oscillating glycolysis of yeast by stochastic, periodic, and steady source of substrate: A model and experimental study. *Proc. Natl. Acad. Sci. U.S.A.* **72**, 3829–3833.

104. BOITEUX, A. and HESS, B. (1974) Oscillation in glycolysis, cellular respiration and communication. *Faraday Symp. Chem. Soc.* **9**, 202–214.

105. BOITEUX, A. and HESS, B. (1978) Visualization of dynamic spatial structures in glycolyzing cell-free extracts of yeast. In *Frontiers in Biological Energetics.* pp. 789–798. (DUTTON, P. L., LEIGH, J. and SCARPA, A., eds), Academic Press, New York.

106. BOITEUX, A., HESS, B. and SEL'KOV, E. E. (1980) Creative functions of instability and oscillations in metabolic systems. *Curr. Top. Cell. Regul.* **14**, 171–203.

107. BOITEUX, A., MARKUS, M., PLESSER, T. and HESS, B. (1981) Applications of progress curves in enzymes kinetics. In *Kinetic Data Analysis: Design and Analysis of Enzyme and Pharmacokinetic Data.* pp. 341–352. (ENDRENYI, L., ed.), Plenum Press, New York.

108. BON, S., CARTAUD, J. and MASSOULIÉ, J. (1978) The dependence of acetylcholinesterase aggregation at low ionic strength upon a polyanionic component. *Eur. J. Biochem.* **85**, 1–14.

109. BON, S. and MASSOULIÉ, J. (1978) Collagenase sensitivity and aggregation properties of *Electrophorus* acetylcholinesterase. *Eur. J. Biochem.* **89**, 89–94.

110. BON, S., VIGNY, M. and MASSOULIÉ, J. (1979) Asymmetric and globular forms of acetylcholinesterase in mammals and birds. *Proc. Natl. Acad. Sci. U.S.A.* **76**, 2546–2550.

111. BONNER, D. M., DeMOSS, J. A. and MILLS, S. E. (1965) The evolution of an enzyme. In *Evolving Genes and Proteins.* pp. 305–318. (BRYSON, V. and VOGEL, H. J., eds), Academic Press, New York.

112. BOSCHETTI, A. (1978) *Biogenese der Chloroplasten und Mitochondrien.* Gustav Fischer Verlag, Stuttgart.

113. BOTHWELL, M. and SCHACHMAN, H. K. (1974) Pathways of assembly of aspartate transcarbamoylase from catalytic and regulatory subunits. *Proc. Natl. Acad. Sci. U.S.A.* **71**, 3221–3225.

114. BOTHWELL, M. A. and SCHACHMAN, H. K. (1980) Equilibrium and kinetic studies of the association of catalytic and regulatory subunits of aspartate transcarbamoylase. *J. Biol. Chem.* **255**, 1962–1970.

115. BOTHWELL, M. A. and SCHACHMAN, H. K. (1980) A model for the assembly of aspartate transcarbamoylase from catalytic and regulatory subunits. *J. Biol. Chem.* **255**, 1971–1977.

116. BOWERS, W. F., CZUBAROFF, V. B. and HASCHEMEYER, R. H. (1970) Subunit structure of L-aspartate β-decarboxylase from *Alcaligenes faecalis. Biochemistry* **9**, 2620–2625.

117. BOWMAN, I. B. R. and FLYNN, I. W. (1976) In *Biology of the Kinetoplastida.* Vol. 1., pp. 435–476. (LUMSDEN, W. H. R. and EVANS, D. A., eds), Academic Press, New York.

118. BOWMAN, B. J. and DAVIS, R. H. (1977) Cellular distribution of ornithine in *Neurospora:* Anabolic and catabolic steady-states. *J. Bacteriol.* **130**, 274–284.

119. BOWMAN, B. J. and DAVIS, R. H. (1977) Arginine catabolism in *Neurospora:* Cycling of ornithine. *J. Bacteriol.* **130**, 285–291.

120. Boy, E., Borne, F. and Patte, J.-C. (1979) Isolation and identification of mutants constitutive for aspartokinase III synthesis in *Escherichia coli* K 12. *Biochimie* **61**, 1151–1160.
121. Boyer, P. D. (1965) In *Oxidases and Related Redox Systems*. Vol. 2., pp. 994–1008. (King, T. E., Mason, H. S. and Morrison, M., eds), Wiley, New York.
122. Boyer, P. D. (1977) Coupling mechanisms in capture, transmission, and use of energy. *Annu. Rev. Biochem.* **46**, 957–966.
123. Boyer, P. D., Chance, B., Ernster, L., Mitchell, P., Racker, E. and Slater, E. C. (1977) Oxidative phosphorylation and photophosphorylation. *Annu. Rev. Biochem.* **46**, 955–1026.
124. Boyer, P. D., Hutton, R. L., Cardon, J. and Araki, M. (1978) Subunit catalytic cooperativity in enzyme catalysis. In *New Trends in the Description of the General Mechanism and Regulation of Enzymes.* pp. 13–33. (Damjanovich, S., Elődi, P. and Somogyi, B., eds), Akadémiai Kiadó, Budapest.
125. Brabson, J. S. and Switzer, R. L. (1975) Purification and properties of *Bacillus subtilis* aspartate transcarbamylase. *J. Biol. Chem.* **250**, 8664–8669.
126. Braun, H., Cogoli, A. and Semenza, G. (1975) Dissociation of small-intestinal sucrase isomaltase complex into enzymatically active subunits. *Eur. J. Biochem.* **52**, 475–480.
127. Braun, T. and Dods, R. F. (1975) Development of a Mn^{2+}-sensitive, "soluble" adenylate cyclase in rat testis. *Proc. Natl. Acad. Sci. U.S.A.* **72**, 1097–1101.
128. Brennan, E. F. and Rocha, V. (1979) Hybrid tryptophan synthase β_2 proteins: Apparent conservation of the β–β binding region of the β monomer among enteric bacteria. *J. Bacteriol.* **140**, 1116–1119.
129. Brew, K., Castellino, F. J., Vanaman, T. C. and Hill, R. L. (1970) The complete amino acid sequence of bovine α-lactalbumin. *J. Biol. Chem.* **245**, 4570–4582.
130. Brew, K. and Hill, R. L. (1975) Lactose biosynthesis. *Rev. Physiol. Biochem. Pharmacol.* **72**, 105–157.
131. Bridgeland, E. S. and Jones, K. M. (1965) L-Serine dehydratase of *Arthrobacter globiformis*. *Biochem. J.* **94**, 29 P.
132. Briehl, R. W. (1963) The relation between the oxygen equilibrium and aggregation of subunits in lamprey hemoglobin. *J. Biol. Chem.* **238**, 2361–2366.
133. Briggs, M. M. and Capaldi, R. A. (1978) Cross-linking studies on a cytochrome c-cytochrome c oxidase complex. *Biochem. Biophys. Res. Commun.* **80**, 553–559.
134. Brimacombe, R., Stöffler, G. and Wittmann, H. G. (1978) Ribosome structure. *Annu. Rev. Biochem.* **47**, 217–249.
135. Brothers, V. M., Tsubota, S. I., Germeraad, S. E. and Fristrom, J. W. (1978) Rudimentary locus of *Drosophila melanogaster*: Partial purification of a carbamyl phosphate synthase-aspartate transcarbamylase-dihydroorotase complex. *Biochem. Gen.* **16**, 321–332.
136. Brown, G. K. and O'Sullivan, W. J. (1977) Subunit structure of the orotate phosphoribosyl-transferase-orotidylate decarboxylase complex from human erythrocytes. *Biochemistry* **16**, 3235–3242.
137. Brown, G. K., Powell, J. F. and Craigh, I. W. (1980) Molecular weight differences between human platelet and placental monoamine oxidase. *Biochem. Pharmacol.* **29**, 2595–2603.
138. Bruch, P., Schnackerz, K. D. and Gracy, R. W. (1976) Matrix-bound phosphoglucose isomerase formation and properties of monomers and hybrids. *Eur. J. Biochem.* **68**, 153–158.
139. Brunner, J., Hauser, H., Braun, H., Wilson, K. J., Wacker, H., O'Neill, B. and Semenza, G. (1979) The mode of association of the enzyme complex sucrase-isomaltase with the intestinal brush border membrane. *J. Biol. Chem.* **254**, 1821–1828.
140. Bruns, H., Krieglstein, J. and Wever, K. (1978) Narkose und intrazelluläre Verteilung der cerebralem Hexokinase. *Anaesthesist* **27**, 557–561.
141. Bryce, C. F. A., Williams, D. C., John, R. A. and Fasella, P. (1976) The anomalous kinetics of coupled aspartate aminotransferase and malate dehydrogenase. *Biochem. J.* **153**, 571–577.

142. BÜCHER, T., BRANSER, B., CONZE, A., KLEIN, F., LAGGUTH, O. and SIES, H. (1972) State of oxidation-reduction and state of binding in the cytosolic NADH-system as disclosed by equilibration with extracellular lactate/pyruvate in hemoglobin-free perfused rat liver. *Eur. J. Biochem.* **27**, 301–317.

143. BÜCHER, T. and SIES, H. (1969) Steady state relaxation of enolase *in vitro* and metabolic throughput *in vitro* of red and white rabbit muscles. *Eur. J. Biochem.* **8**, 273–283.

144. BÜCHER, T. and SIES, H. (1976) In *Use of Isolated Liver Cells and Kidney Tubules in Metabolic Studies.* pp. 41–64. (TAGER, J. M., SÖLING, H. D. and WILLIAMSON, J. R., eds), North-Holland, Amsterdam.

145. BUEHNER, M., FORD, G. C., MORAS, D., OLSEN, K. W. and ROSSMANN, M. G. (1974) Three-dimensional structure of D-glyceraldehyde-3-phosphate dehydrogenase. *J. Mol. Biol.* **90**, 25–49.

146. BURGER, R. M. and GLASER, D. A. (1973) Effect of nalidixic acid on DNA replication by toluene-treated *Escherichia coli. Proc. Natl. Acad. Sci. U.S.A.* **70**, 1955–1958.

147. BURGOYNE, L., CASE, M. E. and GILES, N. H. (1969) Purification and properties of the aromatic (arom) synthetic enzyme aggregate of *Neurospora crassa. Biochim. Biophys. Acta* **191**, 452–462.

148. BURNS, J. A. (1969) Steady states of general multi-enzyme networks and their associated properties. Computational approaches. *FEBS Lett.* **2**, Suppl. S30–S33.

149. BURT, C. T., COHEN, S. M. and BARANY, M. (1979) Analysis of intact tissue with ^{31}P NMR. *Annu. Rev. Biophys. Bioeng.* **8**, 1–25.

150. BUSBY, S. J. W., GADIAN, D. G., RADDA, G. K., RICHARDS, R. E. and SEELEY, P. J. (1978) Phosphorus nuclear-magnetic-resonance studies of compartmentation in muscle. *Biochem. J.* **170**, 103–114.

151. BUSBY, S. J. W. and RADDA, G. K. (1976) Regulation of the glycogen phosphorylase system – From physical measurements to biological speculations. *Curr. Top. Cell. Reg.* **10**, 89–160.

152. BUSSE, D., POHL, B., BARTEL, H. and BUSCHMANN, F. (1980) The Mg^{2+}-dependent adenosine triphosphatase activity in the brush border of rabbit kidney cortex. *Arch. Biochem. Biophys.* **201**, 147–159.

153. BUSSE, D., WAHLE, H. U., BARTEL, H. and POHL, B. (1978) The brush border of rabbit kidney, a cellular compartment free of glycolytic enzymes. *Biochem. J.* **174**, 509–515.

154. BUSTAMANTE, E., MORRIS, H. P. and PEDERSEN, P. L. (1978) Hexokinase – direct link between mitochondrial and glycolytic reactions in rapidly growing cancer cells. *Adv. Exp. Med. Biol.* **92**, 363–380.

155. BUSTAMANTE, E. and PEDERSEN, P. L. (1977) High aerobic glycolysis of rat hepatoma cells in culture: Role of mitochondrial hexokinase. *Proc. Natl. Acad. Sci. U.S.A.* **74**, 3735–3739.

156. BUTLER, J. R., PETTIT, F. H., DAVIS, P. F. and REED, L. J. (1977) Binding of thiamin thiazolone pyrophosphate to mammalian pyruvate dehydrogenase and its effects on kinase and phosphatase activities. *Biochem. Biophys. Res. Commun.* **74**, 1667–1674.

157. CABRAL, F., SOLIOZ, M., RUDIN, Y., SCHATZ, G., CLAVILIER, L. and SLONIMSKY, P. P. (1978) Identification of the structural gene for yeast cytochrome c oxidase subunit II on mitochondrial DNA. *J. Biol. Chem.* **253**, 297–304.

158. CAHN, R. S., INGOLD, C. and PRELOG, V. (1966) Spezifikation der molekularen Chiralität. *Angew. Chem., Int. Edit.* **5**, 385.

159. CALKWAERT, D. M., ROSEMBLATT, M. S. and TCHEN, T. T. (1974) Purification and properties of 4-aminobutanal dehydrogenase from a Pseudomonas species. *J. Biol. Chem.* **249**, 1737–1741.

160. CALLINGHAM, B. A. and PARKINSON, D. (1979) Tritiated pargyline binding to rat liver mitochondrial MAO. In *Monoamine Oxidase: Structure, Function, and Altered Functions.* pp. 81–86. (SINGER, T. P., VON KORFF, R. W. and MURPHY, D. L., eds), Academic Press, New York.

161. CALVO, J. M. and FINK, G. R. (1971) Regulation of biosynthetic pathways in bacteria and fungi. *Annu. Rev. Biochem.* **40**, 943–968.

162. CAPALDI, R. A., BELL, R. L. and BRANCHEK, T. (1977) Changes in order of migration of poly-peptides in complex III and cytochrome c oxidase under different conditions of SDS polyacrylamide gel electrophoresis. *Biochem. Biophys. Res. Commun.* **74**, 425–433.

163. CAPALDI, R. A., SWEETLAND, J. and MERLI, A. (1977) Polypeptides in the succinate–coenzyme Q reductase segment of the respiratory chain. *Biochemistry* **16**, 5707–5710.

164. CARERI, G., GRATTON, E., YANG, P.-H. and RUPLEY, J. A. (1980) Correlation of IR spectro-scopic, heat capacity, dia-magnetic susceptibility and enzymatic measurements on lysozyme powder. *Nature* **284**, 572–573.

165. CARLSON, G. H., BECHTEL, P. J. and GRAVES, D. J. (1979) Chemical and regulatory properties of phosphorylase kinase and cyclic AMP-dependent protein kinase. *Adv. Enzymol.* **50**, 41–115.

166. CARLSON, R. W., WADA, H. G. and SUSSMAN, H. H. (1976) The plasma membrane of human placenta. Isolation of microvillus membrane and characterization of protein and glycoprotein subunit. *J. Biol. Chem.* **251**, 4139–4146.

167. CARON, M. G., LIMBIRD, L. E. and LEFKOWITZ, R. J. (1979) Biochemical characterization of the beta-adrenergic receptor of the frog erythrocyte. *Mol. Cell. Biochem.* **28**, 45–66.

168. CARPENTER, F. H. and HARRINGTON, K. T. (1972) Intermolecular crosslinking of monomeric proteins and cross-linking of oligomeric proteins as a probe of quaternary structure. *J. Biol. Chem.* **247**, 5580–5586.

169. CARRAWAY, K. L. and SHIN, B. C. (1972) Specific modification, isolation, and partial charac-terization of an erythrocyte membrane protein. *J. Biol. Chem.* **247**, 2102–2108.

170. CARTAUD, J., RIEGER, F., BON, S. and MASSOULIÉ, J. (1975) Fine structure of electric eel acetyl-cholinesterase. *Brain Res.* **88**, 127–130.

171. CARVAJAL, N., MARTINEZ, J., DE OCA, F. M., RODRIGUEZ, J. and FERNANDEZ, M. (1978) Subunit interactions and immobilised dimers of human liver arginase. *Biochim. Biophys. Acta* **527**, 1–7.

172. CASPAR, D. L. D. and KLUG, A. (1962) Physical principles in the construction of regular viruses. *Cold Spring Harb. Symp. Quant. Biol.* **27**, 1–24.

173. CASSOLY, R. (1978) Evidence against the binding of native hemoglobin to spectrin of human erythrocytes. *FEBS Lett.* **85**, 357–360.

174. CATE, R. L. and ROCHE, T. E. (1978) A unifying mechanism for stimulation of mammalian pyruvate dehydrogenase kinase by reduced nicotinamide adenine dinucleotide, dihydrolipoamide, acetyl coenzyme A, or pyruvate. *J. Biol. Chem.* **253**, 496–503.

175. CATE, R. L. and ROCHE, T. E. (1979) Function and regulation of mammalian pyruvate de-hydrogenase complex. *J. Biol. Chem.* **254**, 1659–1665.

176. CATE, R. L., ROCHE, T. E. and DAVIS, L. C. (1980) Rapid intersite transfer of acetyl groups and movement of pyruvate dehydrogenase component in the kidney pyruvate dehydrogenase. *J. Biol. Chem.* **255**, 7556–7562.

177. CATTELL, K. J., LINDOP, C. R., KNIGHT, I. G. and BEECHEY, R. B. (1971) The identification of the site of action of *NN'*-dicyclohexylcarbodiimide as a proteolipid in mitochondrial mem-branes. *Biochem. J.* **125**, 169–177.

178. CAWTHON, R. M. and BREAKEFIELD, X. O. (1979) Differences in A and B forms of monoamine oxidase revealed by limited proteolysis and peptide mapping. *Nature* **281**, 692–694.

179. CAWTHON, R. M., PINTAR, J. E., HASELTINE, F. P. and BRAKEFIELD, X. O. (1981) Difference in the structure of A and B forms of human monoamine oxidase. *J. Neurochem.* **37**, 363–372.

180. CELADA, F. and ZABIN, I. (1979) A dimer–dimer binding region in β-galactosidase. *Biochem-istry* **18**, 404–406.

181. CHAIMANEE, P. and YUTHAVONG, Y. (1977) Binding of haemoglobin to spectrin of human erythrocytes. *FEBS Lett.* **78**, 119–123.

182. CHALEFF, R. S. (1974) The inducible quinate-shikimate catabolic pathway in *Neurospora crassa:* Genetic organization. *J. Gen. Microbiol.* **81**, 337–355.

183. CHALEFF, R. S. (1974) Inducible quinate-shikimate catabolic pathway in *Neurospora crassa*. Induction and regulation of enzyme synthesis. *J. Gen. Microbiol.* **81**, 357–372.

184. CHAMBERLIN, M. J. (1974) Bacterial DNA-dependent RNA polymerase. In *The Enzymes*. Vol. 10, pp. 333–374. (BOYER, P. D., ed.), Academic Press, New York.

185. CHAMBON, P. (1974) Eucaryotic RNA polymerases. In *The Enzymes*. Vol. 10, pp. 261–331. (BOYER, P. D., ed.), Academic Press, New York.

186. CHAMBON, P. (1975) Eukaryotic nuclear RNA polymerases. *Annu. Rev. Biochem.* **44**, 613–638.

187. CHAMPOUX, J. J. and DULBECCO, R. (1972) An activity from mammalian cells that untwists superhelical DNA:A possible swivel for DNA replication. *Proc. Natl. Acad. Sci. U.S.A.* **69**, 143–146.

188. CHAN, W. W.-C. (1970) Matrix-bound protein subunits. *Biochem. Biophys. Res. Commun.* **41**, 1198–1204.

189. CHAN, W. W.-C. (1975) Subunit interactions in aspartate transcarbamylase. *J. Biol. Chem.* **250**, 661–667.

190. CHAN, W. W.-C. (1975) Subunit interactions in aspartate transcarbamylase. A model for the allosteric mechanism. *J. Biol. Chem.* **250**, 668–674.

191. CHAN, W. W.-C. (1976) Some experimental approaches for studying subunit interactions in enzymes. *Canad. J. Biochem.* **54**, 521–528.

192. CHAN, W. W.-C. (1978) On the mechanism of assembly of the aspartate transcarbamoylase from *Escherichia coli*. *Eur. J. Biochem.* **90**, 271–281.

193. CHAN, W. W.-C. and ENNS, C. A. (1978) Structure and function of aspartate transcarbamoylase studied using chymotrypsin as a probe. *Can. J. Biochem.* **56**, 654–658.

194. CHAN, W. W.-C. and ENNS, C. A. (1979) Aspartate transcarbamoylase: loss of homotropic but not heterotropic interactions upon modification of the catalytic subunit with a bifunctional reagent. *Can. J. Biochem.* **57**, 798–805.

195. CHAN, W. W.-C., KAISER, C., SALVO, J. M. and LAWFORD, G. R. (1974) Formation of dissociated enzyme subunits by chemical treatment during renaturation. *J. Mol. Biol.* **87**, 847–852.

196. CHAN, W. W.-C. and MORT, J. S. (1973) A complex between the catalytic and regulatory subunits of aspartate transcarbamylase. *J. Biol. Chem.* **248**, 7614–7616.

197. CHAN, W. W.-C., MORT, J. S., CHONG, D. K. K. and MCDONALD, D. M. (1973) Studies on protein subunits. III. Kinetic evidence for the presence of active subunits during the renaturation of muscle aldolase. *J. Biol. Chem.* **248**, 2778–2784.

198. CHAN, W. W.-C. and MOSBACH, K. (1976) Effects of subunit interactions on the activity of lactate dehydrogenase studied in immobilized enzyme systems. *Biochemistry* **15**, 4215–4222.

199. CHAN, W. W.-C., SCHUTT, H. and BRAND, K. (1973) Active subunits of transaldolase bound to Sepharose. *Eur. J. Biochem.* **40**, 533–541.

200. CHAN, W. W.-C. and SHANKS, K. E. (1977) Contribution of subunit interactions to the stability of lactate dehydrogenase. *J. Biol. Chem.* **252**, 6163–6168.

201. CHANCE, B., BARLOW, C., HASELGROVE, J., NAKASE, Y., QUISTORFF, B., MATSCHINSKY, F. and MAYEVSKY, A. (1978) Microheterogeneities of redox states of perfused and intact organs. In *Microenvironments and Metabolic Compartmentation*. pp. 131–148. (SRERE, P. A. and ESTABROOK, R. W., eds), Academic Press, New York–London.

202. CHANCE, B., LEIGH, J. S. and WARING, A. (1977) Structure and function of cytochrome oxidase and its intermediate compounds with oxygen reduction products. In *Structure and Function of Energy-transducing Membranes*. B. B. A. Library, Vol. 14, pp. 1–10. Elsevier, Amsterdam–Oxford–New York.

203. CHANDLER, D. H., WOOLF, C. J. and HEPBURN, R. H. (1978) Gliding edge dislocations in proteins as a mechanism for active ion transport. *Biochem. J.* **169**, 559–565.

204. CHANG, T. W. and GOLDBERG, A. L. (1978) Leucine inhibits oxidation of glucose and pyruvate in skeletal muscle during fasting. *J. Biol. Chem.* **253**, 3696–3701.

205. CHANGEUX, J.-P. (1961) The feedback control mechanism of biosynthetic L-threonine deaminase by L-isoleucine. *Cold Spring Harbor Symp. Quant. Biol.* **26**, 313–318.

206. CHAPPELL, J. B. and CROFTS, A. R. (1965) Gramicidin and ion transport in isolated liver mitochondria. *Biochem. J.* **95**, 393–402.

207. CHENG, S.-C., NARUSE, H. and BRUNNER, E. A. (1978) Effects of sodium thiopental on the tricarboxylic acid cycle metabolism in mouse brain: CO_2 fixation and metabolic compartmentation. *J. Neurochem.* **30**, 1591–1593.

208. CHEREDNIKOVA, T. V., MURONETZ, V. I. and NAGRADOVA, N. K. (1980) Study of subunit interactions in immobilized D-glyceraldehyde-3-phosphate dehydrogenase. *Biochim. Biophys. Acta* **613**, 292–308.

209. CHERRY, R. J. (1979) Rotational and lateral diffusion of membrane proteins. *Biochim. Biophys. Acta* **559**, 289–327.

209a. CHEUNG, W. Y. (1970) Cyclic 3′,5′-nucleotide phospodiesterase. Demonstration of an activator. *Biochem. Res. Commun.* **38**, 533–538.

209b. CHEUNG, W. Y. (1980) Calmodulin plays a pivotal role in cellular regulation. *Science.* **207**, 19–27.

210. CHILLAR, R. K. and BEUTLER, E. (1976) Explanation for the apparent lack of ouabain inhibition of pyruvate production in hemolysates: The "backward" PGK Reaction. *Blood* **47**, 507–512.

211. CHILSON, O. P., COSTELLO, L. A. and KAPLAN, N. O. (1965) Studies on the mechanism of hybridization of lactic dehydrogenases *in vitro*. *Biochemistry* **4**, 271–281.

212. CHOCK, P. B., RHEE, S. G. and STADTMAN, E. R. (1980) Interconvertible enzyme cascades in cellular regulation. *Annu. Rev. Biochem.* **49**, 813–843.

213. CHOTHIA, C. (1974) Hydrophobic bonding and accessible surface area in proteins. *Nature* **248**, 338–339.

214. CHOTHIA, C. (1975) Structural invariants in protein folding. *Nature* **254**, 304–308.

215. CHOTHIA, C. and JANIN, J. (1975) Principles of protein-protein recognition. *Nature* **256**, 705–708.

216. CHOU, A. C. and WILSON, J. E. (1972) Purification and properties of rat brain hexokinase. *Arch. Biochem. Biophys.* **151**, 48–55.

217. CHRISTEN, P., COGOLI-GREUTER, M., HEALY, M. J. and LUBINI, D. (1976) Specific irreversible inhibition of enzymes concomitant to the oxidation of carbanionic enzyme–substrate intermediates by hexacyanoferrate (III.). *Eur. J. Biochem.* **63**, 223–231.

218. CHURCH, W. R., RAWITCH, A. B. and EBNER, K. E. (1981) The fluorescein-mediated interaction of bovine serum albumin with fluorescent derivatives of prolactin and other polypeptides in polarization of fluorescence based assays. *Arch. Biochem. Biophys.* **206**, 285–290.

219. CHURCHICH, J. E. and LEE, Y.-H. (1976) Nanosecond emission anisotropy of interacting enzymes aspartate aminotransferase-glutamate dehydrogenase. *Biochem. Biophys. Res. Commun.* **68**, 409–416.

220. CITRI, N. (1973) Conformational adaptability in enzymes. *Adv. Enzymol.* **37**, 397–648.

221. CLARKE, F. M. and MASTERS, C. J. (1973) Multi-enzyme aggregates: New evidence for an association of glycolytic components. *Biochim. Biophys. Acta* **327**, 223–226.

222. CLARKE, F. M. and MASTERS, C. J. (1974) On the association of glycolytic components in skeletal muscle extracts. *Biochim. Biophys. Acta* **358**, 193–207.

223. CLARKE, F. M. and MASTERS, C. J. (1975) On the association of glycolytic enzymes with structural proteins of skeletal muscle. *Biochim. Biophys. Acta* **381**, 37–46.

224. CLARKE, F. M. and MASTERS, C. J. (1976) Interactions between muscle proteins and glycolytic enzymes. *Int. J. Biochem.* **7**, 359–365.

225. CLARKE, F. M., MASTERS, C. J. and WINZOR, D. J. (1974) Interaction of aldolase with the troponin–tropomyosin complex of bovine muscle. *Biochem. J.* **139**, 785–788.

226. CLARKE, F. M., SHAW, F. D. and MORTON, D. F. (1980) Effect of electrical stimulation post mortem of bovine muscle on the binding of glycolytic enzymes. *Biochem. J.* **186**, 105–109.

227. COBIANCHI, F., RIVA, S., MASTROMEI, G., SPADARI, S., PEDRALI-NOY, G. and FALASCHI, A. (1978) Enhancement of the rate of DNA polymerase-α activity on duplex DNA by a DNA-binding protein and a DNA-dependent ATPase of mammalian cells. *Cold Spring Harbor Symp. Quant. Biol.* **43**, 639–647.

228. COHEN, C. (1979) Cell architecture and morphogenesis: I. The cytoskeletal proteins. *Trends Biochem. Sci.* **4**, 73–77.

229. COHEN, C. (1979) Cell architecture and morphogenesis: II. Examples of embryology. *Trends Biochem. Sci.* **4**, 97–101.

230. COHEN, G. N. (1969) The aspartokinases and homoserine dehydrogenases of *Escherichia coli.* *Curr. Top. Cell. Regul.* **1**, 183–231.

231. COHEN, G. N. and DAUTRYVA, A. (1980) The aspartokinases-homoserine dehydrogenases of *Escherichia coli.* In *Multifunctional Proteins.* pp. 49–121. (BISSWANGER, H. and SCHMINKEOTT, E., eds), Wiley, New York.

232. COHEN, N. D., BEEGEN, H., UTTER, M. F. and WRIGLEY, N. G. (1979) A re-examination of the electron microscopic appearance of pyruvate carboxylase from chicken liver. *J. Biol. Chem.* **254**, 1740–1747.

233. COHEN, N. D., UTTER, M. F., WRIGLEY, N. G. and BARRETT, A. N. (1979) Quaternary structure of yeast pyruvate carboxylase: Biochemical and electron microscopic studies. *Biochemistry* **18**, 2197–2203.

234. COHEN, R. and MIRE, M. (1971) Analytical-band centrifugation of an active enzyme-substrate complex. 1. Principle and practice of the centrifugation. *Eur. J. Biochem.* **23**, 267–275.

235. COHEN, R. J. and BENEDEK, G. B. (1976) The functional relationship between polymerization and catalytic activity of beef liver glutamate dehydrogenase. I. Theory. *J. Mol. Biol.* **108**, 151–178.

236. COHEN, R. J. and BENEDEK, G. B. (1979) The functional relationship between the polymerization and catalytic activity of beef liver glutamate dehydrogenase. III. Analysis of Thusius' critique. *J. Mol. Biol.* **129**, 37–44.

237. COHEN, R. J., JEDZINIAK, J. A. and BENEDEK, G. B. (1976) The functional relationship between polymerization and catalytic activity of beef liver glutamate dehydrogenase. *J. Mol. Biol.* **108**, 179–199.

238. COHLBERG, J. A., PIGIET, V. P. and SCHACHMAN, H. K. (1972) Structure and arrangement of the regulatory subunits in aspartate transcarbamylase. *Biochemistry* **11**, 3396–3411.

239. COLBEAU, A. and MAROUX, S. (1978) Integration of alkaline phosphatase in the intestinal brush border membrane. *Biochim. Biophys. Acta* **511**, 39–51.

240. COHN, W., KIRSCHNER, K. and PAUL, C. (1979) N-(5-Phosphoribosyl)-anthranilate isomerase-indoleglycerolphosphate synthase. 2. Fast-reaction studies show that a fluorescent substrate analogue binds independently to two different sites. *Biochemistry* **18**, 5953–5959.

241. COLBEAU, A., NACHBAUR, J. and VIGNAIS, P. M. (1971) Enzymic characterization and lipid composition of rat liver subcellular membranes. *Biochim. Biophys. Acta* **249**, 462–492.

242. COLEMAN, P. F., SUTTLE, D. P. and STARK, G. R. (1977) Purification from hamster cells of the multifunctional protein that initiates de novo synthesis of pyrimidine nucleotides. *J. Biol. Chem.* **252**, 6379–6385.

243. COLLINS, J. H. and REED, L. J. (1977) Acyl group and electron pair relay system: A network of interacting lipoyl moieties in the pyruvate and α-ketoglutarate dehydrogenase complexes from *Escherichia coli. Proc. Natl. Acad. Sci. U.S.A.* **74**, 4223–4227.

244. CONSTANTINIDES, S. M. and DEAL, W. C., JR. (1969) Reversible dissociation of tetrameric rabbit muscle glyceraldehyde-3-phosphate dehydrogenase into dimers or monomers by adenosine triphosphate. *J. Biol. Chem.* **244**, 5695–5702.

245. CONWAY, A. and KOSHLAND, D. E. JR. (1968) Negative cooperativity in enzyme action. The binding of diphosphopyridine nucleotide to glyceraldehyde 3-phosphate dehydrogenase. *Biochemistry* **7**, 4011–4023.

246. CONWAY, E. J. (1955) Evidence for a redox pump in the active transport of cations. *Int. Rev. Cytol.* **4**, 377–396.

247. COOK, R. A. and KOSHLAND, D. E., JR. (1970) Positive and negative cooperativity in yeast glyceraldehyde 3-phosphate dehydrogenase. *Biochemistry* **9**, 3337–3342.

248. COOK, R. A. and MILNE, J. A. (1977) An effect of enzyme and ligand concentration on the state of aggregation of aspartate transcarbamylase of *E. coli:* I. The binding of CTP and ATP to the enzyme. *Can. J. Biochem.* **55**, 346–358.

249. COOPER, K. D., SHUKLA, J. B. and RENNERT, O. M. (1978) Polyamine compartmentalization in various human disease states. *Clin. Chim. Acta* **82**, 1–7.

250. COOPER, R. H., RANDLE, P. J. and DENTON, R. M. (1974) Regulation of heart muscle pyruvate dehydrogenase kinase. *Biochem. J.* **143**, 625–641.

251. CORNISH-BOWDEN, A. (1975) The physiological significance of negative co-operativity. *J. Theor. Biol.* **51**, 233–235.

252. CORNISH-BOWDEN, A. and KOSHLAND, D. E., JR. (1970) A general method for the quantitative determination of saturation curves for multisubunit proteins. *Biochemistry* **9**, 3325–3336.

253. CORNISH-BOWDEN, A. and KOSHLAND, D. E., JR. (1970) The influence of binding domains on the nature of subunit interactions in oligomeric proteins. Application to unusual kinetic and binding patterns. *J. Biol. Chem.* **245**, 6241–6250.

254. CORNISH-BOWDEN, A. and KOSHLAND, D. E., JR. (1971) The quaternary structure of proteins composed of identical subunits. *J. Biol. Chem.* **246**, 3092–3102.

255. COZZARELLI, N. R. (1980) DNA gyrase and the supercoiling of DNA. *Science* **207**, 953–960.

256. CRABTREE, B. and NEWSHOLME, E. A. (1978) Sensitivity of a near-equilibrium reaction in a metabolic pathway to changes in substrate concentration. *Eur. J. Biochem.* **89**, 19–22.

257. CRAMER, C. L. and DAVIS, R. H. (1979) Screening for amino acid pool mutants of *Neurospora* and yeasts: Replica-printing technique. *J. Bacteriol.* **137**, 1437–1438.

258. CRANE, R. K. (1966) Enzymes and malabsorption: A concept of brush-border-membrane disease. *Gastroenterology* **50**, 254–262.

259. CRANE, R. K. (1980) Brush-border-membrane disease. *Biochem. Soc. Trans.* **8**, 688–690.

260. CRANE, R. K. and SOLS, A. (1953) The association of hexokinase with particulate fractions of brain and other tissue homogenates. *J. Biol. Chem.* **203**, 273–292.

261. CRAWFORD, I. P. (1975) Gene rearrangements in the evolution of the tryptophan synthetic pathway. *Bacteriol. Rev.* **39**, 87–120.

262. CRAWFORD, I. P., DECASTEL, M. and GOLDBERG, M. E. (1978) Assignment of the ends of the β-chain of *E. coli* tryptophan synthase to the F_1 and F_2 domains. *Biochem. Biophys. Res. Commun.* **85**, 309–316.

263. CRAWFORD, I. P. and YANOFSKY, C. (1958) On the separation of the tryptophan synthetase of *Escherichia coli* into two protein components. *Proc. Natl. Acad. Sci. U.S.A.* **44**, 1161–1170.

264. CREIGHTON, T. E. (1970) A steady-state kinetic investigation of the reaction mechanism of the tryptophan synthetase of *Escherichia coli*. *Eur. J. Biochem.* **13**, 1–10.

265. CREIGHTON, T. E. (1977) Conformational restrictions on the pathway of folding and unfolding of the pancreatic trypsin inhibitor. *J. Mol. Biol.* **113**, 275–293.

266. CREIGHTON, T. E. and YANOFSKY, C. (1966) Association of the α and $β_2$ subunits of the tryptophan synthetase of *Escherichia coli*. *J. Biol. Chem.* **241**, 980–990.

267. CREIGHTON, T. E. and YANOFSKY, C. (1970) Chorismate to tryptophan *(Escherichia coli)*-anthranilate synthetase, PR transferase, PRA isomerase, In–GP synthetase, tryptophan synthetase. *Methods Enzymol.* **17a**, 365–380.

268. CRISS, W. E., SAPICO, V. and LITWACK, G. (1970) Rat liver adenosine triphosphate:adenosine monophosphate phosphotransferase activity. I. Purification and physical and kinetic characterization of adenylate kinase III. *J. Biol. Chem.* **245**, 6346–6351.

269. CROSS, R. L. (1981) The mechanism and regulation of ATP synthesis by F_1-ATPases. *Annu. Rev. Biochem.* **50**, 681–714.

270. CROWDER, S. E. and RAGAN, C. I. (1977) Effects of proteolytic digestion by chymotrypsin on the structure and catalytic properties of reduced nicotinamide–adenine dinucleotide–ubiquinone oxidoreductase from bovine heart mitochondria. *Biochem. J.* **165**, 295–301.

271. CSEKE, E. and BOROSS, L. (1970) Factors affecting the reactivity of the activated SH-group of D-glyceraldehyde 3-phosphate dehydrogenase. *Acta Biochim. Biophys. Acad. Sci. Hung.* **5**, 385–397.

272. CSEKE, E., VÁRADI, A., SZABOLCSI, G. and BISZKU, E. (1978) On the molecular sieving property of the human erythrocyte membrane and the localization of some glycolytic enzymes in the cell. *FEBS Lett.* **96**, 15–18.

273. CZECH, M. P. (1977) Molecular basis of insulin action. *Annu. Rev. Biochem.* **46**, 359–384.

274. DAGHER, S. M. and HULTIN, H. O. (1975) Association of glyceraldehyde-3-phosphate dehydrogenase with the particulate fraction of chicken skeletal muscle. *Eur. J. Biochem.* **55**, 185–192.

275. DALY, J. W. (1977) *The Role of Cyclic Nucleotides in the Nervous System.* Plenum Press, New York.

276. DALZIEL, K. (1973) Kinetics of control enzymes. *Symp. Soc. Exp. Biol.* **27**, 21–48.

277. DALZIEL, K., McFERRAN, N. V. and WONACOTT, A. J. (1981) Glyceraldehyde-3-phosphate dehydrogenase. *Phil. Trans. R. Soc. Lond. B* **293**, 105–118.

278. DAMJANOVICH, S. and SOMOGYI, B. (1973) A molecular enzyme model based on oriented energy transfer. *J. Theor. Biol.* **41**, 567–569.

279. DANIELSEN, E. M., SJÖSTRÖM, H., NORÉN, O. and DABELSTEEN, E. (1977) Immunoelectrophoretic studies on pig intestinal brush border proteins. *Biochim. Biophys. Acta* **494**, 332–342.

280. DANNER, D. J., LEMMON, S. K. and ELSAS, L. J. (1978) Substrate specificity and stabilization by thiamin pyrophosphate of rat liver branched chain α-keto acid dehydrogenase. *Biochem. Med.* **19**, 27–38.

281. DANSON, M. J., FERSHT, A. R. and PERHAM, R. N. (1978) Rapid intramolecular coupling of active sites in the pyruvate dehydrogenase complex of *Escherichia coli*: Mechanism for rate enhancement in a multimeric structure. *Proc. Natl. Acad. Sci. U.S.A.* **75**, 5386–5390.

282. DANSON, M. J., HALE, G., JOHNSON, P., PERHAM, R. N., SMITH, J. and SPRAGG, P. (1979) Molecular weight and symmetry of the pyruvate dehydrogenase multienzyme complex of *Escherichia coli. J. Mol. Biol.* **129**, 603–617.

283. DANSON, M. J., HALE, G. and PERHAM, R. N. (1981) The role of lipoic acid residues in the pyruvate dehydrogenase multienzyme complex of *Escherichia coli. Biochem. J.* **199**, 505–511.

284. DANSON, M. J., HOOPER, E. A. and PERHAM, R. N. (1978) Intramolecular coupling of active sites in the pyruvate dehydrogenase multienzyme complex of *Escherichia coli. Biochem. J.* **175**, 193–198.

285. DANSON, M. J. and PERHAM, R. N. (1976) Evidence for two lipoic acid residues per lipoate acetyltransferase chain in the pyruvate dehydrogenase multienzyme complex. *Biochem. J.* **159**, 677–682.

286. DARNALL, D. W. and KLOTZ, I. M. (1975) Subunit constitution of proteins: A table. *Arch. Biochem. Biophys.* **166**, 651–682.

287. DAS GUPTA, U. and RIESKE, J. S. (1973) Identification of a protein component of the antimycin-binding site of the respiratory chain by photoaffinity labeling. *Biochem. Biophys. Res. Commun.* **54**, 1247–1254.

288. DAVID, M. M., SCHEJTER, A., DANIEL, E., AR, A. and BEN-SHAUL, Y. (1977) Subunit structure of hemoglobin from the clam shrimp Cyzicus. *J. Mol. Biol.* **111**, 211–214.

289. DAVIE, E. W., FUJIKAWA, K., KURACHI, K. and KISIEL, W. (1979) The role of serine proteases in the blood coagulation cascade. *Adv. Enzymol.* **48**, 277–318.

290. DAVIES, G. E. and STARK, G. R. (1970) Use of dimethyl suberimidate, a cross-linking reagent, in studying the subunit structure of oligomeric proteins. *Proc. Natl. Acad. Sci. U.S.A.* **66**, 651–656.

291. DAVIS, R. H. (1967) Channeling in *Neurospora* metabolism. In *Organizational Biosynthesis.* pp. 303–322. (VOGEL, H. J., LAMPEN, J. D. and BRYSON, V., eds), Academic Press, New York.

292. DAVIS, R. H. (1980) Arginine metabolism in *Neurospora:* The role of an amino acid compartment. In *Cell Compartmentation and Metabolic Channeling.* pp. 239–243. (NOVER, L., LYNEN, F. and MOTHES, K., eds) VEB Gustav Fischer Verlag, Jena, Elsevier/North-Holland Biomedical Press, Amsterdam–New York–Oxford.

293. DAWSON, R. M. C. and CLARKE, N. G. (1973) A comparison of D-inositol 1:2-cyclic phosphate 2-phosphohydrolase with other phosphodiesterases of kidney. *Biochem. J.* **134,** 59–67.

294. DE, B. K. and KIRTLEY, M. E. (1977) Interaction of phosphoglycerate kinase with human erythrocyte membranes. *J. Biol. Chem.* **252,** 6715–6720.

295. DEAL, W. C., RUTTER, W. J. and VAN HOLDE, K. E. (1963) Reversible dissociation of aldolase into unfolded subunits. *Biochemistry* **2,** 246–251.

296. DE DUVE, C. (1964) Principles of tissue fractionation. *J. Theor. Biol.* **6,** 33–59.

297. DEGANI, C. and DEGANI, Y. (1980) Further evidence for nonsymmetric subunit association and intersubunit cooperativity in creatine kinase. Subunit-selective modifications by 2,4-dinitrophenylthiocyanate. *J. Biol. Chem.* **255,** 8221–8228.

298. DEGANI, Y. and DEGANI, C. (1980) Enzymes with asymmetrically arranged subunits. *Trends Biochem. Sci.* **5,** 337–341.

299. DELANGE, R. J. and HUANG, T.-S. (1971) Egg white avidin: III. Sequence of the 78-residue middle cyanogen bromide peptide. Complete amino acid sequence of the protein subunit. *J. Biol. Chem.* **246,** 698–709.

300. DELONG, R. and POPLIN, L. (1977) On the etiology of ageing. *J. Theor. Biol.* **67,** 111–120.

301. DEMOSS, J. A. (1962) Studies on the mechanism of the tryptophan synthetase reaction. *Biochim. Biophys. Acta* **62,** 279–293.

302. DERIEL, J. K. and PAULUS, H. (1978) Subunit dissociation in the allosteric regulation of glycerol kinase from *Escherichia coli.* 1. Kinetic evidence. *Biochemistry* **17,** 5134–5140.

303. DERIEL, J. K. and PAULUS, H. (1978) Subunit dissociation in the allosteric regulation of glycerol kinase from *Escherichia coli.* 2. Physical evidence. *Biochemistry* **17,** 5141–5145.

304. DERIEL, J. K. and PAULUS, H. (1978) Subunit dissociation in the allosteric regulation of glycerol kinase from *Escherichia coli.* 3. Role in desensitization. *Biochemistry* **17,** 5146–5150.

305. DEROSIER, D. J. and CLIVER, R. M. (1972) A low resolution electron-density map of lipoyl transsuccinylase, the core of the α-ketoglutarate dehydrogenase complex. *Cold Spring Harb. Symp. Quant. Biol.* **36,** 199–203.

306. DEROSIER, D. J., OLIVER, R. M. and REED, L. J. (1971) Crystallization and preliminary structural analysis of dihydrolipoyl transsuccinylase, the core of the 2-oxoglutarate dehydrogenase complex. *Proc. Natl. Acad. Sci. U.S.A.* **68,** 1135–1137.

307. DETWILER, M. and KIRSCHNER, K. (1979) Tryptophan synthase from *Saccharomyces cerevisiae* is a dimer of two polypeptide chains of $M_r \approx 76,000$ each. *Eur. J. Biochem.* **102,** 159–165.

308. DEVIJLDER, J. J. M. and SLATER, E. C. (1968) The reaction between NAD^+ and rabbit-muscle glyceraldehydephosphate dehydrogenase. *Biochem. Biophys. Acta* **167,** 23–34.

309. DICKERSON, R. E. (1972) The structure and history of an ancient protein. *Sci. Amer.* **226** (4), 58–72.

310. DICKINSON, F. M. (1972) Role of the essential thiol groups of yeast alcohol dehydrogenase. *Biochem. J.* **126,** 133–138.

311. DICKINSON, F. M. and MONGER, G. P. (1973) A study of the kinetics and mechanism of yeast alcohol dehydrogenase with a variety of substrates. *Biochem. J.* **131,** 261–270.

312. DIOPOH, J. and OLOMUCKI, M. (1977) New protein reagents. N-(4-chloromercuriphenyl)-4-chloro-3,5-dinitrobenzamide and its use as a probe of the quaternary structure of yeast alcohol dehydrogenase. *Eur. J. Biochem.* **75,** 441–444.

313. DI RIENZO, J. M., NAKAMURA, K. and INOUYE, M. (1978) The outer membrane proteins of

Gram-negative bacteria: biosynthesis, assembly, and functions. *Annu. Rev. Biochem.* **47**, 481–532.

314. DIXON, M. and WEBB, E. C. (1979) *Enzymes*. Third edition. Longman Group Ltd., London.

315. DOCKTER, M. E., STEINEMANN, A. and SCHATZ, G. (1978) Mapping of yeast cytochrome c oxidase by fluorescence resonance energy transfer. Distances between subunit II, heme a, and cytochrome c bound to subunit III. *J. Biol. Chem.* **253**, 311–317.

316. DOI, R. H. (1977) Role of ribonucleic acid polymerase in gene selection in procaryotes. *Bacteriol. Rev.* **41**, 568–594.

317. DOMBRÁDI, V. (1981) Structural aspects of the catalytic and regulatory function of glycogen phosphorylase. *Int. J. Biochem.* **13**, 125–139.

318. DOMBRÁDI, V., HAJDU, J., BOT, G. and FRIEDRICH, P. (1980) Structural changes in glycogen phosphorylase as revealed by crosslinking with bifunctional diimidates: Phospho-dephospho hybrid and phosphorylase *a*. *Biochemistry* **19**, 2295–2299.

319. DOWNIE, J. A., GIBSON, F. and COX, G. B. (1979) Membrane adenosine triphosphatases of prokaryotic cells. *Annu. Rev. Biochem.* **48**, 103–131.

320. DRICKAMER, L. K. (1976) Fragmentation of the 95,000-dalton transmembrane polypeptide in human erythrocyte membranes. *J. Biol. Chem.* **251**, 5115–5123.

321. DRICKAMER, L. K. (1977) Fragmentation of the band 3 polypeptide from human erythrocyte membranes. *J. Biol. Chem.* **252**, 6909–6917.

322. DRICKAMER, L. K. (1978) Orientation of the band 3 polypeptide from human erythrocyte membranes. *J. Biol. Chem.* **253**, 7242–7248.

323. DUBIN, S. B. and CANNELL, D. S. (1975) The effect of succinate on the translational diffusion coefficient of aspartate transcarbamylase. *Biochemistry* **14**, 192–195.

324. DUCHESNE, J. (1977) A unifying biochemical theory of cancer, senescence and maximal life span. *J. Theor. Biol.* **66**, 137–145.

325. DUNTZE, W. and MANNEY, T. R. (1968) Two mechanisms of allelic complementation among tryptophan synthetase mutants of *Saccharomyces cerevisiae*. *J. Bacteriol.* **96**, 2085–2093.

326. DURAND, R., BRIAND, Y., TOURAILLE, S. and ALZIARI, S. (1981) Molecular approaches to phosphate transport in mitochondria. *Trends Biochem. Sci.* **6**, 211–214.

327. DURCHSCHLAG, H. (1975) X-ray small-angle studies of the pyruvate dehydrogenase core complex from *Escherichia coli* K-12. I. Overall structure of the core complex. *Biophys. Struct. Mechanism* **1**, 153–168.

328. DÜRR, M., URECH, K., BOLLER, TH., WIEMKEN, A., SCHWENKE, J. and NAGY, M. (1979) Sequestration of arginine by polyphosphate vacuoles in yeast *(Saccharomyces cerevisiae)*. *Arch. Microbiol.* **121**, 169–175.

329. EASTERBY, J. S. (1973) Coupled enzyme assays: a general expression for the transient. *Biochim. Biophys. Acta* **293**, 552–558.

330. EBY, D. and KIRTLEY, M. E. (1979) Isolation and characterization of glyceraldehyde-3-phosphate dehydrogenase from human erythrocyte membranes. *Arch. Biochem. Biophys.* **198**, 608–613.

331. EDDY, D. and HARMAN, D. (1977) Free radical theory of ageing: Effect of age, sex and dietary precursors on rat brain docosahexanoic acid. *J. Am. Geriat. Soc.* **25**, 220–229.

332. EDWARDS, Y. H., EDWARDS, P. A. and HOPKINSON, D. A. (1973) A trimeric structure for mammalian purine nucleoside phosphorylase. *FEBS Lett.* **32**, 235–237.

333. EGAN, R., MICHEL, H. O., SCHLUETER, R. and JANDORF, B. J. (1975) Physicochemical investigation of the chymotrypsin. II. On the mechanism of dimerization of chymotrypsin. *Arch. Biochem. Biophys.* **66**, 366–373.

334. EHMANN, J. D. and HULTIN, H. O. (1973) Substrate inhibition of soluble and bound lactate dehydrogenase (isoenzyme 5). *Arch. Biochem. Biophys.* **154**, 471–475.

335. EISENBERG, H., JOSEPHS, R. and REISLER, E. (1976) Bovine liver glutamate dehydrogenase. *Adv. Protein Chem.* **30**, 101–181.

336. ELEY, M. H., NAMIHIRA, G., HAMILTON, L., MUNK, P. and REED, L. J. (1972) α-Keto acid de-

hydrogenase complexes. XVIII. Subunit composition of the *Escherichia coli* pyruvate dehydrogenase complex. *Arch. Biochem. Biophys*. **152**, 655–669.

337. ELLENRIEDER, G., KIRSCHNER, K. and SCHUSTER, I. (1972) The binding of oxidized and reduced nicotinamide adenine-dinucleotide to yeast glyceraldehyde-3-phosphate dehydrogenase. *Eur. J. Biochem*. **26**, 220–236.

338. ELORANTA, T. O., MAENPAA, P. H. and RAINA, A. M. (1976) Synthesis of hepatic polyamines, ribonucleic acid and *S*-adenosylmethionine in normal and oestrogen-treated chicks. *Biochem. J*. **154**, 95–103.

339. ENGEL, P. C. (1973) Evolution of enzyme regulator sites: Evidence for partial gene duplication from amino-acid sequence of bovine glutamate dehydrogenase. *Nature* **241**, 118–120.

340. ENGEL, P. C. (1973) Sequence homologies among pyridine nucleotide-linked dehydrogenases: Possible partial gene duplications in glyceraldehyde-3-phosphate dehydrogenase. *FEBS Lett*. **33**, 151–153.

341. ENGELMAN, D. M. and MOORE, P. B. (1972) A new method for the determination of biological quaternary structure by neutron scattering. *Proc. Natl. Acad. Sci. U.S.A*. **69**, 1997–1999.

342. ENGELMAN, D. M., MOORE, P. B. and SCHOENBORN, B. P. (1975) Neutron scattering measurements of separation and shape of proteins in 30S ribosomal subunit of *Escherichia coli:* S2-S5, S5-S8, S3-S7. *Proc. Natl. Acad. Sci. U.S.A*. **72**, 3888–3892.

343. ENNS, C. A. and CHAN, W. W.-C. (1978) Stabilization of the relaxed state of aspartate transcarbamoylase by modification with a bifunctional reagent. *J. Biol. Chem*. **253**, 2511–2513.

344. ERDŐS, E. G. (1977) The angiotensin I converting enzyme. *Fed. Proc*. **36**, 1760–1765.

345. ERECINSKA, M. (1977) A new photoaffinity-labeled derivative of mitochondrial cytochrome *c*. *Biochem. Biophys. Res. Commun*. **76**, 495–501.

346. ERECINSKA, M., BLASIE, J. K. and WILSON, D. F. (1977) Orientation of the hemes of cytochrome c oxidase and cytochrome c in mitochondria. *FEBS Lett*. **76**, 235–239.

347. ERECINSKA, M., OSHINO, R. and WILSON, D. F. (1980) Binding of cytochrome *c* to cytochrome *c*-oxidase in intact mitochondria. A study with radioactive photoaffinity-labeled cytochrome *c*. *Biochem. Biophys. Res. Commun*. **92**, 743–748.

348. ERECINSKA, M. and WILSON, D. F. (1978) Cytochrome c oxidase: A synopsis. *Arch. Biochem. Biophys*. **188**, 1–14.

349. ERNSTER, L., SANDRI, G., HUNDAL, T., CARLSSON, C. and NORDENBRAND, K. (1977) Inhibitors as probes of mitochondrial adenosine triphosphatase. In *Structure and Function of Energy-transducing Membranes*. B.B.A. Library, Vol. 14, pp. 209–222, Elsevier, Amsterdam–Oxford–New York.

350. EVANS, P. R. and HUDSON, P. J. (1979) Structure and control of phosphofructokinase from *Bacillus strearothermophilus*. *Nature* **279**, 500–504.

351. EXPERT-BEZANÇON, A., BARRITAULT, D., MILET, M., GUÉRIN, M.-F. and HAYES, D. H. (1977) Identification of neighbouring proteins in *Escherichia coli* 30S ribosome subunits. *J. Mol. Biol*. **112**, 603–629.

352. FAEDER, E. J. and HAMMES, G. G. (1970) Kinetic studies of tryptophan synthetase. Interaction of substrates with the B subunit. *Biochemistry* **9**, 4043–4049.

353. FAEDER, E. J. and HAMMES, G. G. (1971) Kinetic studies of tryptophan synthetase. Interaction of L-serine, indole, and tryptophan with the native enzyme. *Biochemistry* **10**, 1041–1045.

354. FAHIEN, L. A., HSU, S. L. and KMIOTEK, E. (1977) Effect of aspartate on complexes between glutamate dehydrogenase and various aminotransferases. *J. Biol. Chem*. **252**, 1250–1256.

355. FAHIEN, L. A., KMIOTEK, E. and KAJIWARA, K. (1981) Effect of malate on glutamate dehydrogenase and complexes between glutamate dehydrogenase and mitochondrial aspartate aminotransferase. *Arch. Biochem. Biophys*. **209**, 143–151.

356. FAHIEN, L. A., KMIOTEK, E. and SMITH, L. (1979) Glutamate dehydrogenase-malate dehydrogenase complex. *Arch. Biochem. Biophys*. **192**, 33–46.

357. FAHIEN, L. A., RUOHO, A. E. and KMIOTEK, E. (1978) A study of glutamate dehydrogenase. Aminotransferase complexes with a bifunctional imidate. *J. Biol. Chem.* **253**, 5745–5751.

358. FAHIEN, L. A. and SMITH, S. E. (1974) The enzyme–enzyme complex of transaminase and glutamate dehydrogenase. *J. Biol. Chem.* **249**, 2696–2703.

359. FAHIEN, L. A. and STRMECKI, M. (1969) Studies of gluconeogenic mitochondrial enzymes. II. The conversion of glutamate to α-ketoglutarate by bovine liver mitochondrial glutamate dehydrogenase and glutamate-oxaloacetate transaminase. *Arch. Biochem. Biophys.* **130**, 456–467.

360. FAHIEN, L. A. and STRMECKI, M. (1969) Studies of gluconeogenic mitochondrial enzymes. III. The conversion of α-ketoglutarate to glutamate by bovine liver mitochondrial glutamate dehydrogenase and glutamate-oxaloacetate transaminase. *Arch. Biochem. Biophys.* **130**, 468–477.

361. FAHIEN, L. A. and STRMECKI, M. (1969) Studies of gluconeogenic mitochondrial enzymes. IV. The conversion of oxaloacetate to fumarate by bovine liver mitochondrial malate dehydrogenase and fumarase. *Arch. Biochem. Biophys.* **130**, 478–487.

362. FAHIEN, L. A. and VAN ENGELEN, D. L. (1976) A study of the aminotransferase-glutamate dehydrogenase complex with anilinonaphthalene sulfonate. *Arch. Biochem. Biophys.* **176**, 298–305.

363. FAIRBANKS, G., STECK, T. L. and WALLACH, D. F. H. (1971) Electrophoretic analysis of the major polypeptides of the human erythrocyte membrane. *Biochemistry* **10**, 2606–2617.

364. FEIG, S. A., SEGEL, G. B., SHOHET, S. B. and NATHAN, D. G. (1972) Energy metabolism in human erythrocytes. II. Effects of glucose depletion. *J. Clin. Invest.* **51**, 1547–1554.

365. FEINSTEIN, M. B., FERNANDEZ, S. M. and SHA'AFI, R. I. (1975) Fluidity of natural membranes and phosphatidylserine and ganglioside dispersions. Effects of local anesthetics, cholesterol and protein. *Biochim. Biophys. Acta* **413**, 354–370.

366. FELDHERR, C. M. (1965) The effect of the electron-opaque pore material on exchanges through the nuclear annuli. *J. Cell Biol.* **25**, 43–53.

367. FELDMANN, K., ZEISEL, H. and HELMREICH, E. (1972) Interactions between native and chemically modified subunits of matrix-bound glycogen phosphorylase. *Proc. Natl. Acad. Sci. U.S.A.* **69**, 2278–2282.

368. FELGNER, P. L., MESSER, J. L. and WILSON, J. E. (1979) Purification of a hexokinase-binding protein from the outer mitochondrial membrane. *J. Biol. Chem.* **254**, 4946–4949.

369. FELGNER, P. L. and WILSON, J. E. (1977) Dansylation of tyrosine: Hindrance by *N*-ethylmorpholine and photodegradation of *O*-dansylated derivatives. *Anal. Biochem.* **80**, 601–611.

370. FELGNER, P. L. and WILSON, J. E. (1977) Effect of neutral salts on the interaction of rat brain hexokinase with the outer mitochondrial membrane. *Arch. Biochem. Biophys.* **182**, 282–294.

371. FELLMANN, P., ANDERSEN, J., DEVAUX, P. F., MAIRE, M. and BIENVENUE, A. (1980) Photoaffinity spin-labeling of the Ca^{2+} ATPase in sarcoplasmic reticulum: Evidence for oligomeric structure. *Biochem. Biophys. Res. Commun.* **95**, 289–295.

372. FERGUSON, S. J., LLOYS, W. J., LYONS, M. H. and RADDA, G. K. (1975) The mitochondrial ATPase. Evidence for a single essential tyrosine residue. *Eur. J. Biochem.* **54**, 117–126.

373. FERSHT, A. R. (1971) Conformational equilibria and the salt bridge in chymotrypsin. *Cold Spring Harbor Symp. Quant. Biol.* **36**, 71–73.

374. FILLINGAME, R. H. (1980) The proton-translocating pumps of oxidative phosphorylation. *Annu. Rev. Biochem.* **49**, 1079–1113.

375. FISHBEIN, W. N., ENGLER, W. F., GRIFFIN, J. L., SCURZI, W. and BAHR, G. F. (1977) Electron microscopy of negatively stained jackbean urease at three levels of quaternary structure, and comparison with hydrodynamic studies. *Eur. J. Biochem.* **73**, 185–190.

376. FISHBEIN, W. N., NAGARJAN, K. and SCURZI, W. (1970) Urease catalysis and structure. VI. Correlation of sedimentation coefficients and electrophoretic mobilities for the polymeric urease isozymes. *J. Biol. Chem.* **245**, 5985–5992.

377. FISHBEIN, W. N., NAGARJAN, K. and SCURZI, W. (1973) Urease catalysis and structure. IX. The half-unit and hemipolymers of jack bean urease. *J. Biol. Chem.* **248**, 7870–7877.

378. FISHER, H. F. (1973) Glutamate dehydrogenase-ligand complexes and their relationship to the mechanism of the reaction. *Adv. Enzymol.* **39**, 369–417.

379. FISHER, H. F., CROSS, D. G. and McGREGOR, L. L. (1965) The independence of the substrate specificity of glutamate dehydrogenase on its state of aggregation. *Biochim. Biophys. Acta* **99**, 165–167.

380. FISHER, S., NAGEL, R. L., BOOKCHIN, R. M., ROTH, E. F. and TELLEZ-NAGEL, I. (1975) The binding of hemoglobin to membranes of normal and sickle erythrocytes. *Biochim. Biophys. Acta* **375**, 422–433.

381. FLETTERICK, R. J. and MADSEN, N. B. (1980) The structures and related functions of phosphorylase *a*. *Annu. Rev. Biochem.* **49**, 31–61.

382. FLETTERICK, R. J., SYGUSCH, J., SEMPLE, M. and MADSEN, N. B. (1976) Structure of glycogen phosphorylase *a* at 3.0 Å resolution and its ligand binding sites at 6 Å. *J. Biol. Chem.* **251**, 6142–6146.

383. FLORKIN, M. (1972) A history of biochemistry. Chapter 13: Biocatalysis and the enzymatic theory of metabolism. In *Comprehensive Biochemistry*. Vol. 30, pp. 265–278. (FLORKIN, M. and STOTZ, E. H., eds), Elsevier, Amsterdam–London–New York.

384. FLURI, R., JACKSON, L. E., LEE, W. E. and CRAWFORD, I. P. (1971) Tryptophan synthetase β_2 subunit. Primary structure of the pyridoxyl peptide from the *Escherichia coli* enzyme. *J. Biol. Chem.* **246**, 6620–6624.

385. FÖLDI, J., SZABOLCSI, G. and FRIEDRICH, P. (1973) Interaction of glycolytic enzymes: Increase of the apparent molecular weight of aldolase in rabbit muscle extract. *Acta Biochim. Biophys. Acad. Sci. Hung.* **8**, 263–265.

386. FONT, B., VIAL, C. and GAUTHERON, D. C. (1975) Intracellular and submitochondrial localization of pig heart hexokinase. *FEBS Lett.* **56**, 24–29.

387. FONYÓ, A. and BESSMAN, S. P. (1966) The action of oligomycin and of para-hydroxymercuribenzoate on mitochondrial respiration stimulated by ADP, arsenate and calcium. *Biochem. Biophys. Res. Commun.* **24**, 61–66.

388. FOOTE, A. M., WINKLER, F. K. and MOODY, M. F. (1981) Crystallization and preliminary X-ray study of the catalytic subunit of aspartate transcarbamylase. *J. Mol. Biol.* **146**, 389–391.

389. FORBUSH B., III, KAPLAN, J. H. and HOFFMAN, J. F. (1978) Characterization of a new photoaffinity derivative of ouabain: Labeling of the large polypeptide and of a proteolipid component of the Na, K-ATPase. *Biochemistry* **17**, 3667–3676.

390. FOSSEL, E. T. and SOLOMON, A. K. (1977) Membrane mediated link between ion transport and metabolism in human red cells. *Biochim. Biophys. Acta* **464**, 82–92.

391. FOSSEL, E. T. and SOLOMON, A. K. (1978) Ouabain-sensitive interaction between human red cell membrane and glycolytic enzyme complex in cytosol. *Biochim. Biophys. Acta* **510**, 99–111.

392. FOSSEL, E. T. and SOLOMON, A. K. (1979) Effect of the sodium/potassium ratio on glyceraldehyde-3-phosphate dehydrogenase interaction with red cell vesicles. *Biochim. Biophys. Acta* **553**, 142–153.

393. FOSTER, D. L. and FILLINGAME, R. H. (1979) Energy-transducing H^+-ATPase of *Escherichia coli*. Purification, reconstitution, and subunit composition. *J. Biol. Chem.* **254**, 8230–8236.

394. FRAUENFELDER, H., PETSKO, G. A. and TSERNOGLOU, D. (1979) Temperature-dependent X-ray diffraction as a probe of protein structural dynamics. *Nature* **280**, 558–563.

395. FREY, T. G., CHAN, S. H. P. and SCHATZ, G. (1978) Structure and orientation of cytochrome c oxidase in crystalline membranes. Studies by electron microscopy and by labeling with subunit-specific antibodies. *J. Biol. Chem.* **253**, 4389–4395.

396. FREY, P. A., IKEDA, B. H., CAVINO, G. R., SPECKHARD, D. C. and WONG, S. S. (1978) *Escherichia coli* pyruvate dehydrogenase complex. Site coupling in electron and acetyl group transfer pathways. *J. Biol. Chem.* **253**, 7234–7241.

397. FRIEDEN, C. (1967) Treatment of enzyme kinetic data. II. The multisite case: Comparison of allosteric models and a possible new mechanism. *J. Biol. Chem.* **242**, 4045–4052.

398. FRIEDEN, C. (1970) Kinetic aspects of regulation of metabolic processes. The hysteretic enzyme concept. *J. Biol. Chem.* **245**, 5788–5799.

399. FRIEDEN, C. (1979) Slow transitions and hysteretic behavior in enzymes. *Annu. Rev. Biochem.* **48**, 471–489.

400. FRIEDEN, C. and COLMAN, R. F. (1967) Glutamate dehydrogenase concentration as a determinant in the effect of purine nucleotides on enzymatic activity. *J. Biol. Chem.* **242**, 1705–1715.

401. FRIEDEN, C., GILBERT, H. R. and BOCK, P. E. (1976) Phosphofructokinase. III. Correlation of the regulatory kinetic and molecular properties of the rabbit muscle enzyme. *J. Biol. Chem.* **251**, 5644–5647.

402. FRIEDMAN, F. K. and BEYCHOK, S. (1979) Probes of subunit assembly and reconstitution pathways in multisubunit proteins. *Ann. Rev. Biochem.* **48**, 217–250.

403. FRIEDMAN, M. (1977) Protein crosslinking. In *Adv. Exp. Med. Biol.*, 86 A, 86 B, Plenum Press, New York–London.

404. FRIEDRICH, P. (1974) Dynamic compartmentation in soluble enzyme systems. *Acta Biochim. Biophys. Acad. Sci. Hung.* **9**, 159–173.

405. FRIEDRICH, P. (1979) On the physiological significance of positive and negative cooperativity in enzymes. *J. Theor. Biol.* **81**, 527–532.

406. FRIEDRICH, P., APRÓ-KOVÁCS, V. A. and SOLTI, M. (1977) Study of metabolite compartmentation in erythrocyte glycolysis. *FEBS Lett.* **84**, 183–186.

407. FRIEDRICH, P., ARÁNYI, P. and NAGY, I. (1972) Studies on the relationship between quaternary structure and enzymatic activity of rabbit muscle aldolase. *Acta Biochim. Biophys. Acad. Sci. Hung.* **7**, 11–19.

408. FRIEDRICH, P., FÖLDI, J. and VÁRADI, K. (1974) Freezing-induced alkylation of SH groups and inactivation of rabbit muscle aldolase. *Acta Biochim. Biophys. Acad. Sci. Hung.* **9**, 1–13.

409. FRIEDRICH, P., HAJDU, J. and BARTHA, F. (1979) Subunit arrangements in oligomeric enzymes. *Proceedings of 12th FEBS Meeting, Dresden 1978.* **52**, 239–248.

410. FRIEDRICH, P., NAGY, I. and ARÁNYI, P. (1973) Effect of ultraviolet irradiation on the quaternary structure of rabbit muscle aldolase. *Acta Biochim. Biophys. Acad. Sci. Hung.* **8**, 1–7.

411. FUJII, T. and SATO, M. (1975) Possible adsorption of aldolase and glyceraldehyde-3-phosphate dehydrogenase into the erythrocyte ghosts prepared under a hypotonic condition. *J. Clin. Chem.* **3**, 453–459.

412. FULLER, S. D., CAPALDI, R. A. and HENDERSON, R. (1979) Structure of cytochrome c oxidase in deoxycholate-derived two-dimensional crystals. *J. Mol. Biol.* **134**, 305–327.

413. FULLER, C. C., REED, L. J., OLIVER, R. M. and HACKERT, M. L. (1979) Crystallization of a dihydrolipoyl transacetylase-dihydrolipoyl dehydrogenase subcomplex and its implications regarding the subunit structure of the pyruvate dehydrogenase complex from *Escherichia coli*. *Biochem. Biophys. Res. Commun.* **90**, 431–438.

414. GAERTNER, F. H. (1972) Purification of two multienzyme complexes in the aromatic/tryptophan pathway of *Neurospora*. *Arch. Biochem. Biophys.* **151**, 277–284.

415. GAERTNER, F. H. (1978) Unique catalytic properties of enzyme clusters. *Trends Biochem. Sci.* **3**, 63–65.

416. GAERTNER, F. H. and COLE, K. W. (1976) The protease problem in *Neurospora*. Structural modification of the arom multienzyme system during its extraction and isolation. *Arch. Biochem. Biophys.* **177**, 566–573.

417. GAERTNER, F. H. and COLE, K. W. (1977) A cluster-gene: evidence for one gene, one polypeptide, five enzymes. *Biochem. Biophys. Res. Commun.* **75**, 259–264.

418. GAERTNER, F. H., ERICSON, M. C. and DEMOSS, J. A. (1970) Catalytic facilitation in vitro by two multienzyme complexes from *Neurospora crassa*. *J. Biol. Chem.* **245**, 595–600.

419. GAERTNER, F. H., VITTO, A., ALLISON, D. E., COLE, K. W. and SPADY, G. E. (1978) In *Limited Proteolysis in Microorganisms*. p. 197. (COHEN, G. N. and HOLZER, H., eds), Natl. Inst. Health, Bethesda, Maryland.

419a. GAFNI, A. (1981) Age-related effects in coenzyme binding patterns of rat muscle glyceraldehyde-3-phosphate dehydrogenase. *Biochemistry* **20**, 6041–6046.

420. GARAVITO, R. M., BERGER, D. and ROSSMANN, M. G. (1977) Molecular asymmetry in an abortive ternary complex of lobster glyceraldehyde-3-phosphate dehydrogenase. *Biochemistry* **16**, 4393–4398.

421. GARDNER, J. D., AURBACH, G. D., SPIEGEL, A. M. and BROWN, E. M. (1976) Receptor function and ion transport in turkey erythrocytes. *Recent Progr. Horm. Res.* **32**, 567–595.

422. GARFINKEL, D. (1962) Computer simulation of steady-state glutamate metabolism in rat brain. *J. Theor. Biol.* **3**, 412–422.

423. GARFINKEL, D. (1966) A simulation study of the metabolism and compartmentation in brain of glutamate, aspartate, the Krebs cycle, and related metabolites. *J. Biol. Chem.* **241**, 3918–3929.

424. GARFINKEL, D. (1981) Computer modeling of metabolic pathways. *Trends Biochem. Sci.* **6**, 69–71.

425. GARFINKEL, D. and HESS, B. (1964) Metabolic control mechanisms. *J. Biol. Chem.* **239**, 971–983.

426. GARLID, K. D. (1979) Aqueous-phase structure in cells and organelles. In *Cell-Associated Water*. pp. 293–361. (DROST-HANSEN, W. and CLEGG, J., eds), Academic Press, New York.

427. GAVILANES, F., SALERNO, C. and FASELLA, P. (1981) Heterologous enzyme–enzyme complex between D-fructose-1,6-bisphosphate aldolase and triosephosphate isomerase from *Ceratitis capitata*. *Biochim. Biophys. Acta* **660**, 154–156.

428. GAVILANES, J., LIZARBE, M. A., MUNICIO, A. M., RAMOS, J. A. and RELANO, E. (1979) Lipid requirements for structure and function of the fatty acid synthetase complex from *Ceratitis capitata. J. Biol. Chem.* **254**, 4015–4021.

429. GEIDER, K. and HOFFMANN-BERLING, H. (1981) Proteins controlling the helical structure of DNA. *Annu. Rev. Biochem.* **50**, 233–260.

430. GELLERFORS, P., LUNDÉN, M. and NELSON, B. D. (1976) Evidence for a function of core protein in complex III from beef-heart mitochondria. *Eur. J. Biochem.* **67**, 463–468.

431. GELLERT, M. (1981) DNA topoisomerases. *Annu. Rev. Biochem.* **50**, 879–910.

432. GENNIS, L. S. (1976) Negative homotropic cooperativity and affinity heterogeneity: Preparation of yeast glyceraldehyde-3-phosphate dehydrogenase with maximal affinity homogeneity. *Proc. Natl. Acad. Sci. U.S.A.* **73**, 3928–3932.

433. GEORGE, E. R., BALAKIR, R. A., FILBURN, C. R. and SACKTOR, B. (1977) Cyclic adenosine monophosphate-dependent and -independent protein kinase activity of renal brush border membranes. *Arch. Biochem. Biophys.* **180**, 429–443.

434. GEORGE, S. G. and KENNY, A. J. (1973) Studies on the enzymology of purified preparations of brush border from rabbit kidney. *Biochem. J.* **134**, 43–57.

435. GERHART, J. C. (1970) A discussion of the regulatory properties of aspartate transcarbamylase from *Escherichia coli. Curr. Top. Cell Regul.* **2**, 275–325.

436. GERHART, J. C. and PARDEE, A. B. (1962) The enzymology of control by feedback inhibition. *J. Biol. Chem.* **237**, 891–896.

437. GERHART, J. C. and SCHACHMAN, H. K. (1965) Distinct subunits for the regulation and catalytic activity of aspartate transcarbamylase. *Biochemistry* **4**, 1054–1062.

438. GERHART, J. C. and SCHACHMAN, H. K. (1968) Allosteric interactions in aspartate transcarbamylase. II. Evidence for different conformational states of the protein in the presence and absence of specific ligands. *Biochemistry* **7**, 538–552.

439. GERLACH, E., FLECKENSTEIN, A., CROSS, E. and LÜBBEN, K. (1958) Der intermediäre Phosphat-Stoffwechsel des Menschen-Erythrocyten. *Pflügers Arch.* **266**, 528–555.

440. GERSHON, D. (1979) Current status of age-altered enzymes: Alternative mechanisms. *Mech. Age. Devl.* **9**, 189–196.

441. GHANGAS, G. and REEM, G. H. (1979) Characterization of the subunit structure of human placental nucleoside phosphorylase by immunochemistry. *J. Biol. Chem.* **254**, 4233–4237.

442. GHOSE, T. K., FIECHTER, A. and BLAKEBROUGH, N. (eds) *Advances in Biochemical Engineering.* Vol. 10: *Immobilized Enzymes* I (1978) and Vol. 12: *Immobilized Enzymes* II (1979), Springer-Verlag, Berlin–Heidelberg–New York.

443. GIANNELLI, F. and PAWSEY, S. A. (1976) DNA repair synthesis in human heterokaryons. III. The rapid and slow complementing varieties of *Xeroderma pigmentosum. J. Cell Sci.* **20**, 207–213.

444. GIBSON, F. I., DOWNIE, J. A., COX, G. B. and RADIK, J. (1978) Mu-induced polarity in the *unc* operon of *Escherichia coli. J. Bacteriol.* **134**, 728–736.

445. GIESE, G. and WUNDERLICH, F. (1980) Increased fluidity and loss of temperature and Ca^{2+}/Mg^{2+} sensitivity in nuclear membranes upon removal from the membrane-associated nuclear matrix proteins. *J. Biol. Chem.* **255**, 1716–1721.

446. GILES, N. H. (1978) The organization, function, and evolution of gene clusters in eucaryotes. *Am. Nat.* **112**, 641–657.

447. GILES, N. H., CASE, M. E., PARTRIDGE, C. W. H. and AHMED, S. I. (1967) A gene cluster in *Neurospora crassa* coding for an aggregate of five aromatic synthetic enzymes. *Proc. Natl. Acad. Sci. U.S.A.* **58**, 1453–1460.

448. GILLIES, R. J. (1982) The binding site for aldolase and glyceraldehyde-3-phosphate dehydrogenase in erythrocyte membranes. *Trends Biochem. Sci.* **7**, 41–42.

449. GINSBURG, A. and STADTMAN, E. R. (1970) Multienzyme systems. *Annu. Rev. Biochem.* **39**, 429–472.

450. GIROTTI, A. W. (1976) Glyceraldehyde-3-phosphate dehydrogenase in the isolated human erythrocyte membrane: Selective displacement by bilirubin. *Arch. Biochem. Biophys.* **173**, 210–218.

451. GLENNER, G. G., McMILLAN, P. J. and FOLK, J. E. (1962) A mammalian peptidase specific for the hydrolysis of N-terminal γ-L-glutamyl and aspartyl residues. *Nature* **194**, 867.

452. GNEGY, M. E., COSTA, E. and UZUNOV, P. (1976) Regulation of transsynaptically elicited increase of 3′:5′-cyclic AMP by endogenous phosphodiesterase activator. *Proc. Natl. Acad. Sci. U.S.A.* **73**, 352–355.

453. GOLDBERG, A. L. and CHANG, T. W. (1978) Regulation and significance of amino acid metabolism in skeletal muscle. *Fed. Proc.* **37**, 2301–2307.

454. GOLDBERG, M.-E., CREIGHTON, T. E., BALDWIN, R. L. and YANOFSKY, C. (1966) Subunit structure of the tryptophan synthetase of *Escherichia coli. J. Mol. Biol.* **21**, 71–82.

455. GOLDBERG, M.-E., ORSINI, G., HÖGBERG-RAIBAUD, A. and RAIBAUD, O. (1975) Relation between the quaternary structure and the function of oligomeric enzymes. *Proc. 10th FEBS Meeting* 25–34.

456. GOLDBETER, A. (1974) Kinetic negative cooperativity in the allosteric model of Monod, Wyman and Changeux. *J. Mol. Biol.* **90**, 185–190.

457. GOLDFINE, I. D. (1981) Interaction of insulin, polypeptide hormones, and growth factors with intracellular membranes. *Biochim. Biophys. Acta* **650**, 53–67.

458. GOLDHAMMER, A. R. and PARADIES, H. H. (1979) Phosphofructokinase: Structure and function. *Curr. Top. Cell. Reg.* **15**, 109–141.

459. GOLDIN, S. M. (1977) Active transport of sodium and potassium ions by the sodium and potassium ion-activated adenosine triphosphatase from renal medulla. *J. Biol. Chem.* **252**, 5630–5642.

460. GOLDMAN, R. and KATCHALSKI, E. (1971) Kinetic behavior of a two-enzyme membrane carrying out a consecutive set of reactions. *J. Theor. Biol.* **32**, 243–257.

461. GOLDSTEIN, L. and GARTLER, S. M. (1979) The irreversibility of subunit associations in glucose-6-phosphate dehydrogenase and a suggestion regarding an early step in cellular morphogenesis. *Exp. Cell Res.* **122**, 185–190.

462. GONCHAR, N. A., LVOV, YU, M., SAMSONIDZE, T. G., SEMINA, L. A. and FEIGIN, L. A. (1978) Subunit structure of histidine decarboxylase according to data of small angle X-ray scattering and electron microscopy. *Biofizika,* **23,** 768–774.

463. GORDON, D. J., BOYER, J. L. and KORN, E. D. (1977) Comparative biochemistry of non-muscle actins. *J. Biol. Chem.* **252,** 8300–8309.

464. GORRINGE, D. M. and MOSES, V. (1980) Organization of the glycolytic enzymes in *Escherichia coli. Int. J. Biol. Macromol.* **2,** 161–173.

465. GOSS, N. H., DYER, P. Y., KEECH, D. B. and WALLACE, J. C. (1979) An electron microscopic study of pyruvate carboxylase. *J. Biol. Chem.* **254,** 1734–1739.

466. GOTS, R. E. and BESSMAN, S. P. (1974) The functional compartmentation of mitochondrial hexokinase. *Arch. Biochem. Biophys.* **163,** 7–14.

467. GOTS, R. E., GORIN, F. A. and BESSMAN, S. P. (1972) Kinetic enhancement of bound hexokinase activity by mitochondrial respiration. *Biochem. Biophys. Res. Commun.* **49,** 1249–1255.

468. GOURLEY, D. R. H. (1952) The role of adenosine triphosphate in the transport of phosphate in the human erythrocyte. *Arch. Biochem. Biophys.* **40,** 1–12.

469. GRANDE, H. J., VISSER, A. J. W. G. and VEEGER, C. (1980) Protein mobility inside pyruvate dehydrogenase complexes as reflected by laser-pulse fluorometry. A new approach to multienzyme catalysis. *Eur. J. Biochem.* **106,** 361–369.

470. GRANT, P. T. and SARGENT, J. R. (1960) Properties of L-α-glycerophosphate oxidase and its role in the respiration of *Trypanosoma rhodesiense. Biochem. J.* **76,** 229–237.

471. GRAU, E. M., MARATHE, G. V. and TATE, S. S. (1979) Rapid purification of rat kidney brush borders enriched in γ-glutamyl transpeptidase. *FEBS Lett.* **98,** 91–95.

472. GRAVES, D. J. and WANG, J. H. (1972) α-Glucan phosphorylases: Chemical and physical basis of catalysis and regulation. In *The Enzymes.* pp. 435–482. 3rd edn (BOYER, P. D., ed.), Academic Press, New York–San Francisco–London.

473. GRAZI, E. and TROMBETTA, G. (1980) The aldolase–substrate intermediates and their interaction with glyceraldehyde-3-phosphate dehydrogenase in a reconstructed glycolytic system. *Eur. J. Biochem.* **107,** 369–373.

474. GRAZI, E. and TROMBETTA, G. (1980) Fructose bisphosphate aldolase from rabbit muscle. A new, acid-labile intermediate of the aldolase reaction and the partition of the enzyme among the catalytic intermediates at equilibrium. *Arch. Biochem. Biophys.* **200,** 31–39.

475. GRAY, G. M. and SANTIAGO, N. A. (1977) Intestinal surface amino-oligopeptidases. I. Isolation of two weight isomers and their subunits from rat brush border. *J. Biol. Chem.* **252,** 4922–4928.

476. GREEN, D. E., MURER, E., HULTIN, H. O., RICHARDSON, S. H., SALMON, B., BRIERLEY, G. P. and BAUM, H. (1965) Association of integrated metabolic pathways with membranes. I. Glycolytic enzymes of the red blood corpuscle and yeast. *Arch. Biochem. Biophys.* **112,** 635–647.

477. GREEN, J. R. and HADORN, B. (1977) Glycosidases of the guinea pig brush border membrane. *Biochim. Biophys. Acta* **467,** 86–90.

478. GREEN, J. R. and HAURI, H.-P. (1977) Lactase enzymes in the intestinal brush border membrane of the suckling rat. *FEBS Lett.* **84,** 233–235.

479. GREEN, N. M. (1968) Evidence for a genetic relationship between avidins and lysozymes. *Nature* **217,** 254–256.

480. GREEN, N. M., KONIECZNY, L., TOMS, E. J. and VALENTINE, R. C. (1971) The use of bifunctional biotinyl compounds to determine the arrangement of subunits in avidin. *Biochem. J.* **125,** 781–791.

481. GROSSBARD, L. and SCHIMKE, R. T. (1966) Multiple hexokinases of rat tissues. Purification and comparison of soluble forms. *J. Biol. Chem.* **241,** 3546–3560.

418a. GROSSMAN, S. H., PYLE, J. and STEINER, R. J. (1981) Kinetic evidence for active monomers during the reassembly of denatured creatine kinase. *Biochemistry* **20,** 6122–6128.

482. GUIDOTTI, G. (1979) Coupling on ion transport to enzyme activity. In *The Neurosciences: Fourth Study Program.* pp. 831–840. MIT Press, Cambridge, Mass.

483. GUIDOTTI, G. (1980) The structure of the Band 3 polypeptide, in membrane transport in erythrocytes. *Alfred Benzon Symposium,* Vol. 14., pp. 300–311. (LASSEN, U. V., USSING, M. H. and WIETH, J. O., eds), Munksgaard, Copenhagen.

484. GUPTA, S. K. and ROTHSTEIN, M. (1976) Triosephosphate isomerase from young and old *Turbatrix aceti. Arch. Biochem. Biophys.* **174,** 333–338.

485. GUSEV, N. B., HAJDÚ, J. and FRIEDRICH, P. (1979) Motility of the N-terminal tail of phosphorylase b as revealed by crosslinking. *Biochem. Biophys. Res. Commun.* **90,** 70–77.

486. GUTFREUND, H. (1971) Transients and relaxation kinetics of enzyme reactions. *Annu. Rev. Biochem.* **40,** 315–344.

487. GUY, P., LAW, S. and HARDIE, G. (1978) Mammalian fatty acid synthetase: Evidence for subunit identity and specific removal of the thioesterase component using elastase digestion. *FEBS Lett.* **94,** 33–37.

488. HACKENBROCK, C. R. (1968) Chemical and physical fixation of isolated mitochondria in low-energy and high-energy states. *Proc. Natl. Acad. Sci. U.S.A.* **61,** 598–605.

489. HACKENBROCK, C. R. (1981) Lateral diffusion and electron transfer in the mitochondrial inner membrane. *Trends Biochem. Sci.* **6,** 151–154.

490. HADVARY, P. and KADENBACH, B. (1976) Identification of a membrane protein involved in mitochondrial phosphate transport. *Eur. J. Biochem.* **67,** 573–581.

491. HAGA, T., HAGA, K. and GILMAN, A. G. (1977) Hydrodynamic properties of the β-adrenergic receptor and adenylate cyclase from wild type and variant S49 lymphoma cells. *J. Biol. Chem.* **252,** 5776–5782.

492. HAGGIS, G. H. (1966) *The Electron Microscope In Molecular Biology.* Longmans, Green, London.

493. HAHN, L.-H. E. and HAMMES, G. G. (1978) Structural mapping of aspartate transcarbamoylase by fluorescence energy-transfer measurements: Determination of the distance between catalytic sites of different subunits. *Biochemistry* **17,** 2423–2429.

494. HAJDU, J. (1980) Determination of subunit arrangement and detection of structural changes in oligomeric enzymes by bifunctional reagents. (In Hungarian) Thesis, Budapest.

495. HAJDU, J., BARTHA, F. and FRIEDRICH, P. (1976) Crosslinking with bifunctional reagents as a means for studying the symmetry of oligomeric proteins. *Eur. J. Biochem.* **68,** 373–383.

496. HAJDU, J., DOMBRÁDI, V., BOT, G. and FRIEDRICH, P. (1979) Structural changes in glycogen phosphorylase as revealed by cross-linking with bifunctional diimidates: Phosphorylase b. *Biochemistry* **18,** 4037–4041.

497. HAJDU, J., SOLTI, M. and FRIEDRICH, P. (1975) Cross-linking and coupling of rabbit muscle aldolase and glyceraldehyde-3-phosphate dehydrogenase. *Acta Biochim. Biophys. Acad. Sci. Hung.* **10,** 7–16.

498. HAJDU, J., WYSS, S. R. and AEBI, H. (1977) Properties of human erythrocyte catalases after crosslinking with bifunctional reagents. Symmetry of the quaternary structure. *Eur. J. Biochem.* **80,** 199–207.

499. HALÁSZ, P. and POLGÁR, L. (1982) Lack of asymmetry in the active sites of tetrameric D-glyceraldehyde-3-phosphate dehydrogenase during alkylation in the crystalline state. *FEBS Lett.* **143,** 93–95.

500. HALE, G., BATES, D. L. and PERHAM, R. N. (1979) Subunit exchange in the pyruvate dehydrogenase complex of *Escherichia coli. FEBS Lett.* **104,** 343–346.

501. HALE, G. and PERHAM, R. N. (1979) Polypeptide-chain stoichiometry and lipoic acid content of the pyruvate dehydrogenase complex of *Escherichia coli. Biochem. J.* **177,** 129–137.

502. HALL, J., DAVIS, J. P. and CANTOR, C. R. (1977) Interaction of a fluorescent streptomycin derivative with *Escherichia coli* ribosomes. *Arch. Biochem. Biophys.* **179,** 121–130.

503. HALPER, L. A. and SRERE, P. A. (1977) Interaction between citrate synthase and mitochondrial malate dehydrogenase in the presence of polyethylene glycol. *Arch. Biochem. Biophys.* **184,** 529–534.

504. HALVORSON, H. R. (1979) Relaxation kinetics of glutamate dehydrogenase. Self-association by pressure perturbation. *Biochemistry* **18**, 2480–2487.

505. HAMADA, M., OTSUKA, K.-I., TANAKA, N., OGASHARA, K., KOIKE, K., HIRAOKA, T. and KOIKE, M. (1975) Purification, properties, and subunit composition of pig heart lipoate acetyltransferase. *J. Biochem.* **78.** 187–197.

506. HAMMERSTEDT, R. M., MÖHLER, H., DECKER, K. A. and WOOD, W. A. (1971) Structure of 2-keto-3-deoxy-6-phosphogluconate aldolase. *J. Biol. Chem.* **246**, 2069–2074.

507. HAMMES, G. G. (1964) Mechanism of enzyme catalysis. *Nature* **204**, 342–343.

508. HAMMES, G. G. and ALBERTY, R. A. (1960) The relaxation spectra of simple enzymatic mechanisms. *J. Am. Chem. Soc.* **82**, 1564–1569.

509. *Handbook of Chemistry and Physics* (1959), 41st ed., p. 2523, Chemical Rubber, Cleveland, Ohio.

510. HANKE, D. W., WARDEN, D. A., EVANS, J. O., FANNIN, F. F. and DIEDRICH, D. F. (1980) The kinetic advantage for transport into hamster intestine of glucose generated from phlorizin by brush border β-glucosidase. *Biochim. Biophys. Acta* **599**, 652–663.

511. HANOZET, G., PIRCHER, H.-P., VANNI, P., OESCH, B. and SEMENZA, G. (1981) An example of enzyme hysteresis. The slow and tight interaction of some fully competitive inhibitors with small intestinal sucrase. *J. Biol. Chem.* **256**, 3703–3711.

512. HANSEN, O., JENSEN, J., NØRBY, J. G. and OTTOLENGHI, P. (1979) A new proposal regarding the subunit composition of $(Na^+ + K^+)$ATPase. *Nature* **280**, 410–412.

513. HANSON, K. R. (1966) Symmetry of protein oligomers formed by isologous association. *J. Mol. Biol.* **22**, 405–409.

514. HARRIS, J. I. and PERHAM, R. N. (1965) Glyceraldehyde 3-phosphate dehydrogenases. I. The protein chains in glyceraldehyde 3-phosphate dehydrogenase from pig muscle. *J. Mol. Biol.* **13**, 876–884.

515. HASCHEMEYER, R. H. and HASCHEMEYER, A. E. V. (1973) *Proteins: A guide to study by physical and chemical methods,* John Wiley & Sons, New York–London–Sydney–Toronto.

516. HATANAKA, M., WHITE, E. A., HORIBATA, K. and CRAWFORD, I. P. (1962) A study of the catalytic properties of *Escherichia coli* tryptophan synthetase, a two-component enzyme. *Arch. Biochem. Biophys.* **97**, 596–606.

517. HATEFI, Y., GALANTE, Y. M., STIGGALL, D. L. and RAGAN, C. I. (1979) Proteins, polypeptides prosthetic groups and enzymic properties of Complexes I, II, III, IV and V of the mitochondrial oxidative phosphorylation system. *Methods Enzymol.* **56**, 577–602.

518. HATHAWAY, G. M., KIDA, S. and CRAWFORD, I. P. (1969) Subunit structure of the B component of *Escherichia coli* tryptophan synthetase. *Biochemistry* **8**, 989–997.

519. HAWLEY, S. A. (1978) High-pressure techniques. *Methods Enzymol.* **49**, 14–24.

520. HAYAISHI, O. and UEDA, K. (1977) Poly(ADP-ribose) and ADP-ribosylation of proteins. *Annu. Rev. Biochem.* **46**, 95–116.

521. HAYFLICK, L. (1965) The limited in vitro lifetime of human diploid cell strains. *Exp. Cell Res.* **37**, 614–636.

522. HECK, H. D'A. (1972) Threonine-sensitive aspartokinase-homoserine dehydrogenase of *Escherichia coli* K12. Evidence for a cooperative tetramer. *Biochemistry* **11**, 4421–4427.

523. HEINRICH, R., RAPOPORT, S. M. and RAPOPORT, T. A. (1977) Metabolic regulation and mathematical models. *Prog. Biophys. Molec. Biol.* **32**, 1–82.

524. HEINRICH, R. and RAPOPORT, T. A. (1974) A linear steady-state treatment of enzyme chains. *Eur. J. Biochem.* **42**, 89–95.

525. HEINRICH, R. and RAPOPORT, T. A. (1975) Mathematical analysis of multienzyme systems. II. Steady state and transient control. *Biosystems* **7**, 130–136.

526. HELMREICH, E. (1969) Control of synthesis and breakdown of glycogen, starch and cellulose.

In *Comprehensive Biochemistry*. Vol. 17, pp. 17–92. (FLORKIN, M. and STOTZ, E. H., eds), Elsevier, Amsterdam.

527. HELMREICH, E., MICHAELIDES, M. C. and CORI, C. F. (1967) Effects of substrates and a substrate analog on the binding of 5'-adenylic acid to muscle phosphorylase a. *Biochemistry* 6, 3695–3710.

528. HENDERSON, R., CAPALDI, R. A. and LEIGH, J. S. (1977) Arrangement of cytochrome oxidase molecules in two-dimensional vesicle crystals. *J. Mol. Biol.* 112, 631–648.

529. HENSLEY, P. and SCHACHMAN, H. K. (1979) Communication between dissimilar subunits in aspartate transcarbamoylase: Effect of inhibitor and activator on the conformation of the catalytic polypeptide chains. *Proc. Natl. Acad. Sci. U.S.A.* 76, 3732–3736.

530. HERLAN, G., ECKERT, W. A., KAFFENBERGER, W. and WUNDERLICH, F. (1979) Isolation and characterization of an RNA-containing nuclear matrix from Tetrahymena macronuclei. *Biochemistry* 18, 1782–1788.

531. HERMAN, C. A., ZAHLER, W. L., DOAK, G. A. and CAMPBELL, B. J. (1976) Bull sperm adenylate cyclase: Localization and partial characterization. *Arch. Biochem. Biophys.* 177, 622–629.

532. HERZFELD, J. and STANLEY, H. E. (1974) A general approach to co-operativity and its application to the oxygen equilibrium of hemoglobin and its effectors. *J. Mol. Biol.* 82, 231–265.

533. HESS, B. (1980) Organization of biochemical reactions: From microspace to macroscopic structures. In *Cell Compartmentation and Metabolic Channeling*. pp. 75–92. (NOVER, L., LYNEN, F. and MOTHES, K., eds), VEB Gustav Fischer Verlag, Jena, Elsevier/North-Holland Biomedical Press, Amsterdam–New York–Oxford.

534. HESS, B. and BOITEUX, A. (1972) Heterologous enzyme–enzyme interactions. In *Protein–protein Interaction*, pp. 271–297, Springer-V, Berlin, Heidelberg.

535. HESS, B., BOITEUX, A. and KRÜGER, J. (1969) Cooperation of glycolytic enzymes. *Adv. Enzyme Regul.* 7, 149–167.

536. HESS, B., GOLDBETER, A. and LEFEVER, R. (1978) Temporal, spatial and functional order in regulated biochemical and cellular systems. *Adv. Chem. Physics* 38, 363–413.

537. HESS, B. and WURSTER, B. (1970) Transient time of the pyruvate kinase-lactate dehydrogenase system of rabbit muscle in vitro. *FEBS Lett.* 9, 73–77.

538. HIGASHI, T., RICHARDS, C. S. and UYEDA, K. (1979) The interaction of phosphofructokinase with erythrocyte membranes. *J. Biol. Chem.* 254, 9542–9550.

539. HIGGINS, J. (1963) Analysis of sequential reactions. *Ann. N. Y. Acad. Sci.* 108, 305–321.

540. HILL, C. M., WAIGHT, R. D. and BARDSLEY, W. G. (1977) Does any enzyme follow the Michaelis–Menten equation? *Mol. Cell. Biochem.* 15, 173–178.

541. HILL, R. L. and BREW, K. (1975) Lactose synthetase. *Adv. Enzymol.* 43, 411–490.

542. HILL, T. L. (1976) Steady-state kinetic formalism applied to multienzyme complexes, oxidative phosphorylation, and interacting enzymes. *Proc. Natl. Acad. Sci. U.S.A.* 73, 4432–4436.

543. HILL, T. L. (1977) *Free Energy Transduction in Biology*. Academic Press, New York.

544. HILL, T. L. (1977) Theoretical study of the effect of enzyme–enzyme interactions on steady-state enzyme kinetics. *Proc. Natl. Acad. Sci. U.S.A.* 74, 3632–3636.

545. HILL, T. L. (1977) Further study of the effect of enzyme–enzyme interactions on steady-state enzyme kinetics. *Proc. Natl. Acad. Sci. U.S.A.* 74, 4111–4115.

546. HILL, T. L. (1977) "Virial" expansion of enzyme flux and use of quasi-chemical approximation for two-state enzymes with enzyme–enzyme interactions. *Proc. Natl. Acad. Sci. U.S.A.* 74, 5227–5230.

547. HILL, T. L. (1978) Effect of enzyme–enzyme interactions on steady-state enzyme kinetics. *J. Theor. Biol.* 75, 391–416.

548. HILL, T. L. (1978) Unsymmetrical and concerted examples of the effect of enzyme–enzyme interactions on steady-state enzyme kinetics. *Proc. Natl. Acad. Sci. U.S.A.* 75, 1101–1105.

549. HIRWAY, S. C. and HULTIN, H. O. (1977) A model system for studying chicken lactate dehydrogenase-5 in the particulate phase. *J. Food Sci.* **42**, 1164–1167.

550. HO, M. K. and GUIDOTTI, G. (1975) A membrane protein from human erythrocytes involved in anion exchange. *J. Biol. Chem.* **250**, 675–683.

551. HOAGLAND, V. D. and TELLER, D. C. (1969) Influence of substrates on the dissociation of rabbit muscle D-glyceraldehyde 3-phosphate dehydrogenase. *Biochemistry* **8**, 594–602.

552. HÖCHLI, M. and HACKENBROCK, C. R. (1976) Fluidity in mitochondrial membranes: Thermotropic lateral translational motion of intramembrane particles. *Proc. Natl. Acad. Sci. U.S.A.* **73**, 1636–1640.

553. HÖCHLI, M. and HACKENBROCK, C. R. (1977) Thermotropic lateral translational motion of intramembrane particles in the inner mitochondrial membrane and its inhibition by artificial peripheral proteins. *J. Cell Biol.* **72**, 278–291.

554. HÖCHLI, M. and HACKENBROCK, C. R. (1979) Lateral translational diffusion of cytochrome c oxidase in the mitochondrial energy-transducing membrane. *Proc. Natl. Acad. Sci. U.S.A.* **76**, 1236–1240.

555. HOCMAN, G. (1979) Biochemistry of ageing. *Int. J. Biochem.* **10**, 867–876.

556. HOCMAN, G. (1980) Biochemistry of ageing – II. *Int. J. Biochem.* **12**, 515–522.

557. HOCMAN, G. (1981) Biochemistry of ageing and cancer. *Int. J. Biochem.* **13**, 659–672.

558. HOFFMANN, W., SARZALA, M. G. and CHAPMAN, D. (1979) Rotational motion and evidence for oligomeric structures of sarcoplasmic reticulum Ca^{2+}-activated ATPase. *Proc. Natl. Acad. Sci. U.S.A.* **76**, 3860–3864.

559. HOFMANN, E. C. G. (1955) Abbau und Synthese des DPN in den roten Blutkörperchen des Kaninchens. *Biochem. Z.* **327**, 273–283.

560. HOFMANN, E. and NOLL, F. (1961) Verteilung von DPN und TPN-spezifischen Nukleosidasen in Erythrozyten verschiedener Tierarten. *Acta Biol. Med. Germ.* **6**, 1–6.

561. HOFMANN, E. C. G. and RAPOPORT, S. (1957) DPN- und TPN-spezifische Nucleosidasen in Erythrocyten. II. Eigenschaften und Hemmbarkeit der gereinigten Enzyme. *Biochem. Z.* **329**, 437–448.

562. HOFSTADTER, D. R. (1981) The Magic Cube's cubies are twiddled by cubists and solved by cubemeisters. *Sci. Amer.* **244**/3, 14–26.

563. HOLLAWAY, M. R., OSBORNE, H. H. and SPOTORNO, G. M. L. (1977) The effect of nucleotide binding on subunit interaction in glyceraldehyde-3-phosphate dehydrogenase as determined by the kinetics and thermodynamics of subunit exchange. In *Pyridine Nucleotide-dependent Dehydrogenases.* pp. 101–117. (SUND, H., ed.), Walter de Gruyter, Berlin–New York.

564. HOLLIDAY, R. and TARRANT, G. M. (1972) Altered enzymes in ageing human fibroblasts. *Nature* **238**, 26–30.

565. HOLMES, E. W., JR., MALONE, J. I., WINEGRAD, A. I. and OSKI, F. A. (1967) Hexokinase isoenzymes in human erythrocytes: Association of type II with fetal hemoglobin. *Science* **156**, 646–648.

566. HOLZER, H., BETZ, H. and EBNER, E. (1975) Intracellular proteinases in microorganisms. *Curr. Top. Cell. Regul.* **9**, 103–156.

567. HØRDER, M. and WILKINSON, J. H. (1979) Enzymes. In *Chemical Diagnosis of Disease.* pp. 363–406. (BROWN, S. S., MITCHELL, F. L. and YOUNG, D. S., eds), Elsevier/North-Holland Biomedical Press, Amsterdam–New York–Oxford.

568. HORIUCHI, S., INOUE, M. and MORINO, Y. (1978) γ-Glutamyl transpeptidase: Sidedness of its active site on renal brush-border membrane. *Eur. J. Biochem.* **87**, 429–437.

569. HÖRL, W. H. and HEILMEYER, L. M. G., JR. (1978) Evidence for the participation of a Ca^{2+}-dependent protein kinase and protein phosphatase in the regulation of the Ca^{2+} transport ATPase of the sarcoplasmic reticulum. 2. Effect of phosphorylase kinase and phosphorylase phosphatase. *Biochemistry* **17**, 766–772.

570. HOTTA, Y. and STERN, H. (1978) DNA unwinding proteins from meiotic cells of *Lilium. Biochemistry* **17**, 1872–1880.

571. HOUSLAY, M. D. and MARCHMONT, R. J. (1980) Exposure of mitochondrial outer membranes to neuraminidase selectively destroys monoamine oxidase A activity. *J. Pharm. Pharmacol.* **32**, 65–66.

572. HOWELL, S. H. and WALKER, L. L. (1972) Synthesis of DNA in toluene-treated *Chlamydomonas reinhardi. Proc. Natl. Acad. Sci. U.S.A.* **69**, 490–494.

573. HOWLETT, G. J., BLACKBURN, M. N., COMPTON, J. G. and SCHACHMAN, H. K. (1977) Allosteric regulation of aspartate transcarbamoylase. Analysis of the structural and functional behavior in terms of a two-state model. *Biochemistry* **16**, 5091–5099.

574. HUANG, K., FAIRCLOUGH, H. and CANTOR, C. R. (1975) Singlet energy transfer studies of the arrangement of proteins in the 30S *Escherichia coli* ribosome. *J. Mol. Biol.* **97**, 443–470.

575. HUANG, L., MONTOYA, A. L. and NESTER, E. W. (1974) Characterization of the functional activities of the subunits of 3-deoxy-D-arabinoheptulosonate 7-phosphate synthetase-chorismate mutase from *Bacillus subtilis* 168. *J. Biol. Chem.* **249**, 4473–4479.

576. HUANG, L., MONTOYA, A. L. and NESTER, E. W. (1975) Purification and characterization of shikimate kinase enzyme activity in *Bacillus subtilis. J. Biol. Chem.* **250**, 7675–7681.

577. HÜBSCHER, G., MAYER, R. J. and HANSEN, H. J. M. (1971) Glycolytic enzymes as a multi-enzyme system. *Bioenergetics* **2**, 115–118.

578. HUCHO, F., MÜLLNER, M. and SUND, H. (1975) Investigation of the symmetry of oligomeric enzymes with bifunctional reagents. *Eur. J. Biochem.* **59**, 79–87.

579. IHLER, G. M., GLEW, R. H. and SCHNURE, F. W. (1973) Enzyme loading of erythrocytes. *Proc. Natl. Acad. Sci. U.S.A.* **70**, 2663–2666.

580. IKAI, A. and TANFORD, C. (1971) Kinetic evidence for incorrectly folded intermediate states in the refolding of denatured proteins. *Nature* **230**, 100–102.

581. IKEDA, B. H., SPECKHARD, D. C. and FREY, P. A. (1977) Acetyl group and electron transfer sites in *Escherichia coli* pyruvate dehydrogenase complex. *Fed. Proc.* **36**, 3186.

582. ITO, S. (1965) The enteric surface coat on cat intestinal microvilli. *J. Cell Biol.* **27**, 475–491.

583. JACOB, S. T. and ROSE, K. M. (1980) Basic enzymology of transcription in prokaryotes and eukaryotes: A review. In *Cell Biology: Comprehensive Treatise*. Vol. 3, pp. 114–141. (GOLDSTEIN, L. and PRESCOTT, D. M., eds), Academic Press, New York.

584. JACOBSON, G. R. and STARK, G. R. (1973) Aspartate transcarbamylases. In *The Enzymes*. Vol. 9, pp. 225–308. (BOYER, P. D., ed.), Academic Press, New York.

585. JACOBUS, W. E. and LEHNINGER, A. L. (1973) Creatine kinase of rat heart mitochondria. Coupling of creatine phosphorylation to electron transport. *J. Biol. Chem.* **248**, 4803–4810.

586. JAENICKE, R. and KOBERSTEIN, R. (1971) High pressure dissociation of lactic dehydrogenase. *FEBS Lett.* **17**, 351–354.

587. JAENICKE, R., RUDOLPH, R. and HEIDER, I. (1979) Quaternary structure, subunit activity, and *in vitro* association of porcine mitochondrial malic dehydrogenase. *Biochemistry* **18**, 1217–1222.

588. JAFFÉ, H. H. and ORCHIN, M. (1965) *Symmetry in Chemistry*. Wiley, New York.

589. JAIN, S. K. and HOCHSTEIN, P. (1980) Polymerization of membrane components in aging red blood cells. *Biochem. Biophys. Res. Commun.* **92**, 247–254.

590. JANKOWSKI, J. M. and KLECZKOWSKI, K. (1980) Is protein kinase a subunit of RNA polymerase II, which is responsible for the specificity of transcription? *Biochem. Biophys. Res. Commun.* **96**, 1216–1224.

591. JARRET, H. W. and KYTE, J. (1979) Human erythrocyte calmodulin. Further chemical characterization and the site of its interaction with the membrane. *J. Biol. Chem.* **254**, 8237–8244.

592. JEFFREY, P. D. and ANDREWS, P. R. (1980) Application of calculated sedimentation ratios in the specification of models for protein dimers, trimers, tetramers and pentamers. *Biophys. Chem.* **11**, 61–70.

593. JENKINS, J. A., JOHNSON, L. N., STUART, D. I., STURA, E. A., WILSON, K. S. and ZANOTTI, G. (1981) Phosphorylase: control and activity. *Phil. Trans. R. Soc. Lond. B.* **293**, 23–41.

594. JOHNSON, C. S. and DEAL, W. C., JR. (1978) Low and high temperature asymmetric forms of pig kidney and rabbit muscle phosphofructokinases and reversible, temperature-dependent transitions. *Arch. Biochem. Biophys.* **190**, 560–570.

595. JOHNSON, L. N., STURA, E. A., WILSON, K. S., SANSOM, M. S. P. and WEBER, I. T. (1979) Nucleotide binding to glycogen phosphorylase *b* in the crystal. *J. Mol. Biol.* **134**, 639–653.

596. JOHNSON, R. S. and SCHACHMAN, H. K. (1980) Propagation of conformational changes in Ni(II)-substituted aspartate transcarbamoylase: Effect of active-site ligands on the regulatory chains. *Proc. Natl. Acad. Sci. U.S.A.* **77**, 1995–1999.

597. JONES, J. E. (1924) On the determination of molecular fields. *Proc. Roy. Soc. London, Ser A.* **106**, 441–462.

598. JONES, M. E. (1971) Regulation of pyrimidine and arginine biosynthesis in mammals. *Adv. Enzyme Regul.* **9**, 19–49.

599. JONES, M. E. (1980) Pyrimidine nucleotide biosynthesis in animals: Genes, enzymes, and regulation of UMP biosynthesis. *Annu. Rev. Biochem.* **49**, 253–279.

600. JØRGENSEN, P. L. (1980) Energetics of active transtubular transport; Function of the Na-K-ion pump. *Int. J. Biochem.* **12**, 283–286.

601. JUNGE, W., KRISCH, K. and HOLLANDT, H. (1974) Further investigations on the subunit structure of microsomal carboxylesterases from pig and ox livers. *Eur. J. Biochem.* **43**, 379–389.

602. KACSER, H. (1963) The kinetic structure of organisms. In *Biological Organisation at the Cellular and Supercellular Level.* (HARRIS, R. J. C., ed.), Academic Press, London–New York.

603. KACSER, H., BULFIELD, G. and WALLACE, M. F. (1973) Histidinaemic mutant in the mouse. *Nature* **244**, 77–79.

604. KACSER, H. and BURNS, J. A. (1973) *The Control and Flux Rate Control of Biological Processes.* *Symp. Soc. Exp. Biol.* Vol. 27., pp. 65–104. (DAVIES, D. D., ed.), University Press, Cambridge.

605. KACSER, H. and BURNS, J. A. (1979) Molecular democracy: Who shares the controls? *Biochem. Soc. Trans.* **7**, 1149–1160.

606. KAKIUCHI, S., YAMAZAKI, R., TESHIMA, Y., UENISHI, K. and MIYAMOTO, E. (1975) Multiple cyclic nucleotide phosphodiesterase activities from rat tissues and occurrence of a calcium-plus-magnesium-ion-dependent phosphodiesterase and its protein activator. *Biochem. J.* **146**, 109–120.

607. KÁLMÁN, M., NURIDSÁNY, M. and OVÁDI, J. (1980) Substrate-induced dissociation of glyceraldehyde-phosphate dehydrogenase detected by affinity chromatography. *Biochim. Biophys. Acta* **614**, 285–293.

608. KANNER, B. I., SERRANO, R., KANDRACH, M. A. and RACKER, E. (1976) Preparation and characterization of homogeneous coupling factor 6 from bovine heart mitochondria. *Biochem. Biophys. Res. Commun.* **69**, 1050–1056.

609. KANT, J. A. and STECK, T. L. (1973) Specificity in the association of glyceraldehyde 3-phosphate dehydrogenase with isolated human erythrocyte membranes. *J. Biol. Chem.* **248**, 8457–8464.

610. KANTROWITZ, E. R., FOOTE, J., REED, H. W. and VENSEL, L. A. (1980) Isolation and preliminary characterization of single amino acid substitution mutants of aspartate carbamoyltransferase. *Proc. Natl. Acad. Sci. U.S.A.* **77**, 3249–3253.

611. KANTROWITZ, E. R., PASTRA-LANDIS, S. C. and LIPSCOMB, W. N. (1980) *E. coli* aspartate transcarbamylase: Part I: Catalytic and regulatory functions. *Trends Biochem. Sci.* **5**, 124–128.

612. KARADSHEH, N. S. and UYEDA, K. (1977) Changes in allosteric properties of phosphofructokinase bound to erythrocyte membranes. *J. Biol. Chem.* **252**, 7418–7420.

613. KARADSHEH, N. S., UYEDA, K. and OLIVER, R. M. (1977) Studies on structure of human erythrocyte phosphofructokinase. *J. Biol. Chem.* **252**, 3515–3524.

614. KARLIN, J. N., BOWMAN, B. J. and DAVIS, R. H. (1976) Compartmental behavior of ornithine in *Neurospora crassa. J. Biol. Chem.* **251**, 3948–3955.

615. KARLISH, S. J. D., YATES, D. W. and GLYNN, I. M. (1978) Conformational transitions between Na$^+$-bound and K$^+$-bound forms of (Na$^+$+K$^+$)-ATPase, studied with formycin nucleotides. *Biochim. Biophys. Acta* **525**, 252–264.

616. KASLOW, H. R., JOHNSON, G. L., BROTHERS, V. M. and BOURNE, H. R. (1980) A regulatory component of adenylate cyclase from human erythrocyte membranes. *J. Biol. Chem.* **255**, 3736–3741.

617. KASVINSKY, P. J., SHECHOSKY, S. and FLETTERICK, R. J. (1978) Synergistic regulation of phosphorylase a by glucose and caffeine. *J. Biol. Chem.* **253**, 9102–9106.

618. KATO, N., OMORI, Y., TANI, Y. and OGATA, K. (1976) Alcohol oxidases of *Kloeckera* sp. and *Hansenula polymorpha*. Catalytic properties and subunit structures. *Eur. J. Biochem.* **64**, 341–350.

619. KATZEN, H. M. (1967) The multiple forms of mammalian hexokinase and their significance to the action of insulin. *Adv. Enzyme Reg.* **5**, 335–356.

620. KATZEN, H. M., SODERMANN, D. D. and WILEY, C. E. (1970) Multiple forms of hexokinase. *J. Biol. Chem.* **245**, 4081–4096.

621. KAUFMANN, S. H., COFFEY, D. S. and SHAPER, J. H. (1981) Considerations in the isolation of rat liver nuclear matrix, nuclear envelope, and pore complex lamina. *Exp. Cell Res.* **132**, 105–123.

622. KAUZMANN, W. (1959) Some factors in the interpretation of protein denaturation. *Adv. Protein Chem.* **14**, 1–63.

623. KAVIPURAPU, P. R. and JONES, M. E. (1976) Purification, size, and properties of the complex of orotate phosphoribosyltransferase: Orotidylate decarboxylase from mouse Ehrlich ascites carcinoma. *J. Biol. Chem.* **251**, 5589–5599.

624. KAWAGUCHI, A. and BLOCH, K. (1976) Inhibition of glutamate dehydrogenases by palmitoyl coenzyme A. *J. Biol. Chem.* **251**, 1406–1412.

625. KAWAHARA, K. and TANFORD, C. (1966) The number of polypeptide chains in rabbit muscle aldolase. *Biochemistry* **5**, 1578–1584.

626. KAWATO, S., SIGEL, E., CARAFOLI, E. and CHERRY, R. J. (1980) Cytochrome oxidase rotates in the inner membrane of intact mitochondria and submitochondrial particles. *J. Biol. Chem.* **255**, 5508–5510.

627. KEITH, A. D. (ed.) (1979) *The Aqueous Cytoplasm*. Marcel Dekker, New York.

627a. KELETI, T. (1968) Effect of steric changes in the protein on the kinetics of enzymic reactions. II. Steady-state treament of reactions with one substrate. *Acta Biochim. Biophys. Acad. Sci. Hung.* **3**, 247–258.

628. KELETI, T. (1981) Data analysis from Michaelis–Menten kinetics: Ins and outs. In *Kinetic Data Analysis. Design and Analysis of Enzyme and Pharmacokinetic Experiments*. pp. 353–374. (ENDRÉNYI, L., ed.), Plenum Press, New York.

629. KELETI, T., BATKE, J., OVÁDI, J., JANCSIK, V. and BARTHA, F. (1977) Macromolecular interactions in enzyme regulation. *Adv. Enzyme Regul.* **15**, 233–265.

630. KELETI, T. and SZEGVÁRI, M. (1972) Heat denaturation of D-glyceraldehyde-3-phosphate dehydrogenase holoenzyme. *Acta Biochim. Biophys. Acad. Sci. Hung.* **7**, 115–118.

631. KELLY, J. J. and ALPERS, D. H. (1973) Properties of human intestinal glucoamylase. *Biochim. Biophys. Acta* **315**, 113–120.

632. KEMPE, T. D., SWYRYD, E. A., BRUIST, M. and STARK, G. R. (1976) Stable mutants of mammalian cells that overproduce the first three enzymes of pyrimidine nucleotide biosynthesis. *Cell* **9**, 541–550.

633. KEMPNER, E. S. (1980) Metabolic compartments and their interactions. In *Cell Compartmentation and Metabolic Channeling*, (NOVER, L., LYNEN, F. and MOTHES, K., eds), pp. 211–224, VEB Gustav Fischer Verlag, Jena, Elsevier/North-Holland Biomedical Press, Amsterdam–New York–Oxford.

634. KEMPNER, E. S. and MILLER, J. H. (1968) The molecular biology of *Euglena gracilis* IV. Cellular stratification by centrifuging. *Exp. Cell Res.* **51**, 141–149.

635. KEMPNER, E. S. and MILLER, J. H. (1968) The molecular biology of *Euglena gracilis* V. Enzyme localization. *Exp. Cell Res.* **51**, 150–156.

636. KENNEDY, S. J. (1978) Structure of membrane proteins. *J. Membrane Biol.* **42**, 265–279.

637. KENNY, A. J. and BOOTH, A. G. (1976) Organization of the kidney proximal-tubule plasma membrane. *Biochem. Soc. Trans.* **4**, 1011–1017.

638. KENNY, A. J. and BOOTH, A. G. (1978) Microvilli: Their ultrastructure, enzymology and molecular organization. *Essays Biochem.* **14**, 1–44.

639. KENNY, A. J., BOOTH, A. G. and MACNAIR, R. D. C. (1977) Peptidases of the kidney microvillus membrane. *Acta Biol. Med. Germ.* **36**, 1575–1585.

640. KENNY, J. W., SOMMER, A. and TRAUT, R. R. (1975) Cross-linking studies on the 50S ribosomal subunit of *Escherichia coli* with methyl 4-mercaptobutyrimidate. *J. Biol. Chem.* **250**, 9434–9436.

641. KENT, R. J., LIN, R.-L., SALLACH, H. J. and COHEN, P. P. (1975) Reversible dissociation of a carbamoyl phosphate synthase-aspartate transcarbamoylase–dihydroorotase complex from ovarian eggs of *Rana catesbiana*: Effect of uridine triphosphate and other modifiers. *Proc. Natl. Acad. Sci. U.S.A.* **72**, 1712–1716.

642. KEOKITICHAI, S. and WRIGGLESWORTH, J. M. (1980) Association of glyceraldehyde 3-phosphate dehydrogenase with the membrane of the intact human erythrocyte. *Biochem. J.* **187**, 837–841.

643. KERBEY, A. L., RADCLIFFE, P. M., RANDLE, P. J. and SUDGEN, P. H. (1979) Regulation of kinase reactions in pig heart pyruvate dehydrogenase complex. *Biochem. J.* **181**, 427–433.

644. KERBIRIOU, D. and HERVÉ, G. (1972) Biosynthesis of an aspartate transcarbamylase lacking co-operative interactions. I. Disconnection of homotropic and heterotropic interactions under the influence of 2-thiouracil. *J. Mol. Biol.* **64**, 379–392.

645. KERBIRIOU, D., HERVÉ, G. and GRIFFIN, J. H. (1977) An aspartate transcarbamylase lacking catalytic subunit interactions. Study of conformational changes by ultraviolet absorbance and circular dichroism spectroscopy. *J. Biol. Chem.* **252**, 2881–2890.

646. KERR, M. A. and KENNY, A. J. (1974) The molecular weight and properties of a neutral metallo-endopeptidase from rabbit kidney brush border. *Biochem. J.* **137**, 489–495.

647. KHAILOVA, L. S., FEIGINA, M. M., GEORGIN, S. and SEVERIN, S. E. (1972) Investigation of quaternary structure of muscle pyruvate dehydrogenase. *Biokhimiya*, **37**, 1312–1314.

648. KHAIRALLAH, E. A. and MORTIMORE, G.E. (1976) Assessment of protein turnover in perfused rat liver. Evidence for amino acid compartmentation from differential labeling of free and tRNA-bound valine. *J. Biol. Chem.* **251**, 1375–1384.

649. KIMURA, H. and MURAD, F. (1974) Evidence for two different forms of guanylate cyclase in rat heart. *J. Biol. Chem.* **249**, 6910–6916.

650. KING, N. J. (1967) The glucoamylase of *Coniophora cerebella*. *Biochem. J.* **105**, 577–583.

651. KINNE-SAFFRAN, E. and KINNE, R. (1979) Further evidence for the existence of an intrinsic bicarbonate-stimulated Mg^{2+}-ATPase in brush border membrane isolated from rat kidney cortex. *J. Membrane Biol.* **49**, 235–251.

652. KIRKWOOD, J. G. (1954) The general theory of irreversible processes in solutions of macromolecules. *J. Polymer Sci.* **12**, 1–14.

653. KIRSCHNER, K. (1971) Kinetic analysis of allosteric enzymes. *Curr. Top. Cell. Regul.* **4**, 167–210.

654. KIRSCHNER, K. and BISSWANGER, H. (1976) Multifunctional proteins. *Annu. Rev. Biochem.* **45**, 143–166.

655. KIRSCHNER, K., EIGEN, M., BITTMAN, R. and VOIGT, B. (1966) The binding of nicotinamide-adenine dinucleotide to yeast D-glyceraldehyde-3-phosphate dehydrogenase: Temperature-jump relaxation studies on the mechanism of an allosteric enzyme. *Proc. Natl. Acad. Sci. U.S.A.* **56**, 1661–1667.

656. KIRSCHNER, K. and VOIGT, B. (1968) Reinheitskriterien für das kristallisierbare Isoenzym der D-Glycerinaldehyd-3-phosphat-Dehydrogenase aus Bäckerhefe. *Z. Physiol. Chem.* **349**, 632–644.

657. KIRSCHNER, K., WEISCHET, W. and WISKOCIL, R. L. (1975) Ligand binding to enzyme complexes. In *Protein-Ligand Interaction.* pp. 27–44. (SUND, H. and BLAUER, G., eds), Walter de Gruyter, Berlin.

658. KIRSCHNER, K. and WISKOCIL, R. (1972) Enzyme–enzyme interactions in tryptophan synthetase from *E. coli.* In *Protein–Protein Interactions.* pp. 245–268. (JAENICKE, R. and HELMREICH, E., eds), Springer, New York.

659. KIRSCHNER, K., WISKOCIL, R. L., FOEHN, M. and REZEAU, L. (1975) The tryptophan synthase from *Escherichia coli.* An improved purification procedure for the α-subunit and binding studies with substrate analogues. *Eur. J. Biochem.* **60,** 513–523.

660. KIRTLEY, M. E. and KOSHLAND, D. E., JR. (1967) Models for cooperative effects in proteins containing subunits. Effects of two interacting ligands. *J. Biol. Chem.* **242,** 4192–4205.

661. KISELEV, N. A., LERNER, F. YA. and LIVANOVA, N. B. (1974) Electron microscopy of muscle phosphorylase *a. J. Mol. Biol.* **86,** 587–599.

662. KISELEV, N. A., STELMASCHUK, V. YA. and TSUPRUN, V. L. (1977) Electron microscopy of leucine aminopeptidase. *J. Mol. Biol.* **115,** 33–43.

663. KLEINSCHMIDT, L. J. (1975) Phosphorylation of non-histone proteins in the regulation of chromosome structure and function. *J. Cell Physiol.* **85,** 459–476.

664. KLINGENBERG, M. (1981) Membrane protein oligomeric structure and transport function. *Nature* **290,** 449–454.

665. KLIMAN, H. J. and STECK, T. L. (1980) Association of glyceraldehyde-3-phosphate dehydrogenase with the human red cell membrane. *J. Biol. Chem.* **255,** 6314–6321.

665a. KLINOV, S. V. and KURGANOV, B. I. (1982) The theoretical analysis of kinetic behaviour of "hysteretic" allosteric enzymes. V. Relaxation kinetics of dissociating enzyme systems. *J. Theor. Biol.* **98,** 73–90.

666. KLOTZ, I. M., LANGERMAN, N. R. and DARNALL, D. W. (1970) Quaternary structure of proteins. *Annu. Rev. Biochem.* **39,** 25–62.

667. KNOBLING, A., SCHIFFMANN, D., SICKINGER, H.-D., and SCHWEIZER, E. (1975) Malonyl and palmityl transferase — less mutants of the yeast fatty-acid-synthetase complex. *Eur. J. Biochem.* **56,** 359–367.

668. KNOLL, J. and MAGYAR, K. (1972) Some puzzling pharmacological effects of monoamine oxidase inhibitors. *Adv. Biochem. Psychopharmacol.* **5,** 393–408.

669. KNULL, H. R., TAYLOR, W. F. and WELLS, W. W. (1973) Effects of energy metabolism on *in vivo* distribution of hexokinase in brain. *J. Biol. Chem.* **248,** 5414–5417.

670. KNULL, H. R., TAYLOR, W. F. and WELLS, W. W. (1974) Insulin effects on brain energy metabolism and the related hexokinase distribution. *J. Biol. Chem.* **249,** 6930–6935.

671. KOCH, G. L. E., SHAW, D. C. and GIBSON, F. (1971) The purification and characterization of chorismate mutase-prephenate dehydrogenase from *Escherichia coli* K12. *Biochim. Biophys. Acta* **229,** 795–804.

672. KOCHETOV, G. A., NIKITUSHKINA, L. I. and CHERNOV, N. N. (1970) A complex of functionally-bound enzymes: Transketolase and glyceraldehydephosphate dehydrogenase. *Biochem. Biophys. Res. Commun.* **40,** 873–879.

673. KOCH-SCHMIDT, A. C., MATTIASSON, B. and MOSBACH, K. (1977) Aspects on microenvironmental compartmentation. An evaluation of the influence of restricted diffusion, exclusion effects, and enzyme proximity on the overall efficiency of the sequential two-enzyme system malate dehydrogenase-citrate synthase in its soluble and immobilized form. *Eur. J. Biochem.* **81,** 71–78.

674. KOHN, M. C., MENTEN, L. E. and GARFINKEL, D. (1979) Convenient computer-program for fitting enzymatic rate laws to steady-state data. *Computers Biomed. Res.* **12,** 461–469.

675. KOPPERSCHLÄGER, G., BÄR, J., NISSLER, K. and HOFMANN, E. (1977) Physicochemical parameters and subunit composition of yeast phosphofructokinase. *Eur. J. Biochem.* **81,** 317–325.

676. KORN, E. D. (1978) Biochemistry of actomyosin-dependent cell motility (A review). *Proc. Natl. Acad. Sci. U.S.A.* **75**, 588–599.

677. KORNBLATT, J. A. and LAKE, D. F. (1980) Cross-linking of cytochrome oxidase subunits with difluorodinitrobenzene. *Can. J. Biochem.* **58**, 219–224.

678. KOSHLAND, D. E., JR. (1963) The role of flexibility in enzyme action. *Cold Spring Harbor Symp. Quant. Biol.* **28**, 473–482.

679. KOSHLAND, D. E., Jr., NÉMETHY, G. and FILMER, D. (1966) Comparison of experimental binding data and theoretical models in proteins containing subunits. *Biochemistry* **5**, 365–385.

680. KOSTYUCHENKO, V. I., KOSTYUCHENKO, D. A. and KOLOMIJTSEVA, G. YA. (1981) Dependence of RNA synthesis in the cell cycle of *Physarum polycephalum* on concentration and activity of RNA-polymerase II. *Biokhimiya* **46**, 327–332.

681. KOWALSKY, A. and BOYER, P. D. (1960) A carboxypeptidase-H_2O^{18} procedure for determination of COOH-terminal residues and its application to aldolase. *J. Biol. Chem.* **235**, 604–608.

682. KRAKOW, J. S. and KUMAR, S. A. (1977) Biosynthesis of ribonucleic acid. In *Comprehensive Biochemistry*. Vol. 24, pp. 105–184. (FLORKIN, M., NEUBERGER, A. and VAN DEENEN, L. L. M., eds), Elsevier, Amsterdam–London–New York.

683. KRATKY, O. (1963) X-ray small angle scattering with substances of biological interest in diluted solutions. *Progr. Biophys. Biophys. Chem.* **13**, 105–173.

684. KREBS, E. G. and BEAVO, J. A. (1979) Phosphorylation-dephosphorylation of enzymes. *Annu. Rev. Biochem.* **48**, 923–959.

685. KREIL, G. (1981) Transfer of proteins across membranes. *Annu. Rev. Biochem.* **50**, 317–348.

686. KRESHECK, G. C. and KLOTZ, I. M. (1969) The thermodynamics of transfer of amides from an apolar to an aqueous solution. *Biochemistry* **8**, 8–12.

687. KRESZE, G. B., DIETL, B. and RONFT, H. (1980) Mammalian lipoate acetyltransferase: Molecular weight determination by gel filtration in the presence of guanidinium chloride. *FEBS Lett.* **112**, 48–50.

688. KRESZE, G. B., OESTERHELT, D., LYNEN, F., CASTORPH, H. and SCHWEIZER, E. (1976) Localization of the central and peripheral SH-groups on the same polypeptide chain of yeast fatty acid synthetase. *Biochem. Biophys. Res. Commun.* **69**, 893–899.

689. KUCHEL, P. W., NICHOL, L. W. and JEFFREY, P. D. (1974) Steady state kinetics of consecutive enzyme catalysed reactions involving single substrates: Procedures for the interpretation of coupled assays. *J. Theor. Biol.* **48**, 39–49.

690. KUEBBING, D. and WERNER, R. (1975) A model for compartmentation of de novo and salvage thymidine nucleotide pools in mammalian cells. *Proc. Natl. Acad. Sci. U.S.A.* **72**, 3333–3336.

691. KUEHN, G. D., AFFOLTER, H.-U., ATMAR, V. J., SEEBECK, T., GUBLER, U. and BRAUN, R. (1979) Polyamine-mediated phosphorylation of a nucleolar protein from *Physarum polycephalum* that stimulates rRNA synthesis. *Proc. Natl. Acad. Sci. U.S.A.* **76**, 2541–2545.

692. KUMAGAI, H. and MILES, E. W. (1971) The B protein of *Escherichia coli* tryptophan synthetase. II. New β-elimination and β-replacement reactions. *Biochem. Biophys. Res. Commun.* **44**, 1271–1278.

693. KUNTZ, G., STÖCKEL, P. and HEIDRICH, H.-G. (1978) The conformer nature of the multiple forms of beef liver catalase as obtained by biochemical and small-angle X-ray scattering experiments. A model for the quaternary structure of the beef liver catalase molecule. *Hoppe-Seyler's Z. Physiol. Chem.* **359**, 959–973.

694. KURGANOV, B. I. (1967) A kinetic method for the calculation of protein molecules' association constants. *Mol. Biol.* **1**, 17–27.

695. KURGANOV, B. I. (1968) Kinetic analysis of dissociating enzymatic systems. *Mol. Biol.* **2**, 430–447.

696. KURGANOV, B. I. (1973) The kinetic criteria for validity of the regulatory enzyme models with indirect cooperativity. *Acta Biol. Med. Germ.* **31**, 181–201.

697. KURGANOV, B. I. (1974) Kinetic behaviour of associating enzyme systems of the types $M \rightleftharpoons$ $\rightleftharpoons M_2 \rightleftharpoons M_3 \ldots$ and $2M \rightleftharpoons D \rightleftharpoons D_2 \rightleftharpoons D_3 \rightleftharpoons \ldots$ *Mol. Biol.* **8**, 525–535.

698. KURGANOV, B. I. (1975) Regulatory properties of slowly equilibrating association–dissociation enzyme systems. *Proc. 9th FEBS Meeting* **32**, 29–42.

699. KURGANOV, B. I. (1977) The theoretical analysis of kinetic behaviour of "hysteretic" allosteric enzymes. IV. Kinetics of dissociation–association processes of allosteric enzymes. *J. Theor. Biol.* **68**, 521–543.

700. KURGANOV, B. I., DOROZHKO, A. I., KAGAN, Z. S. and YAKOVLEV, V. A. (1975) The features of kinetic behavior of slowly equilibrating association–dissociation systems. *Mol. Biol.* **9**, 533–542.

701. KURGANOV, B. I., DOROZHKO, A. I., KAGAN, Z. S. and YAKOVLEV, V. A. (1975) Deviations from hyperbolic kinetics in slowly dissociating allosteric enzyme systems. *Biokhimiya* **40**, 793–801.

702. KURGANOV, B. I., DOROZHKO, A. I., KAGAN, Z. S. and YAKOVLEV, V. A. (1976) The theoretical analysis of kinetic behaviour of "hysteretic" allosteric enzymes. I. The kinetic manifestations of slow conformational change of an oligomeric enzyme in the Monod, Wyman and Changeux model. *J. Theor. Biol.* **60**, 247–269.

703. KURGANOV, B. I., KLINOV, S. V. and SUGROBOVA, N. P. (1978) Regulation of enzyme activity in adsorptive enzyme systems. In *Symp. Biol. Hung.* Vol. 21, pp. 81–106. (DAMJANOVICH, S., ELŐDI, P. and SOMOGYI, B., eds), Akadémiai Kiadó, Budapest.

704. KURGANOV, B. I. and LOBODA, N. I. (1979) Regulation of enzyme activity in adsorptive enzyme systems. *J. Theor. Biol.* **79**, 281–301.

705. KURGANOV, B. I. and YAKOVLEV, V. A. (1973) Kinetic analysis of the dissociating enzyme systems of the type monomer $\rightleftharpoons$ dimer $\rightleftharpoons$ tetramer and monomer $\rightleftharpoons$ tetramer. *Mol. Biol.* **7**, 429–447.

706. KURLAND, C. G. (1977) Structure and function of the bacterial ribosome. *Annu. Rev. Biochem.* **46**, 173–200.

706a. KUTER, M. R., MASTERS, C. J., WALSH, T. P. and WINZOR, D. J. (1981) Effect of ionic strength on the interaction between aldolase and actin-containing filaments. *Arch. Biochem. Biophys.* **212**, 306–310.

707. KWON, T.-W. and OLCOTT, H. S. (1965) Augmentation of aldolase activity by glyceraldehyde-3-phosphate dehydrogenase. *Biochem. Biophys. Res. Commun.* **19**, 300–305.

708. LAD, P. M., HILL, D. E. and HAMMES, G. G. (1973) Influence of allosteric ligands on the activity and aggregation of rabbit muscle phosphofructokinase. *Biochemistry* **12**, 4303–4309.

709. LAKATOS, S. and ZÁVODSZKY, P. (1976) The effect of substrates on the association equilibrium of mammalian D-glyceraldehyde 3-phosphate dehydrogenase. *FEBS Lett.* **63**, 145–148.

710. LAKE, J. A. (1981) The ribosome. *Sci. Amer.* **245** (2), 56–69.

711. LAKE, J. A. and KAHAN, L. (1975) Ribosomal proteins S5, S11, S13 and S19 localized by electron microscopy of antibody-labeled subunits. *J. Mol. Biol.* **99**, 631–644.

712. LAMPKIN, S. L., COLE, K. W., VITTO, A. and GAERTNER, F. H. (1976) The protease problem in *Neurospora*. Variable stability of enzymes in aromatic amino acid metabolism. *Arch. Biochem. Biophys.* **177**, 561–565.

713. LANGLEY, K. E., VILLAREJO, M. R., FOWLER, A. V. and ZAMENHOF, P. J. (1975) Molecular basis of β-galactosidase α-complementation. *Proc. Natl. Acad. Sci. U.S.A.* **72**, 1254–1257.

714. LANGLOIS, R., LEE, C. C., CANTOR, C. R., VINCE, R. and PESTKA, S. (1976) The distance between two functionally significant regions of the 50S *Escherichia coli* ribosome: the Erythromycin binding site and proteins L7/L12. *J. Mol. Biol.* **106**, 297–313.

715. LARDY, H. A., JOHNSON, D. and McMURRAY, W. C. (1958) Antibiotics as tools for metabolic studies. I. A survey of toxic antibiotics in respiratory, phosphorylative and glycolytic systems. *Arch. Biochem. Biophys.* **78**, 587–597.

716. LARNER, J., GALASKO, G., CHENG, K., DePADI-ROACH, A. A., HUANG, L., DAGGY, P. and KELLOGG, J. (1979) Generation by insulin of a chemical mediator that controls protein phosphorylation and dephosphorylation. *Science* **206**, 1408–1410.

717. LARSEN, F. L. and VINCENZI, F. F. (1979) Calcium transport across the plasma membrane: Stimulation by calmodulin. *Science* **204**, 306–309.

718. LASZLO, E. (1972) *The Systems View of the World: The Natural Philosophy of the New Developments in the Sciences,* Braziller, G., New York.

719. LATZKOVITS, L., FAJSZI, Cs. and SZENTISTVÁNYI, I. (1972) Tracer kinetic analysis of phosphate incorporation of erythrocytes *in vitro.* II. Model analysis of the system with the ATP pool not in steady state. *Acta Biochim. Biophys. Acad. Sci. Hung.* **7**, 307–314.

720. LATZKOVITS, L., SZENTISTVÁNYI, I. and FAJSZI, Cs. (1972) Tracer kinetic analysis of phosphate incorporation into erythrocytes *in vitro. Acta Biochim. Biophys. Acad. Sci. Hung.* **7**, 55–66.

721. LAURENT, M., SEYDOUX, F. J. and DESSEN, P. (1979) Allosteric regulation of yeast phosphofructokinase. Correlation between equilibrium binding, spectroscopic and kinetic data. *J. Biol. Chem.* **254**, 7515–7520.

722. LEBHERZ, H. G. (1972) Stability of quaternary structure of mammalian and avian fructose diphosphate aldolases. *Biochemistry* **11**, 2243–2250.

723. LECOQ, D., HERVAGAULT, J. F., BROUN, G., JOLY, G., KERNEVEZ, J. P. and THOMAS, D. (1975) The kinetic behavior of an artificial bienzyme membrane. *J. Biol. Chem.* **250**, 5496–5500.

724. LEE, B. and RICHARDS, F. M. (1971) The interpretation of protein structures: Estimation of static accessibility. *J. Mol. Biol.* **55**, 379–400.

725. LEESE, H. J. and SEMENZA, G. (1973) On the identity between the small intestinal enzymes phlorizin hydrolase and glycosylceramidase. *J. Biol. Chem.* **248**, 8170–8173.

726. LEFKOWITZ, R. J. (1978) Identification and regulation of alpha- and beta-adrenergic receptors. *Fed. Proc.* **37**, 123–129.

727. LEFKOWITZ, R. J. and WILLIAMS, L. T. (1977) Catecholamine binding to the β-adrenergic receptor. *Proc. Natl. Acad. Sci. U.S.A.* **74**, 515–519.

728. LEGRAIN, C. and STALON, V. (1976) Ornithine carbamoyltransferase from *Escherichia coli* W. Purification, structure and steady-state kinetic analysis. *Eur. J. Biochem.* **63**, 289–301.

729. LEIGH, J. S. and HARMON, H. J. (1977) Heme plane orientation in mitochondrial membranes. *Biophys. J.* **17**, 251a.

730. LEONARD, K. R., ARAD, T., WINGFIELD, P. and WEISS, H. (1980) *EBEC Reports* **1**, 83–84.

731. LETKO, G. and BOHNENSACK, R. (1975) Further characterization of the association of glyceraldehyde-3-phosphate dehydrogenase with reticulocyte membranes. *Acta Biol. Med. Germ.* **34**, 1145–1151.

732. LEVINE, S., GILLETT, T. A., HAGEMAN, E. and HANSEN, R. G. (1969) Uridine diphosphate glucose pyrophosphorylase. II. Polymeric and subunit structure. *J. Biol. Chem.* **244**, 5729–5734.

733. LEVINSON, B. B., ULLMAN, B. and MARTIN, D. W. (1979) Pyrimidine pathway variants of cultured mouse lymphoma cells with altered levels of both orotate phosphoribosyltransferase and orotidylate decarboxylase. *J. Biol. Chem.* **254**, 4396–4401.

734. LEVITZKI, A. (1974) Negative co-operativity in clustered receptors as a possible basis for membrane action. *J. Theor. Biol.* **44**, 367–372.

735. LEVITZKI, A. and KOSHLAND, D. E., JR. (1976) The role of negative cooperativity and half-of-the-sites reactivity in enzyme regulation. *Curr. Top. Cell. Regul.* **10**, 1–40.

736. LIANG, C. T. and SACKTOR, B. (1977) Preparation of renal cortex basal-lateral and brush border membranes. Localization of adenylate cyclase and guanylate cyclase activities. *Biochim. Biophys. Acta* **466**, 474–487.

737. LIFSON, S. and WARSHEL, A. (1968) Consistent force field for calculations of conformations, vibrational spectra, and enthalpies of cycloalkane and n-alkane molecules. *J. Chem. Phys.* **49**, 5116–5129.

738. LILJAS, A. and ROSSMANN, M. G. (1974) X-ray studies of protein interactions. *Annu. Rev. Biochem.* **43**, 475–507.

739. LIMBIRD, L. E. and LEFKOWITZ, R. J. (1978) Agonist-induced increase in apparent β-adrenergic receptor size. *Proc. Natl. Acad. Sci. U.S.A.* **75**, 228–232.

740. LIN, D. C. and LIN, S. (1978) High affinity binding of [³H] dihydrocytochalasin B to peripheral membrane proteins related to the control of cell shape in the human red cell. *J. Biol. Chem.* **253**, 1415–1419.

741. LINN, T. C., PELLEY, J. W., PETTIT, F. H., HUCHO, F., RANDALL, D. D. and REED, L. J. (1972) α-Keto acid dehydrogenase complexes. XV. Purification and properties of the component enzymes of the pyruvate dehydrogenase complexes from bovine kidney and heart. *Arch. Biochem. Biophys.* **148**, 327–342.

742. LINN, T. C., PETTIT, F. H. and REED, L. J. (1969) α-Keto acid dehydrogenase complexes. X. Regulation of the activity of the pyruvate dehydrogenase complex from beef kidney mitochondria by phosphorylation and dephosphorylation. *Proc. Natl. Acad. Sci. U.S.A.* **62**, 234–241.

743. LIOU, R.-S. and ANDERSON, S. (1980) Activation of rabbit muscle phosphofructokinase by F-actin and reconstituted thin filaments. *Biochemistry* **19**, 2684–2688.

744. LIU, L. F., LIU, C.-C. and ALBERTS, B. M. (1980) Type II DNA topoisomerases: Enzymes that can unknot a topologically knotted DNA molecule via a reversible double-stranded break. *Cell* **19**, 697–707.

745. LLEWELLYN, D. J. and SMITH, G. D. (1979) Study of chorismate mutase-prephenate dehydrogenase in crude cell extracts of *Escherichia coli*. *Biochemistry* **18**, 4707–4714.

746. LODISH, H. F. and ROTHMAN, J. E. (1979) The assembly of cell membranes. *Sci. Amer.* **240/1**, 38–53.

747. LOEWENSTEIN, W. R., KANNO, Y. and SOCOLAR, S. J. (1978) The cell-to-cell channel. *Fed. Proc.* **37**, 2645–2650.

748. LONDESBOROUGH, J. (1977) Characterization of an adenosine 3':5'-cyclic monophosphate phosphodiesterase from baker's yeast. *Biochem. J.* **163**, 467–476.

749. LONDESBOROUGH, J. and JÖNKÄRRI, L. (1982) Low K_m cyclic AMP phosphodiesterase of yeast may be bound to ribosomes associated with the nucleus. *Mol. Cell. Biochem.* **46**, 65–71.

750. LONDON, F. (1930) Über einige Eigenschaften und Anwendungen der Molekularkräfte. *Z. Phys. Chem. Abt. Bx* **11**, 222–251.

751. LOSICK, R. and CHAMBERLIN, M. J. (eds.) (1976) *RNA Polymerase,* Cold Spring Harbor Laboratory, New York.

752. LUDWIG, B., DOWNER, N. W. and CAPALDI, R. A. (1979) Labeling of cytochrome c oxidase with [³⁵S] diazobenzenesulfonate. Orientation of this electron transfer complex in the inner mitochondrial membrane. *Biochemistry* **18**, 1401–1407.

753. LUE, P. F. and KAPLAN, J. G. (1969) The aspartate transcarbamylase and carbamoyl phosphate synthetase of yeast: A multi-functional enzyme complex. *Biochem. Biophys. Res. Commun.* **34**, 426–433.

754. LUMRY, R. (1959) Some aspects of the thermodynamics and mechanism of enzymic catalysis. In *The Enzymes* 2nd ed., Vol 1, pp. 157–231. (BOYER, P. D., LARDY, H. and MYRBÄCK, K., eds), Academic Press, New York.

755. LUMSDEN, J. and COGGINS, J. R. (1978) The subunit structure of the arom multienzyme complex of *Neurospora crassa*. Evidence from peptide 'maps' for the identity of the subunits. *Biochem. J.* **169**, 441–444.

756. LYNEN, A., SEDLACZEK, E. and WIELAND, O. H. (1978) Partial purification and characterization of a pyruvate dehydrogenase-complex-inactivating enzyme from rat liver. *Biochem. J.* **169**, 321–328.

757. LYNEN, F. (1964) Coordination of metabolic processes by multienzyme complexes. In *New Perspectives in Biology*. Vol. 4, pp. 132–146. (SELA, M., ed.), Elsevier, Amsterdam.

758. LYNEN, F. (1967) In *Progress in Biochemical Pharmacology*. Vol. 3, pp. 1–31. (KRITCHEVSKY, D. PAOLETTI, R. and STEINBERG, M. D., eds), Karger, Basel.

759. LYNEN, F. (1969) Yeast fatty acid synthase. *Methods Enzymol.* **14,** 17–33.

760. MACGREGOR, J. S., SINGH, V. N., DAVOUST, S., MELLONI, E., PONTREMOLI, S. and HORECKER, B. L. (1980) Evidence for formation of a rabbit liver aldolase–rabbit liver fructose-1,6-bisphosphatase complex. *Proc. Natl. Acad. Sci. U. S. A.* **77,** 3889–3892.

761. MACGREGOR, R. D. and TOBIAS, C. A. (1972) Molecular sieving of red cell membranes during gradual osmotic hemolysis. *J. Membr. Biol.* **10,** 345–356.

762. MACHICAO, F. and WIELAND, O. H. (1980) Subunit structure of dihydrolipoamide acetyltransferase component of pyruvate dehydrogenase complex from bovine kidney. *Z. Physiol. Chem.* **361,** 1093–1106.

763. MACLENNAN, D. H. (1970) Purification and properties of an adenosine triphosphatase from sarcoplasmic reticulum. *J. Biol. Chem.* **245,** 4508–4518.

764. MACLANNAN, D. H. and TZAGOLOFF, A. (1968) Studies on the mitochondrial adenosine triphosphatase system. IV. Purification and characterization of the oligomycin sensitivity conferring protein. *Biochemistry* **7,** 1603–1610.

765. MADSEN, N. B., KASVINSKY, P. J. and FLETTERICK, R. J. (1978) Allosteric transitions of phosphorylase a and the regulation of glycogen metabolism. *J. Biol. Chem.* **253,** 9097–9101.

766. MAGUIRE, M. E., ROSS, E. M. and GILMAN, A. G. (1977) β-Adrenergic receptor: Ligand binding properties and the interaction with adenylyl cyclase. *Adv. in Cycl. Nucl. Res.* **8,** 1–83.

767. MAIRE, M., MØLLER, J. V. and TANFORD, C. (1976) Retention of enzyme activity by detergent-solubilized sarcoplasmic Ca^{2+}-ATPase. *Biochemistry* **15,** 2336–2342.

768. MAKOWSKI, L. D., CASPAR, L. D., PHILIPS, W. C. and GOODENOUGH, D. A. (1977) Gap junction structures. II. Analysis of the X-ray diffraction data. *J. Cell Biol.* **74,** 629–645.

769. MALATHI, P. and CRANE, R. K. (1969) Phlorizin hydrolase: A β-glucosidase of hamster intestinal brush border membrane. *Biochim. Biophys. Acta* **173,** 245–256.

770. MANEN, C.-A. and RUSSELL, D. H. (1977) Ornithine decarboxylase may function as an initiation factor for RNA polymerase I. *Science* **195,** 505–506.

771. MANLEY, E. R., WEBSTER, T. A. and SPIVEY, H. O. (1980) Kinetics of coupled aspartate aminotransferase-malate dehydrogenase reactions and instability of oxaloacetate on anion-exchange resin. *Arch. Biochem. Biophys.* **205,** 380–387.

772. MANNEY, T. R. (1970) Physiological advantage of the mechanism of the tryptophan synthetase reaction. *J. Bacteriol.* **102,** 483–488.

773. MANSOUR, T. E. and AHLFORS, C. E. (1968) Studies on heart phosphofructokinase. Some kinetic and physical properties of the crystalline enzyme. *J. Biol. Chem.* **243,** 2523–2533.

774. MARCHAND, A., CHAPOUTHIER, G. and MASSOULIÉ, J. (1977) Developmental aspects of acetylcholinesterase activity in chick brain. *FEBS Lett.* **78,** 233–236.

775. MARCHESI, V. T. (1979) Spectrin: Present status of a putative cyto-skeletal protein of the red cell membrane. *J. Membrane Biol.* **51,** 101–131.

776. MARETZKI, D., GROTH, J., TSAMALOUKAS, A. G., GRÜNDEL, M., KRÜGER, S. and RAPOPORT, S. (1974) The membrane association and dissociation of human glyceraldehyde-3-phosphate dehydrogenase under various conditions of hemolysis. Immunochemical evidence for the lack of binding under cellular conditions. *FEBS Lett.* **39,** 83–87.

777. MARETZKI, D. and RAPOPORT, S. (1972) Glyzerinaldehyd-3-phosphat-Dehydrogenase aus Erythrozyten des Menschen. I. Isolierung und einige Eigenschaften. *Acta Biol. Med. Germ.* **29,** 207–221.

778. MARKERT, C. L. (1963) Lactate dehydrogenase isozymes: Dissociation and recombination of subunits. *Science* **140,** 1329–1330.

779. MARKHAM, R., FREY, S. and HILLS, G. J. (1963) Methods for the enhancement of image detail and accentuation of structure in electron microscopy. *Virology* **20,** 88–102.

780. MARKUS, M., PLESSER, T., BOITEUX, A., HESS, B. and MALCORATI, M., (1980) Analysis of progress curves. *Biochem. J.* **189,** 421–433.

781. MAROUX, S., LOUVARND, D. and BARATTI, J. (1973) The aminopeptidase from hog intestinal brush border. *Biochim. Biophys. Acta* **321**, 282–295.

782. MARTIN, G. M. (1979) Genetic and evolutionary aspects of aging. *Fed. Proc.* **38**, 1962–1967.

783. MARTONOSI, A. and HALPIN, R. A. (1971) Sarcoplasmic reticulum X. The protein composition of sarcoplasmic reticulum membranes. *Arch. Biochem. Biophys.* **144**, 66–77.

784. MASTERS, C. J. (1967) Characteristics of aldolase variformity. *Biochem. Biophys. Res. Commun.* **28**, 978–984.

785. MASTERS, C. J. (1977) Metabolic control and the microenvironment. *Curr. Top. Cell. Regul.* **12**, 75–105.

786. MASTERS, C. J. (1978) Interactions between soluble enzymes and subcellular structure. *Trends Biochem. Sci.* **3**, 206–208.

787. MASTERS, C. J. and WINZOR, D. J. (1981) Physicochemical evidence against the concept of an interaction between aldolase and glyceraldehyde-3-phosphate dehydrogenase. *Arch. Biochem. Biophys.* **209**, 185–190.

788. MATCHETT, W. H. and DE MOSS, J. A. (1964) Physiological channeling of tryptophan in *Neurospora crassa*. *Biochim. Biophys. Acta* **86**, 91–99.

789. MATLIB, M. A., BOESMAN-FINKELSTEIN, M. and SRERE, P. A. (1978) The kinetics of rat liver citrate synthase *in situ*. *Arch. Biochem. Biophys.* **191**, 426–430.

790. MATCHETT, W. H. (1974) Indole channeling by tryptophan synthase of *Neurospora*. *J. Biol. Chem.* **249**, 4041–4049.

791. MATCHETT, W. H. and DE MOSS, J. A. (1975) The subunit structure of tryptophan synthase from *Neurospora crassa*. *J. Biol. Chem.* **250**, 2941–2946.

792. MATTHEWS, D. M. (1972) Rates of peptide uptake by small intestine in peptide transport in bacteria and mammalian gut. *Ciba Foundation Symposium,* pp. 71–92. Elsevier, North-Holland, Amsterdam.

793. MATTIASSON, B., JOHANSSON, A.-C. and MOSBACH, K. (1974) Preparation of a soluble, bifunctional enzyme aggregate and studies on its kinetic behaviour in polymer media. *Eur. J. Biochem.* **46**, 341–349.

794. MATTIASSON, B. and MOSBACH, K. (1971) Studies on a matrix-bound three-enzyme system. *Biochim. Biophys. Acta* **235**, 253–257.

795. MAVRIDIS, I. M. and TULINSKY, A. (1976) The folding and quaternary structure of trimeric 2-keto-3-deoxy-6-phosphogluconic aldolase at 3.5 Å resolution. *Biochemistry* **15**, 4410–4417.

796. MAYER, F., WALLACE, J. C. and KEECH, D. B. (1980) Further electron microscopic studies on pyruvate carboxylase. *Eur. J. Biochem.* **112**, 265–272.

797. MAYER, R. J. and HÜBSCHER, G. (1971) Mitochondrial hexokinase from small-intestinal mucosa and brain. *Biochem. J.* **124**, 491–500.

798. MAZLIAK, P. (1977) Glyco- and phospholipids of biomembranes in higher plants. In *Lipids and Lipid Polymers in Higher Plants*. pp. 48–74. (TEVINI, M. and LICHTENHALTER, H. K., eds), Springer, Berlin.

799. MCCLURE, W. R. (1969) A kinetic analysis of coupled enzyme assays. *Biochemistry* **8**, 2782–2786.

800. MCCRACKEN, S. and MEIGHEN, E. (1979) Elucidation of the quaternary structure of reversibly immobilized alkaline phosphatase derivatives. *Can. J. Biochem.* **57**, 834–842.

801. MCDANIEL, C. F., KIRTLEY, M. E. and TANNER, M. J. A. (1974) The interaction of glyceraldehyde 3-phosphate dehydrogenase with human erythrocyte membranes. *J. Biol. Chem.* **249**, 6478–6485.

802. MCELVAIN, S. M. and SCHROEDER, J. P. (1949) Orthoesters and related compounds from malono- and succinonitriles. *J. Am. Chem. Soc.* **71**, 40–46.

803. MCFADDEN, B. A., LORD, J. M., ROWE, A. and DILKS, S. (1975) Composition, quaternary structure, and catalytic properties of D-ribulose-1,5-bisphosphate carboxylase from *Euglena gracilis*. *Eur. J. Biochem.* **54,** 195–206.

804. McGHEE, J. D. and FELSENFELD, F. (1980) Nucleosome structure. *Annu. Rev. Biochem.* **49,** 1115–1156.

805. McKEEHAN, W. L. and HARDESTY, B. (1969) Purification and partial characterization of the aminoacyl transfer ribonucleic acid binding enzyme from rabbit reticulocytes. *J. Biol. Chem.* **244,** 4330–4339.

806. McMINN, C. L. and OTTAWAY, J. H. (1976) On the control of enzyme pathways. *J. Theor. Biol.* **56,** 57–73.

807. MEIGHEN, E. A., PIGIET, V. and SCHACHMAN, H. K. (1970) Hybridization of native and chemically modified enzymes. III. The catalytic subunits of aspartate transcarbamylase. *Proc. Natl. Acad. Sci. U.S.A.* **65,** 234–241.

808. MEIGHEN, E. A. and SCHACHMAN, H. K. (1970) Hybridization of native and chemically modified enzymes. I. Development of a general method and its application to the study of the subunit structure of aldolase. *Biochemistry* **9,** 1163–1176.

809. MEIGHEN, E. A. and SCHACHMAN, H. K. (1970) Hybridization of native and chemically modified enzymes. II. Native and succinylated glyceraldehyde 3-phosphate dehydrogenase. *Biochemistry* **9,** 1177–1184.

810. MEISTER, A. and TATE, S. S. (1976) Glutathione and related γ-glutamyl compounds: Biosynthesis and utilization. *Annu. Rev. Biochem.* **45,** 559–604.

811. MELNICK, R. L. and HULTIN, H. O. (1973) Studies on the nature of the subcellular localization of lactate dehydrogenase and glyceraldehyde-3-phosphate dehydrogenase in chicken skeletal muscle. *J. Cell. Physiol.* **81,** 139–148.

812. MENNECIER, F., WEBER, A., TUDURY, C. and DREYFUS, J. C. (1979) Modifications of aldolase during in vivo aging of rabbit red cells. *Biochimie* **61,** 79–85.

813. METZLER, D. E. (1977) *Biochemistry, The Chemical Reactions of Living Cells*, pp. 853–855. Academic Press, New York.

814. MEUNIER, J.-C. and DALZIEL, K. (1978) Kinetic studies of glyceraldehyde-3-phosphate dehydrogenase from rabbit muscle. *Eur. J. Biochem.* **82,** 483–492.

815. MEYER, F., HEILMEYER, L. M. G., HASCHKE, R. H. and FISCHER, E. H. (1970) Control of phosphorylase activity in a muscle glycogen particle. I. Isolation and characterization of the protein–glycogen complex. *J. Biol. Chem.* **245,** 6642–6648.

816. MICHAELIS, L. and MENTEN, M. L. (1913) Die Kinetik der Invertinwirkung. *Biochem. Z.* **49,** 333–369.

817. MIDELFORT, C. F. and MEHLER, A. H. (1972) Deamidation in vivo of an asparagine residue of rabbit muscle aldolase. *Proc. Natl. Acad. Sci. U.S.A.* **69,** 1816–1819.

818. MIEKKA, S. I. and INGHAM, K. C. (1980) Influence of hetero-association on the precipitation of proteins by poly(ethyleneglycol). *Arch. Biochem. Biophys.* **203,** 630–641.

819. MILES, E. W. (1979) Tryptophan synthase: Structure, function, and subunit interaction. *Adv. Enzymol.* **49,** 127–186.

820. MILES, E. W. and MORIGUCHI, M. (1977) Tryptophan synthase of *Escherichia coli*. Removal of pyridoxal 5'-phosphate and separation of the α and β_2 subunits. *J. Biol. Chem.* **252,** 6594–6599.

821. MILLS, G. C. and HILL, F. L. (1971) Metabolic control mechanisms in human erythrocytes. The role of glyceraldehyde phosphate dehydrogenase. *Arch. Biochem. Biophys.* **146,** 306–311.

822. MILSTEIN, J. and BREMERMANN, H. J. (1979) Parameter identification of the Calvin photosynthesis cycle. *J. Math. Biol.* **7,** 99–116.

823. MINTON, A. P. (1981) Excluded volume as a determinant of macromolecular structure and reactivity. *Biopolymers* **20,** 2093–2120.

824. MINTON, A. P. and WILF, J. (1981) Effect of macromolecular crowding upon the structure and function of an enzyme: Glyceraldehyde-3-phosphate dehydrogenase. *Biochemistry* **20,** 4821–4826.

825. MITCHELL, C. D., MITCHELL, W. B. and HANAHAN, D. J. (1965) Enzyme and hemoglobin retention in human erythrocyte stroma. *Biochim. Biophys. Acta* **104**, 348–358.

826. MITCHELL, P. (1977) Vectorial chemiosmotic processes. *Annu. Rev. Biochem.* **46**, 996–1005.

827. MITTELBACH, P. (1964) Zur Röntgen-kleinwinkelstreuung verdünnter kolloider Systeme. *Acta Phys. Austriaca* **19**, 53–102.

828. MOCKRIN, S. C., BYERS, L. D. and KOSHLAND, D. E. (1975) Subunit interactions in yeast glyceraldehyde-3-phosphate dehydrogenase. *Biochemistry* **14**, 5428–5437.

829. MOMANY, F. A., CARRUTHERS, L. M., MCGUIRE, R. F. and SCHERAGA, M. A. (1974) Intermolecular potentials from crystal data. III. Determination of empirical potentials and application to the packing configurations and lattice energies in crystals of hydrocarbons, carboxylic acids, amines, and amides. *J. Phys. Chem.* **78**, 1595–1620.

830. MOMSEN, G., ROSE, Z. B. and GUPTA, R. K. (1979) A reappraisal of [^{31}P] NMR studies indicating enzyme complexation in red blood cells. *Biochem. Biophys. Res. Commun.* **91**, 651–657.

831. MONACO, H. L., CRAWFORD, J. L. and LIPSCOMB, W. N. (1978) Three-dimensional structures of aspartate carbamoyltransferase from *Escherichia coli* and of its complex with cytidine triphosphate. *Proc. Natl. Acad. Sci. U.S.A.* **75**, 5276–5280.

832. MONOD, J., CHANGEUX, J.-P. and JACOB, F. (1963) Allosteric proteins and cellular control systems. *J. Mol. Biol.* **6**, 306–329.

833. MONOD, J., COHEN-BAZIRE, G. and COHN, M. (1951) Sur la biosynthèse de la β-galactosidase (lactase) chez *Escherichia coli*. La specificité de l'induction. *Biochim. Biophys. Acta* **7**, 585–599.

834. MONOD, J., WYMAN, J. and CHANGEUX, J.-P. (1965) On the nature of allosteric transitions: A plausible model. *J. Mol. Biol.* **12**, 88–118.

835. MOODY, M. F., VACHETTE, P. and FOOTE, A. M. (1979) Changes in the X-ray solution scattering of aspartate transcarbamylase following the allosteric transition. *J. Mol. Biol.* **133**, 517–532.

836. MOODY, M. F., VACHETTE, P., FOOTE, A. M., TARDIEU, A., KOCH, M. H. J. and BORDAS, J. (1980) Stopped-flow X-ray scattering: The dissociation of aspartate transcarbamylase. *Proc. Natl. Acad. Sci. U.S.A.* **77**, 4040–4043.

837. MOORE, P. B., LANGER, J. A., SCHOENBORN, B. P. and ENGELMAN, D. M. (1977) Triangulation of proteins in the 30S ribosomal subunit of *Escherichia coli*. *J. Mol. Biol.* **112**, 199–234.

838. MOOSEKER, M. S. and TILNEY, L. G. (1975) Organization of an actin filament-membrane complex. Filament polarity and membrane attachment in the microvilli of intestinal epithelial cells. *J. Cell Biol.* **67**, 725–743.

839. MORAS, D., OLSEN, K. W., SABESAN, M. N., BUEHNER, M., FORD, G. C. and ROSSMANN, M. G. (1975) Studies of asymmetry in the three-dimensional structure of lobster D-glyceraldehyde-3-phosphate dehydrogenase. *J. Biol. Chem.* **250**, 9137–9162.

840. MORGAN, R. S., MILLER, S. L. and MCADON, J. M. (1979) The symmetry of self-complementary surfaces. *J. Mol. Biol.* **127**, 31–39.

841. MORI, M. and TATIBANA, M. (1978) Multi-enzyme complex of glutamine-dependent carbamoylphosphate synthetase with aspartate carbamoyltransferase and dihydroorotase from rat ascites-hepatoma cells. *Eur. J. Biochem.* **86**, 381–388.

842. MORIMOTO, H., LEHMANN, H. and PERUTZ, M. F. (1971) Molecular pathology of human haemoglobin: Stereochemical interpretation of abnormal oxygen affinities. *Nature* **232**, 408–413.

843. MORINO, Y. and SNELL, E. E. (1967) The subunit structure of tryptophanase. I. The effect of pyridoxal phosphate on the subunit structure and physical properties of tryptophanase. *J. Biol. Chem.* **242**, 5591–5601.

844. MORT, J. S. and CHAN, W. W.-C. (1975) Subunit interactions in aspartate transcarbamylase. Characterization of a complex between the catalytic and the regulatory subunits. *J. Biol. Chem.* **250**, 653–660.

845. MORTON, D. J., CLARKE, F. M. and MASTERS, C. J. (1977) An electron microscope study of the interaction between fructose diphosphate aldolase and actin-containing filaments. *J. Cell Biol.* **74**, 1016–1023.

846. MOSBACH, K. and MATTIASSON, B. (1970) Matrix-bound enzymes. Part II: Studies on a matrix-bound two-enzyme-system. *Acta Chem. Scand.* **24**, 2093–2100.

847. MOSBACH, K. and MATTIASSON, B. (1976) Multistep enzyme systems. *Methods Enzymol.* **44**, 453–478.

848. MOSBACH, K. and MATTIASSON, B. (1978) Immobilized model systems of enzyme sequences. *Curr. Top. Cell. Regul.* **14**, 197–241.

849. MOSES, R. E. (1972) Replicative deoxyribonucleic acid synthesis in a system diffusible for macromolecules. *J. Biol. Chem.* **247**, 6031–6038.

850. MOSES, R. E. and RICHARDSON, C. C. (1970) Replication and repair of DNA in cells of *Escherichia coli* treated with toluene. *Proc. Natl. Acad. Sci. U.S.A.* **67**, 674–681.

851. MOWBRAY, J. and MOSES, V. (1976) The tentative identification in *Escherichia coli* of a multi-enzyme complex with glycolytic activity. *Eur. J. Biochem.* **66**, 25–36.

852. MUNSON, K. B. (1981) Light-dependent inactivation of $(Na^+ + K^+)$-ATPase with a new photo-affinity reagent, chromium arylazido-β-alanyl ATP. *J. Biol. Chem.* **256**, 3223–3230.

853. MURPHY, A. J. (1976) Cross-linking of the sarcoplasmic reticulum ATPase protein. *Biochem. Biophys. Res. Commun.* **70**, 160–166.

854. MURTHY, M. R. N., GARAVITO, R. M., JOHNSON, J. E. and ROSSMANN, M. G. (1980) Structure of apo-D-glyceraldehyde-3-phosphate dehydrogenase at 3.0 Å resolution. *J. Mol. Biol.* **138**, 859–872.

855. MURTHY, S. N. P., LIU, T., KAUL, R. K., KÖHLER, H. and STECK, T. L. (1981) The aldolase-binding site of the human erythrocyte membrane is at the NH_2 terminus of band 3. *J. Biol. Chem.* **256**, 11203–11208.

856. NAGRADOVA, N. K., GOLOVINA, T. O. and MEVKH, A. T. (1974) Immobilized dimers of D-glyceraldehyde-3-phosphate dehydrogenase. *FEBS Lett.* **49**, 242–245.

857. NAKANO, M., SUMI, Y. and MIYAKAWA, M. (1977) Purification and properties of trehalase from rat intestinal mucosal cells. *J. Biochem.* **81**, 1041–1049.

858. NAKAZAWA, K., SANO, M. and SAITO, T. (1976) Subcellular distribution and properties of guanylate cyclase in rat cerebellum. *Biochim. Biophys. Acta* **444**, 563–570.

859. NEER, E. J. (1974) The size of adenylate cyclase. *J. Biol. Chem.* **249**, 6527–6531.

860. NEER, E. J. (1978) Size and detergent binding of adenylate cyclase from bovine cerebral cortex. *J. Biol. Chem.* **253**, 1498–1502.

861. NEFF, N. H. and YANG, H.-Y. (1974) Another look at the monoamine oxidases and the mono-amine oxidase inhibitor drugs. *Life Sci.* **14**, 2061–2074.

862. NELBACH, M. E., PIGIET, V. P., GERHART, J. C. and SCHACHMAN, H. K. (1972) A role for zinc in the quaternary structure of aspartate transcarbamylase from *Escherichia coli*. *Biochemistry* **11**, 315–327.

863. NEMAT-GORGANI, M. and WILSON, J. E. (1980) Ambiquitous behavior—a biological phenomenon of general significance. *Curr. Top. Cell. Regul.* **16**, 45–54.

864. NEWSHOLME, E. A. (1980) A possible metabolic basis for the control of body weight. *N. Engl. J. Med.* **302**, 400–405.

865. NEWSHOLME, E. A. and CRABTREE, B. (1973) Metabolic aspects of enzyme activity regulation. *Symp. Soc. Exp. Biol.* **27**, 429–460.

866. NEWSHOLME, E. A. and CRABTREE, B. (1976) Substrate cycles in metabolic regulation and in heat generation. *Biochem. Soc. Symp.* **41**, 61–109.

867. NEWSHOLME, E. A. and CRABTREE, B. (1979) Theoretical principles in the approaches to control of metabolic pathways and their application to glycolysis in muscle. *J. Mol. Cell. Cardiol.* **11**, 839–856.

868. NEWSHOLME, E. A. and CRABTREE, B. (1981) Flux-generating and regulatory steps in metabolic control. *Trends Biochem. Sci.* **6**, 53–56.

869. NG, S. Y., PARKER, C. S. and ROEDER, R. G. (1979) Transcription of cloned *Xenopus* 5S RNA genes by *X. laevis* RNA polymerase III in reconstituted system. *Proc. Natl. Acad. Sci. U.S.A.* **76**, 136–140.

870. NICHOL, L. W., JACKSON, W. J. H. and WINZOR, D. J. (1967) A theoretical study of the binding of small molecules to a polymerizing protein system. A model for allosteric effects. *Biochemistry* **6**, 2449–2456.

871. NICHOL, L. W., KUCHEL, P. W. and JEFFREY, P. D. (1974) The detection and consequences of association of two enzymes involved in catalysing consecutive reactions. *Biophys. Chem.* **2**, 354–358.

872. NICOLIS, G. and PRIGOGINE, I. (1977) *Self-Organization in Nonequilibrium Systems.* Wiley, New York.

873. NIEHAUS, W. G. and HAMMERSTEDT, R. H. (1976) Mode of orthophosphate uptake and ATP labeling by mammalian cells. *Biochim. Biophys. Acta* **443**, 515–524.

874. NISSELBAUM, J. S. and GREEN, S. (1969) A simple ultramicro method for determination of pyridine nucleotides in tissues. *Anal. Biochem.* **27**, 212–217.

875. NORÉN, O., SJÖSTRÖM, H., DANIELSEN, E. M., STAUN, M., JEPPESEN, L. and SVENSSON, B. (1979) Comparison of two pig intestinal brush border peptidases with the corresponding renal enzymes. *Z. Physiol. Chem.* **360**, 151–157.

876. NOVER, L., LYNEN, F. and MOTHES, K. (eds) (1980) *Cell Compartmentation and Metabolic Channeling.* VEB Gustav Fischer Verlag, Jena, Elsevier/North-Holland Biomedical Press, Amsterdam–New York–Oxford.

877. O'BRIEN, W. J. and FRERMAN, F. E. (1977) Evidence for a complex of three beta-oxidation enzymes in *Escherichia coli:* induction and localization. *J. Bacteriol.* **132**, 532–540.

878. OHNISHI, M. and URRY, D. W. (1970) Solution conformation of valinomycin–potassium ion complex. *Science* **168**, 1091–1092.

879. O'KEEFFE, E. T., HILL, R. L. and BELL, J. E. (1980) Active site of bovine galactosyltransferase: Kinetic and fluorescence studies. *Biochemistry* **19**, 4954–4962.

880. O'KEEFFE, E. T., MORDICK, T. and BELL, J. E. (1980) Bovine galactosyltransferase: Interaction with α-lactalbumin and the role of α-lactalbumin in lactose synthase. *Biochemistry* **19**, 4962–4966.

881. OLSEN, K. W., GARAVITO, R. M., SABESAN, M. N. and ROSSMANN, M. G. (1976) Anion binding sites in the active center of D-glyceraldehyde-3-phosphate dehydrogenase. *J. Mol. Biol.* **107**, 571–576.

882. OLSEN, K. W., GARAVITO, R. M., SABESAN, M. N. and ROSSMANN, M. G. (1976) Studies of coenzyme binding to glyceraldehyde-3-phosphate dehydrogenase. *J. Mol. Biol.* **107**, 577–584.

883. OLSEN, B. R., JIMENEZ, S. A., KIVIRIKKO, K. I. and PROCKOP, D. J. (1970) Electron microscopy of protocollagen proline hydroxylase from chick embryos. *J. Biol. Chem.* **245**, 2649–2655.

884. OLSON, J. W. and RUSSEL, D. H. (1980) Prolonged ornithine decarboxylase induction in regenerating carcinogen-treated liver. *Cancer Research* **40**, 4373–4380.

885. ONCLEY, J. L. (1941) *Ann. N. Y. Acad. Sci.* **41**, 121.

886. OPPERDOES, F. R. and BORST, P. (1977) Localization of nine glycolytic enzymes in a microbody-like organelle in *Trypanosoma brucei:* The glycosome. *FEBS Lett.* **80**, 360–364.

887. OPPERDOES, F. R., BORST, P., BAKHER, S. and LEENE, W. (1977) Localization of glycerol-3-phosphate oxidase in the mitochondrion and particulate NAD$^+$-linked glycerol-3-phosphate dehydrogenase in the microbodies of the blood-stream form of *Trypanosoma brucei. Eur. J. Biochem.* **76**, 29–39.

888. OPPERDOES, F. R., BORST, P. and SPITS, H. (1977) Particle-bound enzymes in the bloodstream form of *Trypanosoma brucei. Eur. J. Biochem.* **76**, 21–28.

889. ORGEL, L. E. (1963) The maintenance of the accuracy of protein synthesis and its relevance to ageing. *Proc. Natl. Acad. Sci. U.S.A.* **49**, 517–521.

890. OSBORNE, H. H. and HOLLAWAY, M. R. (1975) The investigation of substrate-induced changes in subunit interactions in glyceraldehyde 3-phosphate dehydrogenase by measurement of the kinetics and thermodynamics of subunit exchange. *Biochem. J.* **151**, 37–45.

891. OSHINO, N., JAMIESON, D., SUGANO, T. and CHANCE, B. (1975) Optical measurement of the catalase–hydrogen peroxide intermediate (Compound I) in the liver of anaesthetized rats and its implication to hydrogen peroxide production *in situ. Biochem. J.* **146**, 67–77.

892. OTTAWAY, J. H. (1979) Disequilibrium among triose phosphates: The role of aldolase. *Biochem. Soc. Trans.* **7**, 400–402.

893. OTTAWAY, J. H. (1979) Sequestration of metabolites: Insights into metabolic control. *Biochem. Soc. Trans.* **7**, 1161–1167.

894. OTTAWAY, J. H. (1979) Simulation of metabolic events. *Techn. Life Sci.* B2/II: *Techn. Metabol. Res.* **B 219**, 1–27.

895. OTTAWAY, J. H. and McMINN, C. L. (1979) The regulation of acetoacetate metabolism in heart. *Biochem. Soc. Trans.* **7**, 411–412.

896. OTTAWAY, J. H. and MOWBRAY, J. (1977) The role of compartmentation in the control of glycolysis. *Curr. Top. Cell. Regul.* **12**, 107–208.

897. OTTOLENGHI, P. (1975) The reversible delipidation of a solubilized sodium-plus-potassium ion-dependent adenosine triphosphatase from the salt gland of the spiny dogfish. *Biochem. J.* **151**, 61–66.

898. OVÁDI, J., BATKE, J., BARTHA, F. and KELETI, T. (1979) Effect of association–dissociation on the catalytic properties of glyceraldehyde 3-phosphate dehydrogenase. *Arch. Biochem. Biophys.* **193**, 28–33.

899. OVÁDI, J. and KELETI, T. (1978) Kinetic evidence for interaction between aldolase and D-glyceraldehyde-3-phosphate dehydrogenase. *Eur. J. Biochem.* **85**, 157–161.

900. OVÁDI, J., SALERNO, C., KELETI, T. and FASELLA, P. (1978) Physico-chemical evidence for the interaction between aldolase and glyceraldehyde-3-phosphate dehydrogenase. *Eur. J. Biochem.* **90**, 499–503.

901. OVÁDI, J., TELEGDI, M., BATKE, J. and KELETI, T. (1971) Functional non-identity of subunits and isolation of active dimers of D-glyceraldehyde-3-phosphate dehydrogenase. *Eur. J. Biochem.* **22**, 430–438.

902. OVERFIELD, R. E. and WRAIGHT, C. A. (1980) Oxidation of cytochromes c and c_2 by bacterial photosynthetic reaction centers in phospholipid vesicles. 2. Studies with negative membranes. *Biochemistry* **19**, 3328–3334.

903. PACKMAN, P. M. and JAKOBY, W. B. (1967) Crystalline quinolinate phosphoribosyltransferase. II. Properties of the enzyme. *J. Biol. Chem.* **242**, 2075–2079.

904. PAETKAU, V. and LARDY, H. A. (1967) Phosphofructokinase: Correlation of physical and enzymatic properties. *J. Biol. Chem.* **242**, 2035–2042.

905. PAIK, W. K. and KIM, S. (1980) *Protein Methylation.* Wiley, New York.

906. PAIK, W. K., POLASTRO, E. and KIM, S. (1980) Cytochrome C methylation: Enzymology and biologic significance. *Curr. Top. Cell. Regul.* **16**, 87–111.

907. PAINE, P. L., MOORE, L. C. and HOROWITZ, S. B. (1975) Nuclear envelope permeability. *Nature* **254**, 109–114.

908. PANDIT, M. W. and RAO, M. S. N. (1975) Studies on self-association of proteins. Self-association of α-chymotrypsin at its isoelectric point in buffer solutions of ionic strength 0.1. *Biochemistry* **14**, 4106–4110.

909. PARADIES, H. H. and VETTERMANN, W. (1979) The structure of a hexamer of phosphofructokinase: size and shape of a crosslinked hexamer of phosphofructokinase. *Arch. Biochem. Biophys.* **194**, 88–100.

910. PARKER, J. C. and HOFFMAN, J. F. (1967) The role of membrane phosphoglycerate kinase in the control of glycolytic rate by active cation transport in human red blood cells. *J. Gen. Physiol.* **50,** 893–916.

911. PARKER, P. J. and RANDLE, P. J. (1978) Branched chain 2-oxo-acid dehydrogenase complex of rat liver. *FEBS Lett.* **90,** 183–186.

912. PATEL, V. B. and GILES, N. H. (1979) Purification of the arom multienzyme aggregate from *Euglena gracilis. Biochim. Biophys. Acta* **567,** 24–34.

913. PATTHY, L. (1978) Role of nascent α-ketoaldehyde in substrate-dependent oxidative inactivation of aldolase. *Eur. J. Biochem.* **88,** 191–196.

914. PATTHY, L. and THÉSZ, J. (1980) Origin of the selectivity of α-dicarbonyl reagents for arginyl residues of anion-binding sites. *Eur. J. Biochem.* **105,** 387–393.

915. PATTHY, L., VÁRADI, A., THÉSZ, J. and KOVÁCS, K. (1979) Identification of the C-1-phosphate-binding arginine residue of rabbit-muscle aldolase. Isolation of 1,2-cyclohexanedione-labeled peptide by chemisorption chromatography. *Eur. J. Biochem.* **99,** 309–313.

916. PATTHY, L. and VAS, M. (1978) Aldolase-catalysed inactivation of glyceraldehyde-3-phosphate dehydrogenase. *Nature* **276,** 94–95.

917. PAUKERT, J. L., STRAUS, L. D'ARI and RABINOWITZ, J. C. (1976) Formyl-methenyl-methylene-tetrahydrofolate synthetase- (combined). An ovine protein with multiple catalytic activities. *J. Biol. Chem.* **251,** 5104–5111.

918. PAULING, L. (1935) The oxygen equilibrium of hemoglobin and its structural interpretation. *Proc. Natl. Acad. Sci. U.S.A.* **21,** 186–191.

919. PAULING, L. and PRESSMAN, D. (1945) The serological properties of simple substances. IX. Hapten inhibition of precipitation of antisera homologous to the *o-, m-,* and *p*-azophenylarsonic acid groups. *J. Am. Chem. Soc.* **67,** 1003–1012.

920. PAULS, H., BREDENBRÖCKER, B. and SCHONER, W. (1980) Inactivation of $(Na^+ + K^+)$-ATPase by chromium (III) complexes of nucleotide triphosphates. *Eur. J. Biochem.* **109,** 523–533.

921. PAULUS, H. and DeRIEL, J. K. (1975) Absence of kinetic cooperativity in the allosteric model of Monod, Wyman and Changeux. *J. Mol. Biol.* **97,** 667–671.

922. PEDERSEN, P. L. (1975) Mitochondrial adenosine triphosphatase. *J. Bioenerg.* **6,** 243–275.

923. PEKAR, A. H. and FRANK, B. H. (1972) Conformation of proinsulin. A comparison of insulin and proinsulin self-association at neutral pH. *Biochemistry* **11,** 4013–4016.

924. PELLEFIGUE, F., BUTLER, J. B., SPIELBERG, S. P., HOLLENBERG, M. D., GOODMAN, S. I. and SCHULMAN, J. D. (1976) Normal amino acid uptake by cultured human fibroblasts does not require gamma-glutamyl transpeptidase. *Biochem. Biophys. Res. Commun.* **73,** 997–1002.

925. PENEFSKY, H. S. (1979) Mitochondrial ATPase. *Adv. Enzymol.* **49,** 223–280.

926. PENNINCKX, M., SIMON, J.-P. and WIAME, J.-M. (1974) Interaction between arginase and L-ornithine carbamoyltransferase in *Saccharomyces cerevisiae.* Purification of *S. cerevisiae* enzymes and evidence that these enzymes as well as rat-liver arginase are trimers. *Eur. J. Biochem.* **49,** 429–442.

927. PERBAL, B., GUEGUEN, P. and HERVÉ, G. (1977) Biosynthesis of *Escherichia coli* aspartate transcarbamylase. II. Correlated biosynthesis of the catalytic and regulatory chains and cytoplasmic association of the subunits. *J. Mol. Biol.* **110,** 319–340.

928. PERBAL, B. and HERVÉ, G. (1972) Biosynthesis of *Escherichia coli* aspartate transcarbamylase. I. Parameters of gene expression and sequential biosynthesis of the subunits. *J. Mol. Biol.* **70,** 511–529.

929. PERHAM, R. N. (1975) Self-assembly of biological macromolecules. *Philos. Trans. R. Soc. London, B,* **272,** 123–136.

930. PERHAM, R. N., DUCKWORTH, H. W. and ROBERTS, G. C. (1981) Mobility of polypeptide chain, in the pyruvate dehydrogenase complex revealed by proton NMR. *Nature* **292,** 474–477.

931. PERHAM, R. N. and HOOPER, E. A. (1977) Polypeptide chain stoicheiometry in the self-assembly of the pyruvate dehydrogenase multienzyme complex of *Escherichia coli. FEBS Lett.* **73,** 137–140.

932. PERHAM, R. N. and ROBERTS, G. C. K. (1981) Limited proteolysis and proton N.M.R. spectroscopy of the 2-oxoglutarate dehydrogenase multienzyme complex of *Escherichia coli*. *Biochem. J.* 199, 733–740.

933. PERUTZ, M. F. (1970) Stereochemistry of cooperative effects of haemoglobin. *Nature* 228, 726–739.

934. PERUTZ, M. F. (1978) Electrostatic effects in proteins. *Science* 201, 1187–1191.

935. PETERS, K. and RICHARDS, F. M. (1977) Chemical cross-linking: Reagents and problems in studies of membrane structure. *Annu. Rev. Biochem.* 46, 523–551.

936. PETTE, D. (1978) Cytosolic organization of carbohydrate-metabolism enzymes in cross-striated muscle. *Biochem. Soc. Trans.* 6, 9–11.

937. PETTE, D. and BÜCHER, T. (1963) Proportionskonstante Gruppen in Beziehung zur Differenzierung der Enzymaktivitätsmuster von Skelett-Muskeln des Kaninchens. *Z. Physiol. Chem.* 331, 180–195.

938. PETTE, D., SMITH, M. E., STAUDTE, H. W. and VRBOVÁ, G. (1973) Effects of long-term electrical stimulation on some contractile and metabolic characteristics of fast rabbit muscles. *Pflügers Arch.* 338, 257–272.

939. PETTIT, F. H., YEAMAN, S. J. and REED, L. J. (1978) Purification and characterization of branched chain α-keto acid dehydrogenase complex of bovine kidney. *Proc. Natl. Acad. Sci. U.S.A.* 75, 4881–4885.

940. PHAN, S. H. and MAHLER, H. R. (1976) Studies on cytochrome oxidase. Partial resolution of enzymes containing seven or six subunits, from yeast and beef heart, respectively. *J. Biol. Chem.* 251, 257–269.

941. PHAN, S. H. and MAHLER, H. R. (1976) Studies on cytochrome oxidase. Preliminary characterization of an enzyme containing only four subunits. *J. Biol. Chem.* 251, 270–276.

942. PHILLIPS, A. T. (1974) Ligand-induced oligomerization and regulatory mechanism. *Crit. Rev. Biochem.* 2, 343–378.

943. PILLAI, S. and BACHHAWAT, B. K. (1979) Affinity immobilization and "negative" crosslinking: A probe for tertiary and quaternary protein structure. *J. Mol. Biol.* 131, 877–881.

944. PINDER, J. C., UNGEWICKELL, E., BRAY, D. and GRATZER, W. B. (1978) The spectrin–actin complex and erythrocyte shape. *J. Supramol. Struct.* 8, 439–445.

945. PINNER, A. (1892) *Die Imidoäter und ihre Derivative*, Berlin; cf. *Chem. Revs.* (1944) 35,424/R 109.

946. PLEDGER, W. J., THOMPSON, W. J. and STRADA, S. J. (1975) Isolation of an activator of multiple forms of cyclic nucleotide phosphodiesterase of rat cerebrum by isoelectric focusing. *Biochim. Biophys. Acta* 391, 334–340.

947. PLIETZ, P., DAMASCHUN, G., KOPPERSCHLÄGER, G. and MÜLLER, I. I. (1978) Small-angle X-ray scattering studies on the quaternary structure of phosphofructokinase from baker's yeast. *FEBS Lett.* 91, 230–232.

948. POLGÁR, L. (1964) The mechanism of action of glyceraldehyde-3-phosphate dehydrogenase. *Experientia* 20, 408, 1–14.

949. POLITZ, S. M. and GLITZ, D. G. (1977) Ribosome structure: Localization of N^6,N^6-dimethyladenosine by electron microscopy of a ribosome–antibody complex. *Proc. Natl. Acad. Sci. U.S.A.* 74, 1468–1472.

950. PONTREMOLI, S., MELLONI, E., SALAMINO, F., MICHETTI, M., BOTELHO, L. H., EL-DORRY, H. A., CHU, D. K., ISAACS, C. E. and HORECKER, B. L. (1978) Origin of tryptophan in fructose 1,6-bisphosphatase purified from livers of fed rabbits. *Arch. Biochem. Biophys.* 191, 825–827.

951. PONTREMOLI, S., MELLONI, E., SALAMINO, F., SPARATORE, B. and HORECKER, B. L. (1978) Interactions of Zn^{2+} and Mg^{2+} with rabbit liver fructose 1,6-bisphosphatase. *Arch. Biochem. Biophys.* 188, 90–97.

952. PONTREMOLI, S., MELLONI, E., SALAMINO, F., SPARATORE, B., MICHETTI, M., SINGH, V. N.

and HORECKER, B. L. (1979) Evidence for an interaction between fructose 1,6-bisphosphatase and fructose 1,6-bisphosphate aldolase. *Arch. Biochem. Biophys.* **197**, 356–363.

953. POPP, D. A., KIECHLE, F. L., KOTAGAL, N. and JARETT, L. (1980) Insulin stimulation of pyruvate dehydrogenase in an isolated plasma membrane–mitochondrial mixture occurs by activation of pyruvate dehydrogenase phosphatase. *J. Biol. Chem.* **255**, 7540–7543.

954. PORTER, D. H. and CARDENAS, J. M. (1981) Single subunits of Sepharose-bound pyruvate kinase are inactive. *Biochemistry* **20**, 2532–2537.

955. POWELL, L. D. and CANTLEY, L. C. (1980) Structural changes in $(Na^+ + K^+)$-ATPase accompanying detergent inactivation. *Biochim. Biophys. Acta* **599**, 436–447.

956. POYTON, R. O. and SCHATZ, G. (1975) Cytochrome c oxidase from bakers' yeast. III. Physical characterization of isolated subunits and chemical evidence for two different classes of polypeptides. *J. Biol. Chem.* **250**, 752–761.

957. PRAMANIK, A., PAWAR, S., ANTONIAN, E. and SCHULZ, H. (1979) Five different enzymatic activities are associated with the multienzyme complex of fatty acid oxidation from *Escherichia coli*. *J. Bacteriol.* **137**, 469–473.

958. PRANKERD, T. A. J. and ALTMAN, K. I. (1954) A study of the metabolism of phosphorus in mammalian red cells. *Biochem. J.* **58**, 622–633.

959. PRIEELS, J.-P., BELL, J. E., SCHINDLER, M., CASTELLINO, F. J. and HILL, R. L. (1979) Involvement of histidine-32 in the biological activity of α-lactalbumin. *Biochemistry* **18**, 1771–1776.

960. PROVERBIO, F. and HOFFMAN, J. F. (1977) Membrane compartmentalized ATP and its preferential use by the Na, K-ATPase of human red cell ghosts. *J. Gen. Physiol.* **69**, 605–632.

961. PURICH, D. L. and FROMM, H. J. (1971) The kinetics and regulation of rat brain hexokinase. *J. Biol. Chem.* **246**, 3456–3463.

962. PURICH, D. L., FROMM, H. J. and RUDOLPH, F. B. (1973) The hexokinases: Kinetic, physical and regulatory properties. *Adv. Enzymol.* **39**, 249–326.

963. QUIGLEY, J. P. and GOTTERER, G. S. (1969) Distribution of $(Na^+ - K^+)$-stimulated ATPase activity in rat intestinal mucosa. *Biochim. Biophys. Acta* **173**, 456–468.

964. RACKER, E. (1977) Mechanisms of energy transformations. *Annu. Rev. Biochem.* **46**, 1006–1014.

965. RACKER, E. and HORSTMAN, L. L. (1967) Partial resolution of the enzymes catalyzing oxidative phosphorylation. XIII. Structure and function of submitochondrial particles completely resolved with respect to coupling factor 1. *J. Biol. Chem.* **242**, 2547–2551.

966. RACKER, E. and KRIMSKY, I. (1958) Approaches to the mechanism of action of double-headed enzymes. *Fed. Proc.* **17**, 1135–1141.

967. RACKER, E., TYLER, D. D., ESTABROOK, R. W., CONOVER, T. E., PARSONS, D. F. and CHANCE, B. (1965) In *Oxidases and Related Redox Systems*. pp. 1077–1094. (KING, T. E., MASON, H. S. and MORRISON, M. eds), Wiley, New York.

968. RÁCZ, O., BISZKU, E. and STRAUB, F. B. (1979) Artifacts imitating aging of glucose-6-phosphate dehydrogenase in human erythrocytes. *Eur. J. Biochem.* **96**, 503–507.

969. RADDA, G. K. and SEELEY, P. J. (1979) Recent studies on cellular metabolism by nuclear magnetic resonance. *Annu. Rev. Physiol.* **41**, 749–769.

970. RAGHAVAN, N. V. and TULINSKY, A. (1979) The structure of α-chymotrypsin. II. Fourier phase refinement and extension of the dimeric structure at 1.8 Å resolution by density modification. *Acta Cryst.* B **35**, 1776–1785.

971. RAIBAUD, O. and GOLDBERG, M. E. (1976) The dissociated tryptophanase subunit is inactive. *J. Biol. Chem.* **251**, 2820–2824.

972. RAIBAUD, O. and GOLDBERG, M. E. (1977) The reactivity of one essential cysteine as a conformational probe in *Escherichia coli* tryptophanase. Application to the study of the structural influence of subunit interactions. *Eur. J. Biochem.* **73**, 591–599.

973. RAIJMAN, L. and BARTULIS, T. (1979) Effect of ATP translocation on citrulline and oxalo-acetate synthesis by isolated rat liver mitochondria. *Arch. Biochem. Biophys.* **195,** 188–197.

974. RAINE, D. N. (1979) Inborn errors of metabolism, In *Chemical Diagnosis of Disease.* pp. 927–1008. (BROWN, S. S., MITCHELL, F. L. and YOUNG, D. S., eds.), Elsevier/North-Holland Biomedical Press, Amsterdam–New York–London.

975. RALSTON, G. B. (1978) Physical-chemical studies on spectrin. *J. Supramol. Struct.* **8,** 361–373.

976. RAMASWAMY, K., MALATHI, P., CASPARY, W. F. and CRANE, R. K. (1974) Studies on the transport of glucose from disaccharides by hamster small intestine in vitro. II. Characteristics of the desaccharidase-related transport system. *Biochim. Biophys. Acta* **345,** 39–48.

977. RAMASWAMY, S. and RADHAKRISHNAN, A. N. (1975) Lactase-phlorizin hydrolase complex from monkey small intestine. Purification, properties and evidence for two catalytic sites. *Biochim. Biophys. Acta* **403,** 446–455.

978. RANDALL, J., STARLING, D., BALDWIN, J. P. and IBEL, K. (1975) *Brookhaven Symp. Biol.* **27,** IV. 78–IV. 96.

979. RANKIN, B. B., McINTYRE, T. M. and CURTHOYS, N. P. (1980) Brush border membrane hydrolysis of *S*-benzyl-cysteine-*P*-nitroanilide, an activity of aminopeptidase M. *Biochem. Biophys. Res. Commun.* **96,** 991–996.

980. RAPOPORT, S. (1968) Regulation of glycolysis in mammalian erythrocytes. *Essays in Biochem.,* Vol. 4., pp. 69–103. Academic Press, London–New York.

981. RAPOPORT, S. (1973) Control mechanisms of red cell glycolysis. In *The Human Red Cell in Vitro.* (GREENWALT, T. J. and JAMIESON, G. A. eds), Grune, Stratton, New York–London.

982. RAPOPORT, S. and LUEBERING, J. (1950) The formation of 2,3-diphosphoglycerate in rabbit erythrocytes: The existence of a diphosphoglycerate mutase. *J. Biol. Chem.* **183,** 507–516.

983. RAPOPORT, T. A., HEINRICH, R. and RAPOPORT, S. M. (1976) The regulatory principles of glycolysis in erythrocytes *in vivo* and *in vitro. Biochem. J.* **154,** 449–469.

984. RASCHED, I. R., BOHN, A. and SUND, H. (1977) Studies of glutamate dehydrogenase. Analysis of quaternary structure and contact areas between the polypeptide chains. *Eur. J. Biochem.* **74,** 365–377.

985. RASULOV, A. S., EVSTIGNEYEVA, T. G., KRETOVICH, W. L., STELMASCHUK, V. YA., SAMSONIDZE, T. G. and KISELEV, N. A. (1977) Purification, properties and quaternary structure of glutamine synthetase from *Chlorella. Biokhimiya* **42,** 350–358.

986. RECSEI, P. A. and SNELL, E. E. (1973) Prohistidine decarboxylase from Lactobacillus 30a. A new type of zymogen. *Biochemistry* **12,** 365–371.

987. REDDY, G. P. V. and MATHEWS, C. (1978) Functional compartmentation of DNA precursors in T4 phage-infected bacteria. *J. Biol. Chem.* **253,** 3461–3467.

988. REED, L. J. (1969) Pyruvate dehydrogenase complex. *Curr. Top. Cell. Regul.* **1,** 233–251.

989. REED, L. J. (1974) Multi enzyme complexes. *Acc. Chem. Res.* **7,** 40–46.

990. REED, L. J. and COX, D. J. (1970) Multienzyme complexes. In *The Enzymes.* Vol. 1., pp. 213–240. (BOYER, P. D., ed.), Academic Press, New York–London.

991. REED, L. J., LINN, T. C., HUCHO, F., NAMIHIRA, G., BARRERA, C. R., ROCHE, T. E., PELLEY, J. W. and RANDALL, D. D. (1972) Molecular aspects of the regulation of the mammalian pyruvate dehydrogenase complex. In *Metabolic Interconversion of Enzymes.* pp. 281–291. (WIELAND, O., HELMREICH, E. and HOLZER, H., eds), Springer, Berlin.

992. REED, L. J., PETTIT, F. H., ELEY, M. H., HAMILTON, L., COLLINS, J. H. and OLIVER, R. M. (1975) Reconstitution of the *Escherichia coli* pyruvate dehydrogenase. *Proc. Natl. Acad. Sci. U.S.A.* **72,** 3068–3072.

993. REED, L. J., PETTIT, F. H., YEAMAN, S. J., TEAGUE, W. M. and BLEILE, D. M. (1980) Structure, function and regulation of the mammalian pyruvate dehydrogenase complex. In *Trends in Enzymology.* Vol. 1. *Enzyme Regulation and Mechanism of Action.* pp. 47–56. (MILDNER, P. and RIES, B. eds). Pergamon Press, Oxford.

994. REEVES, R. E. and SOLS, A. (1973) Regulation of *Escherichia coli* phosphofructokinase *in situ*. *Biochem. Biophys. Res. Commun.* **50**, 459–466.

995. REINHART, G. D. and LARDY, H. A. (1980) Rat liver phosphofructokinase: kinetic and physiological ramifications of the aggregation behavior. *Biochemistry* **19**, 1491–1495.

996. REISLER, E. and EISENBERG, H. (1970) Studies on the viscosity of solutions of bovine liver glutamate dehydrogenase and on related hydrodynamic models; Effect on toluene on enzyme association. *Biopolymers* **9**, 877–889.

997. REISS, U. and GERSHON, D. (1976) Comparison of cytoplasmic superoxide dismutase in liver, heart and brain of aging rats and mice. *Biochem. Biophys. Res. Commun.* **73**, 255–262.

998. REITHMEIER, R. A. F. (1979) Fragmentation of the band 3 polypeptide from human erythrocyte membranes. *J. Biol. Chem.* **254**, 3054–3060.

999. REUTER, R., ESCHRICH, K., SCHELLENBERGER, W. and HOFMANN, E. (1979) Kinetic modelling of yeast phosphofructokinase. *Acta Biol. Med. Germ.* **38**, 1067–1079.

1000. RICARD, J., MOUTTET, C. and NARI, J. (1974) Subunit interactions in enzyme catalysis. Kinetic models for one-substrate polymeric enzymes. *Eur. J. Biochem.* **41**, 479–497.

1001. RICHARDS, F. M. (1974) The interpretation of protein structures: Total volume, group volume distributions and packing density. *J. Mol. Biol.* **82**, 1–14.

1002. RICHARDS, F. M. (1977) Areas, volumes, packing and protein structure. *Annu. Rev. Biophys. Bioeng.* **6**, 151–176.

1003. RICHARDS, K. E. and WILLIAMS, R. C. (1972) Electron microscopy of aspartate transcarbamylase and its catalytic subunit. *Biochemistry* **11**, 3393–3395.

1004. RICHARDSON, J. C. W. and AGUTTER, P. S. (1980) The relationship between the nuclear membranes and the endoplasmic reticulum in interphase cells. *Biochem. Soc. Trans.* **8**, 459–465.

1005. RICHARDSON, R. M. and BREW, K. (1980) Lactose synthase. An investigation of the interaction site α-lactalbumin for galactosyl transferase by differential kinetic labeling. *J. Biol. Chem.* **255**, 3377–3385.

1006. RICHAUD, F., PHUC, N. H., CASSAN, M. and PATTE, J.-C. (1980) Regulation of aspartokinase-3 synthesis in *Escherichia coli*. Isolation of mutants containing LYSC-LAC fusions. *J. Bacteriol.* **143**, 513–515.

1007. RICHTER, J. J., SUNDERLAND, E., JUHL, U. and KORNGUTH, S. (1978) Extraction of mitochondrial proteins by volatile anesthetics. *Biochim. Biophys. Acta* **543**, 106–115.

1008. RIEGER, F. and VIGNY, M. (1976) Solubilization and physical-chemical characterization of rat brain acetylcholinesterase: Development and maturation of its molecular forms. *J. Neurochem.* **27**, 121–129.

1009. ROBERTS, C. M. and SOKATCH, J. R. (1978) Branched chain amino acids as activators of branched chain ketoacid dehydrogenase. *Biochem. Biophys. Res. Commun.* **82**, 828–833.

1010. ROBERTS, H. and HESS, B. (1977) Kinetics of cytochrome c oxidase from yeast. Membrane-facilitated electrostatic binding of cytochrome c showing a specific interaction with cytochrome c oxidase and inhibition by ATP. *Biochim. Biophys. Acta* **462**, 215–234.

1011. ROBINSON, J. D. and FLASHNER, M. S. (1979) The $(Na^+ + K^+)$-activated ATPase. Enzymatic and transport properties. *Biochim. Biophys. Acta* **549**, 145–176.

1012. ROBINSON, G. A., BUTCHER, R. W. and SUTHERLAND, E. W. (1971) *Cyclic AMP*. pp. 150–151. Academic Press, New York–London.

1013. ROCHE, T. E. and CATE, R. L. (1976) Evidence for lipoic acid mediated NADH and acetyl-CoA stimulation of liver and kidney pyruvate dehydrogenase kinase. *Biochem. Biophys. Res. Commun.* **72**, 1375–1383.

1014. ROCHE, T. E. and CATE, R. L. (1977) Purification of porcine liver pyruvate dehydrogenase complex and characterization of its catalytic and regulatory properties. *Arch. Biochem. Biophys.* **183**, 664–677.

1015. RODBELL, M., KRANS, H. M. J., POHL, S. L. and BIRNBAUMER, L. (1971) The glucagon-sensitive adenyl cyclase system in plasma membranes of rat liver. IV. Effects of guanyl nucleotides on binding of ^{125}I-glucagon. *J. Biol. Chem.* **246**, 1872–1876.

1016. ROLBIN, J. A., KAYUSHINA, R. L., FEIGIN, L. A. and SCHEDRIN, B. M. (1973) On the computing of X-ray small angle scattering by models of macromolecules. *Kristallographiya* **18**, 701–705.

1017. ROME, E. M., HIRABAYSHI, T. and PERRY, S. V. (1973) X-ray diffraction of muscle labelled with antibody to troponin C. *Nature New. Biol.* **244**, 154–155.

1018. ROSE, I. A. and WARMS, J. V. B. (1967) Mitochondrial hexokinase. *J. Biol. Chem.* **242**, 1635–1645.

1019. ROSE, G. D. and ROY, S. (1980) Hydrophobic basis of packing in globular proteins. *Proc. Natl. Acad. Sci. U.S.A.* **77**, 4643–4647.

1020. ROSE, I. A. and WARMS, J. V. B. (1970) Control of red cell glycolysis. *J. Biol. Chem.* **245**, 4009–4015.

1021. ROSE, K. M., STETLER, D. A. and JACOB, S. T. (1981) Protein kinase activity of RNA polymerase I purified from a rat hepatoma: Probable function of $Mr = 42,000$ and 24,600 polypeptides. *Proc. Natl. Acad. Sci. U.S.A.* **78**, 2833–2837.

1022. ROSEMBLATT, M. S., CALLEWAERT, D. M. and TCHEN, T. T. (1973) Succinic semialdehyde dehydrogenase from a *Pseudomonas* species. II. Physical and immunochemical properties of the enzyme. *J. Biol. Chem.* **248**, 6014–6018.

1023. ROSENBUSCH, J. P. and WEBER, K. (1971) Localization of the zinc binding site of aspartate transcarbamylase in the regulatory subunit. *Proc. Natl. Acad. Sci. U.S.A.* **68**, 1019–1023.

1024. ROSS, A. H. and MCCONNELL, H. M. (1977) Reconstitution of band 3, the erythrocyte anion exchange protein. *Biochem. Biophys. Res. Commun.* **74**, 1318–1325.

1025. ROSS, P. D. and SUBRAMANIAN, S. (1981) Thermodynamics of protein association reactions: Forces contributing to stability. *Biochemistry* **20**, 3096–3102.

1026. ROSSI, A., MENEZES, L. C. and PUDLES, J. (1975) Yeast hexokinase A: Succinylation and properties of the active subunit. *Eur. J. Biochem.* **59**, 423–432.

1027. ROSSMANN, M. G., ADAMS, M. J., BUEHNER, M., FORD, G. C., HACHERT, M. L., LILJAS, A., RAO, S. T., BANASZAK, L. J., HILL, E., TSERNOGLOU, D. and WEBB, L. (1973) Molecular symmetry axes and subunit interfaces in certain dehydrogenases. *J. Mol. Biol.* **76**, 533–537.

1028. ROTHMAN, J. E. and LENARD, J. (1977) Membrane asymmetry. The nature of membrane asymmetry provides clues to the puzzle of how membranes are assembled. *Science* **195**, 743–753.

1029. ROTHSTEIN, A. (1979) The role of protein channels in anion transport across the red cell membrane. In *Function and Molecular Aspects of Biomembrane Transport.* pp. 15–24. (QUAGLIA-RIELLO, E., PALMIERI, F., PAPA, S., KLINGENBERG, M., eds), Elsevier/North-Holland, Amsterdam–New York–Oxford.

1030. ROTHSTEIN, M. (1979) Formation of altered enzymes in aging animals. *Mech. Age. Devl.* **9**, 197–202.

1031. ROTHSTEIN, M., COPPENS, M. and SHARMA, H. K. (1980) Effect of aging on enolase from rat muscle, liver and heart. *Biochim. Biophys. Acta* **614**, 591–600.

1032. ROTHSTEIN, M. and SHARMA, H. K. (1978) Altered enzymes in the free-living nematode *Turbatrix aceti*, aged in the absence of fluoro-deoxyuridine. *Mech. Age. Devl.* **8**, 175–180.

1033. RÜDIGER, H. W., LANGENBECK, U. and GOEDDE, H. W. (1972) Oxidation of branched chain α-ketoacids in *Streptococcus faecalis* and its dependence on lipoic acid. *Z. Physiol. Chem.* **353**, 875–882.

1034. RUDOLPH, R., HEIDER, I. and JAENICKE, R. (1977) Mechanism of reactivation and refolding of glyceraldehyde-3-phosphate dehydrogenase from yeast after denaturation and dissociation. *Eur. J. Biochem.* **81**, 563–570.

1035. RUDOLPH, R., HEIDER, I., WESTHOF, E. and JAENICKE, R. (1977) Mechanism of refolding and reactivation of lactic dehydrogenase from pig heart after dissociation in various solvent media. *Biochemistry* **16**, 3384–3390.

1036. RUSSELL, L. K., KIRKLEY, S. A., KLEYMAN, T. R. and CHAN, S. H. P. (1976) Isolation and properties of OSCP and an F_1-ATPase binding protein from rat liver mitochondria. Evidence against OSCP as the linking "stalk" between F_1 and the membrane. *Biochem. Biophys. Res. Commun.* **73**, 434–443.

1037. RUTTER, W. J., WOODFIN, B. H. and BLOSTEIN, R. E. (1963) Enzymic homology. Structural and catalytic differentiation of fructose diphosphate aldolase. *Acta Chem. Scand.* **17**, S 226–S 232.

1038. SACHS, J. R. (1977) Kinetics of inhibition of Na^+-K^+ pump by external sodium. *J. Physiol.* **264**, 449–470.

1039. SALEEMUDDIN, M. and ZIMMERMANN, U. (1978) Use of glyceraldehyde-3-phosphate dehydrogenase-depleted human erythrocyte ghosts as specific high affinity adsorbents for the purification of glyceraldehyde-3-phosphate dehydrogenase from various tissue. *Biochim. Biophys. Acta* **527**, 182–192.

1040. SALEMME, F. R. (1977) Structure and function of cytochrome c. *Annu. Rev. Biochem.* **46**, 299–329.

1041. SALERNO, C., OVÁDI, J., CHURCHICH, J. and FASELLA, P. (1975) Interaction between transaminases and dehydrogenases. In *Mechanism of Action and Regulation of Enzymes. FEBS Symposia* Vol. 32., pp. 147–160. (KELETI, T., ed.), Akadémiai Kiadó, Budapest.

1042. SALERNO, C., OVÁDI, J., KELETI, T. and FASELLA, P. (1982) Kinetics of coupled reactions catalyzed by aspartate aminotransferase and glutamate dehydrogenase. *Eur. J. Biochem.* **121**, 511–517.

1043. SALHANY, J. M., CORDES, K. A. and GAINES, E. D. (1980) Light-scattering measurements of hemoglobin binding to the erythrocyte membrane. Evidence for transmembrane effects related to a disulfonic stilbene binding to band 3. *Biochemistry* **19**, 1447–1454.

1044. SALHANY, J. M. and GAINES, K. C. (1981) Connections between cytoplasm proteins and the erythrocyte membrane. *Trends Biochem. Sci.* **6**, 13–15.

1045. SARKADI, B. (1980) Active calcium transport in human red cells. *Biochim. Biophys. Acta* **604**, 159–190.

1046. SARZALA, M. G. and MICHALAK, M. (1978) Studies on the heterogeneity of sarcoplasmic reticulum vesicles. *Biochim. Biophys. Acta* **513**, 221–235.

1047. SASAJIMA, K., KAWACHI, T., SATO, S. and SUGIMURA, T. (1975) Purification and properties of α,α-trehalase from the mucosa of rat small intestine. *Biochim. Biophys. Acta* **403**, 139–146.

1048. SATRE, M., KLEIN, G. and VIGNAIS, P. V. (1976) Structure of beef heart mitochondrial F_1-ATPase. Arrangement of subunits as disclosed by cross-linking reagents and selective labeling by radioactive ligands. *Biochim. Biophys. Acta* **453**, 111–120.

1049. SAUERMANN, G. (1976) Studies on ribonucleic acid metabolism using nuclear columns release of rapidly labeled RNA from rat liver nuclei. *Z. Physiol. Chem.* **357**, 1117–1124.

1050. SAVAGEAU, M. A. (1971) Parameter sensitivity as a criterion for evaluating and comparing the performance of biochemical systems. *Nature* **229**, 542–544.

1051. SAVAGEAU, M. A. (1971) Concepts relating the behavior of biochemical systems to their underlying molecular properties. *Arch. Biochem. Biophys.* **145**, 612–621.

1052. SAVAGEAU, M. A. (1972) The behavior of intact biochemical control systems. *Curr. Top. Cell. Regul.* **6**, 63–130.

1053. SAWADOGO, M., HUET, J. and FROMAGEOT, P. (1980) Similar binding site for P_{37} factor on yeast RNA polymerases A and B. *Biochem. Biophys. Res. Commun.* **96**, 258–264.

1054. SCALES, D. and INESI, G. (1976) Assembly of ATPase protein in sarcoplasmic reticulum membranes. *Biophys. J.* **16**, 735–751.

1055. SCHACHMAN, H. K. and EDELSTEIN, S. J. (1966) Ultracentrifuge studies with absorption optics. IV. Molecular weight determinations at the microgram level. *Biochemistry* **5**, 2681–2705.

1056. SCHADE, B. C., RUDOLPH, R., LÜDEMANN, H.-D. and JAENICKE, R. (1980) Reversible high-pressure dissociation of lactic dehydrogenase from pig muscle. *Biochemistry* **19**, 1121–1126.

1057. SCHATZ, G. and MASON, T. L. (1974) The biosynthesis of mitochondrial proteins. *Annu. Rev. Biochem.* **43**, 51–87.

1058. SCHEEK, R. M., KALKMAN, M. L., BERDEN, J. A. and SLATER, E. C. (1980) Subunit interactions in glyceraldehyde-3-phosphate dehydrogenases. Their involvement in nucleotide binding and cooperativity. *Biochim. Biophys. Acta* **613**, 275–291.

1059. SCHEEK, R. M. and SLATER, E. C. (1978) Preparation and properties of rabbit muscle glyceraldehyde-phosphate dehydrogenase with equal binding parameters for the third and fourth NAD$^+$ molecules. *Biochim. Biophys. Acta* **526**, 13–24.

1060. SCHIMMEL, P. R. and SÖLL, D. (1979) Aminoacyl-tRNA synthetases: general features and recognition of transfer RNAs. *Annu. Rev. Biochem.* **48**, 601–648.

1061. SCHLEGEL-HAUETER, S., HORE, P., KERRY, K. R. and SEMENZA, G. (1972) The preparation of lactase and glucoamylase of rat small intestine. *Biochim. Biophys. Acta* **258**, 506–519.

1062. SCHLESSINGER, J., SHECHTER, Y., WILLINGHAM, M. C. and PASTAN, I. (1978) Direct visualization of the binding, aggregation, and internalization of insulin and epidermal growth factor on fibroblastic cells. *Proc. Natl. Acad. Sci. U.S.A.* **75**, 2659–2663.

1063. SCHNEIDER, H., LEMASTER, J. J. and HACKENBROCK, C. R. (1981) In *Functions of Quinones in Energy Conserving Systems.* (TRUMPOWER, B. L. ed.), Academic Press, New York.

1064. SCHNEIDER, H., LEMASTERS, J. J., HÖCHLI, M. and HACKENBROCK, C. R. (1980) Fusion of liposomes with mitochondrial inner membranes. *Proc. Natl. Acad. Sci. U.S.A.* **77**, 442–446.

1065. SCHNEIDER, H., LEMASTERS, J. J., HÖCHLI, M. and HACKENBROCK, C. R. (1980) Liposome-mitochondrial inner membrane fusion. Lateral diffusion of integral electron transfer components. *J. Biol. Chem.* **255**, 3748–3756.

1066. SCHNEPF, E. (1964) Zur Feinstruktur von *Geosiphon pyriforme.* Ein Versuch zur Deutung cytoplasmatischer Membranen und Kompartimente. *Arch. Mikrobiol.* **49**, 112–131.

1067. SCHRIER, S. L. (1966) Organization of enzymes in human erythrocyte membranes. *Am. J. Physiol.* **210**, 139–143.

1068. SCHRIER, S. L. (1967) ATP synthesis in human erythrocyte membranes. *Biochim. Biophys. Acta* **135**, 591–598.

1069. SCHRIER, S. L. (1970) Transfer of inorganic phosphate across human erythrocyte membranes. *J. Lab. Clin. Med.* **75**, 422–434.

1070. SCHRIER, S. L., BEN-BASSAT, I., JUNGA, I., SEEGER, M. and GRUMET, F. C. (1975) Characterization of erythrocyte membrane associated enzymes (glyceraldehyde-3-phosphate dehydrogenase and phosphoglyceric kinase). *J. Lab. Clin. Med.* **85**, 797–810.

1071. SCHULMAN, J. D., GOODMAN, S. I., MACE, J. W., PATRICK, A. D., TIETZE, F. and BUTLER E. J. (1975) Glutathionuria: Inborn error of metabolism due to tissue deficiency of gamma-glutamyl transpeptidase. *Biochem. Biophys. Res. Commun.* **65**, 68–74.

1072. SCHULZ, G. E. and SCHIRMER, R. H. (1979) *Principles of Protein Structure,* Springer Verlag, New York–Heidelberg–Berlin.

1073. SCHUMM, D. E. and WEBB, T. E. (1975) Differential effect of ATP on RNA and DNA release from nuclei of normal and neoplastic liver. *Biochem. Biophys. Res. Commun.* **67**, 706–713.

1074. SCHWARTZ, A., ADAMS, R. J., BALL, W. J., COLLINS, J. H., GUPTE, S. S., LANE, L. K., REEVES, A. S. and WALLICK, E. T. (1980) Structure, function and regulation of Na-K-ATPase. *Int. J. Biochem.* **12**, 287–293.

1075. SCHWARTZ, A., ENTMAN, M. L., KANIIKE, K., LANE, L. K., VAN WINKLE, W. B. and BORNET, E. P. (1976) The rate of calcium uptake into sarcoplasmic reticulum of cardiac muscle and skeletal muscle. Effects of cyclic AMP-dependent protein kinase and phosphorylase *b* kinase. *Biochim. Biophys. Acta* **426**, 57–72.

1076. SCHWEIZER, E., KNIEP, B., CASTORPH, H. and HOLZNER, U. (1973) Pantetheine-free mutants of the yeast fatty-acid-synthetase complex. *Eur. J. Biochem.* **39**, 353–362.

1077. Scopes, R. K. (1973) Studies with a reconstituted muscle glycolytic system. The rate and extent of creatine phosphorylation by anaerobic glycolysis. *Biochem. J.* **134**, 197–208.

1078. Sebald, W., Weiss, H. and Jackl, G. (1972) Inhibition of the assembly of cytochrome oxidase in *Neurospora crassa* by chloramphenicol. *Eur. J. Biochem.* **30**, 413–417.

1079. Seeley, P. J., Busby, S. J. W., Gadian, D. G., Radda, G. K. and Richards, R. E. (1976) A new approach to metabolite compartmentation in muscle. *Biochem. Soc. Trans.* **4**, 62–64.

1080. Segal, H. L. (1976) Mechanism and regulation of protein turnover in animal cells. *Curr. Top. Cell. Regul.* **11**, 183–201.

1081. Segel, G. B., Feig, S. A., Glader, B. E., Müller, A., Dutcher, P. and Nathan, D. G. (1975) Energy metabolism in human erythrocytes: The role of phosphoglycerate kinase in cation transport. *Blood,* **46**, 271–278.

1082. Senior, A. E. (1973) The structure of mitochondrial ATPase. *Biochim. Biophys. Acta* **301**, 249–277.

1083. Senior, A. E. (1975) Mitochondrial adenosine triphosphatase. Location of sulfhydryl groups and disulfide bonds in the soluble enzyme from beef heart. *Biochemistry* **14**, 660–664.

1084. Senior, A. E. (1979) Mitochondrial ATPase: A review. In *Membrane Proteins in Energy Transduction.* pp. 233–278. (Capaldi, R. A., ed.), Dekker, New York.

1085. Serrano, R., Gancedo, J. M. and Gancedo, C. (1973) Assay of yeast enzymes *in situ*. A potential tool in regulation studies. *Eur. J. Biochem.* **34**, 479–482.

1086. Serrano, R., Kanner, B. I. and Racker, E. (1976) Purification and properties of the proton-translocating adenosine triphosphatase complex of bovine heart mitochondria. *J. Biol. Chem.* **251**, 2453–2461.

1087. Severin, S. E., Gomazkova, V. S., Krasovskaya, O. E. and Stafeeva, O. A. (1978) The kinetic properties of the α-ketoglutarate dehydrogenase complex from pigeon breast muscle. *Biokhimiya* **43**, 2241–2248.

1088. Seydoux, F., Malhotra, O. P. and Bernhard, S. A. (1974) Half-site reactivity. *CRC Crit. Rev. Biochem.* **2**, 227–257.

1089. Sha'afi, R. I. (1977) Water and small nonelectrolyte permeation in red cells, In *Membrane Transport in Red Cells.* pp. 221–256. (Ellory, J. C. and Lew, V. L. eds), Academic Press, London.

1090. Shaklai, N., Yguerabide, J. and Ranney, H. M. (1977) Interaction of hemoglobin with red blood cell membranes as shown by a fluorescent chromophore. *Biochemistry* **16**, 5585–5592.

1091. Shaklai, N., Yguerabide, J. and Ranney, H. M. (1977) Classification and localization of hemoglobin binding sites on the red blood cell membrane. *Biochemistry* **16**, 5593–5597.

1092. Shakoori, A. R., Romen, W., Oelschläger, W., Schlatterer, B. and Siebert, G. (1972) A new technique for the isolation of nucleoli from animal cells. *Z. Physiol. Chem.* **353**, 1735–1748.

1093. Sharma, H. K. and Rothstein, M. (1978) Serological evidence for the alteration of enolase during ageing. *Mech. Age. Devl.* **8**, 341–354.

1094. Sharma, H. K. and Rothstein, M. (1978) Age-related changes in the properties of enolase from *Turbatrix aceti. Biochemistry* **17**, 2869–2876.

1095. Sharma, H. K. and Rothstein, M. (1980) Altered enolase in aged *Turbatrix aceti* results from conformational changes in the enzyme. *Proc. Natl. Acad. Sci. U.S.A.* **77**, 5865–5868.

1096. Shepherd, G. B. and Hammes, G. G. (1976) Fluorescence energy transfer measurements between ligand binding sites of the pyruvate dehydrogenase multienzyme complex. *Biochemistry* **15**, 311–317.

1097. Shepherd, G. B. and Hammes, G. G. (1977) Fluorescence energy transfer measurements in the pyruvate dehydrogenase multienzyme complex from *Escherichia coli* with chemically modified lipoic acid. *Biochemistry* **16**, 5234–5241.

1098. Sherwood, P., Kelly, P., Kelleher, J. K. and Wright, B. E. (1979) TFLUX: A general purpose program for the interpretation of radioactive tracer experiments. *Comput. Progr. Biomed.* **10**, 66–74.

1099. SHIAO, D. D. F. and STURTEVANT, J. M. (1969) Calorimetric investigations of the binding of inhibitors to α-chymotrypsin. I. The enthalpy of dilution of α-chymotrypsin and of proflavin, and the enthalpy of binding of indole, N-acetyl-D-tryptophan, and proflavin to α-chymotrypsin. *Biochemistry* **12**, 4910–4917.

1100. SHIGA, T., MAEDA, N., SUDA, T., KON, K. and SEKIYA, M. (1979) The decreased membrane fluidity of in vivo aged human erythrocytes. A spin label study. *Biochim. Biophys. Acta* **553**, 84–95.

1101. SHILL, J. P. and NEET, K. E. (1975) Allosteric properties and the slow transition of yeast hexokinase. *J. Biol. Chem.* **250**, 2259–2268.

1102. SHILL, J. P., PETERS, B. A. and NEET, K. E. (1974) Monomer-dimer equilibria of yeast hexokinase during reacting enzyme sedimentation. *Biochemistry* **13**, 3864–3871.

1103. SHIN, B. C. and CARRAWAY, K. L. (1973) Association of glyceraldehyde 3-phosphate dehydrogenase with the human erythrocyte membrane. *J. Biol. Chem.* **248**, 1436–1444.

1104. SHINSHI, H. and KATO, K. (1978) Dissociated active subunits of tobacco phosphodiesterase. *Biochim. Biophys. Acta* **524**, 357–361.

1105. SHIPLEY, R. A. and CLARK, R. E. (1972) *Tracer Methods for in vivo Kinetics: Theory and Applications*, Academic Press, New York–London.

1106. SIA, C. L. and HORECKER, B. L. (1968) Dissociation of protein subunits by maleylation. *Biochem. Biophys. Res. Commun.* **31**, 731–737.

1107. SIEBERT, G. (1978) The limited contribution of the nuclear envelope to metabolic compartmentation. *Biochem. Soc. Trans.* **6**, 5–9.

1108. SIES, H. (1980) From enzymology *in vitro* to physiological chemistry *in vivo*. *Trends Biochem. Sci.* **5**, 182–185.

1109. SIES, H. (ed.) (1982) *Metabolic Compartmentation*, Academic Press, London–New York–Paris–San Diego–San Francisco–Sao Paolo–Sydney–Tokyo–Toronto.

1110. SIES, H. and BRAUSER, B. (1980) Analysis of cellular electron transport systems in liver and other organs by absorbance and fluorescence techniques. *Methods Biochem. Anal.* **26**, 285–325.

1111. SIES, H., BÜCHER, T., OSKINO, N. and CHANCE, B. (1973) Heme occupancy of catalase in hemoglobin-free perfused rat liver and of isolated rat liver catalase. *Arch. Biochem. Biophys.* **154**, 106–116.

1112. SIES, H., HÄUSSINGER, D. and GROSSKOPF, M. (1974) Mitochondrial nicotinamide nucleotide systems: Ammonium chloride responses and associated metabolic transitions in hemoglobin-free perfused rat liver. *Z. Physiol. Chem.* **355**, 305–320.

1113. SIEZEN, R. J. and BRUGGEN, E. F. J. (1974) Structure and properties of hemocyanins. XII. Electron microscopy of dissociation products of *Helix pomatia* α-hemocyanin: Quaternary structure. *J. Mol. Biol.* **90**, 77–89.

1114. SIGEL, P. and PETTE, D. (1969) Intracellular localization of glycogenolytic and glycolytic enzymes in white and red rabbit skeletal muscle. A gel film method for coupled enzyme reactions in histochemistry. *J. Histochem. Cytochem.* **17**, 225–237.

1115. SIGRIST, H., SIGRIST-NELSON, K. and GITLER, C. (1977) Single-phase butanol extraction: A new tool for proteolipid isolation. *Biochem. Biophys. Res. Commun.* **74**, 178–184.

1116. SIMON, I. (1972) Study of the position of NAD and its effect on the conformation of D-glyceraldehyde-3-phosphate dehydrogenase by small-angle X-ray scattering. *Eur. J. Biochem.* **30**, 184–189.

1117. SIMOONS, F. J. (1969) Primary adult lactose intolerance and the milking habit: A problem in biological and cultural interrelations. I. Review of the medical research. *Am. J. Digest. Dis.* **14**, 819–836.

1118. SIMOONS, F. J. (1970) Primary adult lactose intolerance and the milking habit: A problem in biological and cultural interrelations. II. A culture historical hypothesis. *Am. J. Digest. Dis.* **15**, 695–710.

1119. SITTE, P. (1977) Functional organization of biomembranes. In *Lipids and Lipid Polymers in Higher Plants*. pp. 1–28. (TEVINI, M. and LICHTENTHALTER, H. K., eds), Springer, Berlin.

1120. SITTE, P. (1981) Die Endosymbionten-Hypothese — eine kritische Betrachtung zur Zell-Evolution. *Nova Acta Leopoldina* N. F. **56**, 41–58.

1121. SITTE, P. (1980) General principles of cellular compartmentation. In *Cell Compartmentation and Metabolic Channeling.* pp. 17–32. (NOVER, L., LYNEN, F. and MOTHES, K. eds), VEB Gustav Fischer Verlag, Jena, Elsevier/North-Holland Biomedical Press, Amsterdam–New York–Oxford.

1122. SKOU, J. C. (1957) The influence of some cations on an adenosine triphosphatase from peripheral nerves. *Biochim. Biophys. Acta* **23**, 394–401.

1123. SKRINSKA, V. A., MESSINEO, L., TOWNS, R. L. R. and PEARSON, K. H. (1978) The effects of EDTA and contaminating metals on Fe, Cu, and Zn in deoxyribonucleoprotein. *Int. J. Biochem.* **9**, 637–638.

1124. SLATER, E. C. (1953) Mechanism of phosphorylation in the respiratory chain. *Nature* **172**, 975–978.

1125. SLATER, E. C. (1977) Mechanism of oxidative phosphorylation. *Annu. Rev. Biochem.* **46**, 1015–1026.

1126. SLOAN, D. L. and VELICK, S. F. (1973) Protein hydration changes in the formation of the nicotinamide adenine dinucleotide complexes of glyceraldehyde 3-phosphate dehydrogenase of yeast. *J. Biol. Chem.* **248**, 5419–5423.

1127. SMITH, G. D. and SCHACHMAN, H. K. (1971) A disproportionation mechanism for the all-or-none dissociation of mercurial-treated glyceraldehyde phosphate dehydrogenase. *Biochemistry* **10**, 4576–4588.

1128. SMITH, J. B., STERNWEIS, P. C. and HEPPEL, L. A. (1975) Partial purification of active delta and epsilon subunits of the membrane ATPase from *Escherichia coli. J. Supramol. Struct.* **3**, 248–255.

1129. SMITH, L. H. JR., HUGULEY, C. M. JR. and BAIN, J. A. (1972) Hereditary orotic aciduria. In *The Metabolic Basis of Inherited Disease.* pp. 1003–1021. (STANBURY, J. B., WYNGARDEN, J. B. and FREDERICKSON, D. S., eds), McGraw-Hill, New York.

1130. SMITH, S. and STERN, A. (1979) Subunit structure of the mammalian fatty acid synthetase: Further evidence for a homodimer. *Arch. Biochem. Biophys.* **197**, 379–387.

1131. SOKOLOFF, L., FITZGERALD, G. G. and KAUFMAN, E. E. (1977) Cerebral nutrition and energy metabolism. In *Nutrition and the Brain.* Vol. 1, pp. 87–139. (WURTMAN, R. J. and WURTMAN, J. J. eds), Raven, New York.

1132. SOLS, A., FELIÚ, J. E. and ARAGÓN, J. J. (1976) In *Metabolic Interconversion of Enzymes.* pp. 191–197. (SHALTIEL, S., ed.), Springer, Berlin–Heidelberg.

1133. SOLS, A. and MARCO, R. (1970) Concentrations of metabolites and binding sites. Implications in metabolic regulation. *Curr. Top. Cell. Regul.* **2**, 227–273.

1134. SOLTI, M. (1979) Supramolecular organization of glycolytic enzymes and metabolite compartmentation in human red cells. (In Hungarian.) Thesis, Budapest.

1135. SOLTI, M., BARTHA, F., HALÁSZ, N., TÓTH, G., SIROKMÁN, F. and FRIEDRICH, P. (1981) Localization of glyceraldehyde-3-phosphate dehydrogenase in intact human erythrocytes. Evaluation of membrane adherence on autoradiographs at low grain density. *J. Biol. Chem.* **256**, 9260–9265.

1136. SOLTI, M. and FRIEDRICH, P. (1976) Partial reversible inactivation of enzymes due to binding to the human erythrocyte membrane. *Molec. Cell. Biochem.* **10**, 145–152.

1137. SOLTI, M. and FRIEDRICH, P. (1979) The "enzyme-probe" method for characterizating metabolite pools. The use of NAD-glycohydrolase in human erythrocyte sonicate as a model system. *Eur. J. Biochem.* **95**, 551–559.

1138. SOMMER, A. and TRAUT, R. R. (1976) Identification of neighbouring protein pairs in the *Escherichia coli* 30S ribosomal subunit by crosslinking with methyl-4-mercaptobutyrimidate. *J. Mol. Biol.* **106**, 995–1015.

1139. SOMOGYI, B. and DAMJANOVICH, S. (1971) A molecular enzyme kinetic model. *Acta Biochim. Biophys. Acad. Sci. Hung.* **6**, 353–364.

1140. SOMOGYI, B. and DAMJANOVICH, S. (1973) A theoretical model for calculation of the rate constant of enzyme–substrate complex formation. *Acta Biochim. Biophys. Acad. Sci. Hung.* **8**, 153–160.

1141. SOMOGYI, B. and DAMJANOVICH, S. (1975) Relationship between the lifetime of an enzyme–substrate complex and the properties of the molecular environment. *J. Theor. Biol.* **51**, 393–401.

1142. SONE, N., YOSHIDA, M., HIRATA, H. and KAGAWA, Y. (1975) Purification and properties of a dicyclohexylcarbodiimide-sensitive adenosine triphosphatase from a thermophilic bacterium. *J. Biol. Chem.* **250**, 7917–7923.

1143. SOPER, J. W., DECKER, G. L. and PEDERSEN, P. L. (1979) Mitochondrial ATPase complex. A dispersed, cytochrome-deficient, oligomycin-sensitive preparation from rat liver containing molecules with a tripartite structural arrangement. *J. Biol. Chem.* **254**, 11170–11176.

1144. SOWERS, A. E. and HACKENBROCK, C. R. (1980) Electric field displacement and free lateral diffusion of intramembrane particles in the mitochondrial inner membrane. *Fed. Proc.* **39**, 1055.

1145. SOWERS, A. E. and HACKENBROCK, C. R. (1980) In *38th Ann. Proc. Electron Microscopy Soc. Amer.* pp. 620–621. (BAILEY, G. W. ed.).

1146. SPECKHARD, D. C., IKEDA, B. H., WONG, S. S. and FREY, P. A. (1977) Acetylation stoichiometry of *Escherichia coli* pyruvate dehydrogenase complex. *Biochem. Biophys. Res. Commun.* **77**, 708–713.

1147. SPECTOR, M., O'NEAL, S. and RACKER, E. (1980) Phosphorylation of the β-subunit of Na^+K^+-ATPase in Ehrlich ascites tumor by a membrane-bound protein kinase. *J. Biol. Chem.* **255**, 8370–8373.

1148. SPECTOR, M., O'NEAL, S. and RACKER, E. (1981) Regulation of phosphorylation of the β-subunit of the Ehrlich ascites tumor Na^+K^+-ATPase by a protein kinase cascade. *J. Biol. Chem.* **256**, 4219–4227.

1149. SPERANZA, M. L. and GOZZER, C. (1978) Purification and properties of NAD^+-dependent glyceraldehyde-3-phosphate dehydrogenase from spinach leaves. *Biochim. Biophys. Acta* **522**, 32–42.

1150. SPIEGEL, A. M., DOWNS, R. W., LEVINE, M. A. JR., SINGER, M. J., KRAWIETZ, W. JR., MARX, S. J., WOODARD, C. J., REEN, S. A. and AURBACH, G. D. (1981) The role of guanine nucleotides in regulation of adenylate cyclase activity. *Recent Progr. Horm. Res.* **37**, 635–665.

1151. SPIRIN, A. S., SERDYUK, I. N., SHPUNGIN, J. L. and VASILIEV, V. D. (1979) Quaternary structure of the ribosomal 30S subunit: Model and its experimental testing. *Proc. Natl. Acad. Sci. U.S.A.* **76**, 4867–4871.

1152. SPRANG, S. and FLETTERICK, R. J. (1979) The structure of glycogen phosphorylase *a* at 2.5 Å resolution. *J. Mol. Biol.* **131**, 523–551.

1153. SRERE, P. A. (1967) Enzyme concentrations in tissues. *Science* **158**, 936–937.

1154. SRERE, P. A. (1972) Is there an organization of Krebs cycle enzymes in the mitochondrial matrix? In *Energy Metabolism and the Regulation of Metabolic Processes in Mitochondria.* pp. 79–91. (MEHLMAN, M. A. and HANSON, R. W., eds). Academic Press, New York.

1155. SRERE, P. A. (1976) Apparent K_m's and apparent concentrations: An apparent conundrum. In *Glucaneogenesis: Its Regulation in Mammalian Species.* pp. 153–161. (HANSON, R. W. and MEHLMAN, W. A., eds), Wiley, New York.

1156. SRERE, P. A. (1980) The infrastructure of the mitochondrial matrix. *Trends Biochem. Sci.* **5**, 120–121.

1157. SRERE, P. A. (1981) Protein crystals as a model for mitochondrial matrix proteins. *Trends Biochem. Sci.* **6**, 4–7.

1158. SRERE, P. A. and ESTABROOK, R. W. (eds) (1978) *Microenvironments and Metabolic Compartmentation.* Academic Press, New York.

1159. SRERE, P. A. and HENSLEE, J. G. (1980) Is there an infrastructure in the mitochondrial matrix? In *Cell Compartmentation and Metabolic Channelling.* pp. 159–168. (NOVER, L., LYNEN, F. and Mo-

THES, K., eds), **VEB** Gustav Fischer Verlag, Jena, Elsevier/North-Holland Biomedical Press, Amsterdam–New York–Oxford.

1160. SRERE, P. A., MATTIASSON, B. and MOSBACH, K. (1973) An immobilized three-enzyme system: A model for microenvironmental compartmentation in mitochondria. *Proc. Natl. Acad. Sci. U.S.A.* **70**, 2534–2538.

1161. SRERE, P. A. and MOSBACH, K. (1974) Metabolic compartmentation: Symbiotic, organellar, multienzymic and microenviromental. *Annu. Rev. Microbiol.* **28**, 61–83.

1162. STADTMAN, E. R. (1966) Allosteric regulation of enzyme activity. *Adv. Enzymol.* **28**, 41–154.

1163. STADTMAN, E. R. (1970) Mechanism of enzyme regulation in metabolism. In *The Enzymes* (BOYER, P. D., ed.), Vol. **1**, pp. 397–459. Academic Press, New York–London.

1164. STALLCUP, W. B. and KOSHLAND, D. E., JR. (1973) Half-of-the-sites reactivity and negative co-operativity: The case of yeast glyceraldehyde 3-phosphate dehydrogenase. *J. Mol. Biol.* **80**, 41–62.

1165. STANLEY, C. J., PACKMAN, L. C., DANSON, M. J., HENDERSON, C. E. and PERHAM, R. N. (1981) Intramolecular coupling of active sites in the pyruvate dehydrogenase multienzyme complexes from bacterial and mammalian sources. *Biochem. J.* **195**, 715–721.

1166. STEBBING, N. (1980) Evolution of compartmentation, metabolic channeling and control of biosynthetic pathways. In *Cell Compartmentation and Metabolic Channeling.* pp. 93–105. (NOVER, L., LYNEN, F. and MOTHES, K., eds), VEB Gustav Fischer Verlag, Jena, Elsevier/North-Holland Biomedical Press, Amsterdam–New York–Oxford.

1167. STECK, T. L. (1974) Organization of proteins in the human red blood cell membrane. *J. Cell. Biol.* **62**, 1–19.

1168. STECK, T. L. (1978) The band 3 protein of the human red cell membrane: A review. *J. Supramol. Struct.* **8**, 311–324.

1169. STECK, T. L. and KANT, J. A. (1973) Preparation of impermeable ghosts and inside-out vesicles from human erythrocyte membranes. *Methods Enzymol.* **31**, 172–180.

1170. STECK, T. L., KOZIARZ, J. J., SINGH, M. K., REDDY, G. and KÖHLER, H. (1978) Preparation and analysis of seven major, topographically defined fragments of band 3, the predominant transmembrane polypeptide of human erythrocyte membranes. *Biochemistry* **17**, 1216–1222.

1171. STECK, T. L., RAMOS, B. and STRAPAZON, E. (1976) Proteolytic dissection of band 3, the predominant transmembrane polypeptide of the human erythrocyte membrane. *Biochemistry* **15**, 1154–1161.

1172. STEIGER, R. F., OPPERDOES, F. R. and BONTEMPS, J. (1980) Subcellular fractionation of *Trypanosoma brucei* bloodstream forms with special reference to hydrolases. *Eur. J. Biochem.* **105**, 163–175.

1173. STEIN, R. B. and BLUM, J. J. (1979) Quantitative analysis of intermediary metabolism in tetrahymena. *J. Biol. Chem.* **254**, 10385–10395.

1174. STEIN, R. B. and BLUM, J. J. (1980) Quantitative analysis of intermediary metabolism in tetrahymena. *J. Biol. Chem.* **255**, 4198–4205.

1175. STEINHAGEN-THIESSEN, E. and HILZ, H. (1976) Age-dependent decrease in creatine kinase and aldolase activities in human striated muscle is not caused by an accumulation of faulty proteins. *Mech. Age. Devl.* **5**, 447–457.

1176. STEITZ, T. A., ANDERSON, W. F., FLETTERICK, R. J. and ANDERSON, C. M. (1977) High resolution crystal structures of yeast hexokinase complexes with substrates, activators, and inhibitors. *J. Biol. Chem.* **252**, 4494–4500.

1177. STEITZ, T. A., FLETTERICK, R. J., ANDERSON, W. F. and ANDERSON, C. M. (1976) High resolution X-ray structure of yeast hexokinase, an allosteric protein exhibiting a non-symmetric arrangement of subunits. *J. Mol. Biol.* **104**, 197–222.

1178. STEITZ, T. A., SHOHAM, M. and BENNETT, W. S. JR. (1981) Structural dynamics of yeast hexokinase during catalysis. *Phil. Trans. R. Soc. Lond.* **B 293**, 43–52.

1179. STELLWAGEN, E., CRONLUND, M. M. and BARNES, L. D. (1973) A thermostable enolase from the extreme thermophile *Thermus aquaticus* YT-1. *Biochemistry* 12, 1552–1559.

1180. STELLWAGEN, E. and SCHACHMAN, H. K. (1962) The dissociation and reconstitution of aldolase. *Biochemistry* 1, 1056–1069.

1181. STENDER, W. (1981) Inhibition of *E. coli* RNA polymerase by Fab fragments from subunit specific antibodies. *Biochem. Biophys. Res. Commun.* 100, 198–204.

1182. STEWART, M., MORTON, D. J. and CLARKE, F. M. (1980) Interaction of aldolase with actin-containing filaments. *Biochem. J.* 186, 99–104.

1183. STINSON, R. A. and HOLBROOK, J. J. (1973) Equilibrium binding of nicotinamide nucleotides to lactate dehydrogenases. *Biochem. J.* 131, 719–728.

1184. STOOPS, J. K., AWAD, E. S., ARSLANIAN, M. J., GUNSBERG, S., WAKIL, S. J. and OLIVER, R. M. (1978) Studies on the yeast fatty acid synthetase. Subunit composition and structural organization of a large multifunctional enzyme complex. *J. Biol. Chem.* 253, 4464–4475.

1185. STÖCKEL, P., MAYER, A. and KELLER, R. (1973) X-ray small-angle-scattering investigation of a giant respiratory protein: Hemoglobin *Tubifex tubifex*. *Eur. J. Biochem.* 37, 193–200.

1186. STOOPS, J. K. and WAKIL, S. J. (1980) Yeast fatty acid synthetase: Structure-function relationship and nature of the β-ketoacyl synthetase site. *Proc. Natl. Acad. Sci. U.S.A.* 77, 4544–4548.

1187. STRAPAZON, E. and STECK, T. L. (1976) Binding of rabbit muscle aldolase to band 3, the predominant polypeptide of the human erythrocyte membrane. *Biochemistry* 15, 1421–1424.

1188. STRAPAZON, E. and STECK, T. L. (1977) Interaction of the aldolase and the membrane of human erythrocytes. *Biochemistry* 16, 2966–2971.

1189. SUBRAMANI, S., BOTHWELL, M. A., GIBBONS, I., YANG, Y. R. and SCHACHMAN, H. K. (1977) Ligand-promoted weakening of intersubunit bonding domains in aspartate transcarbamoylase. *Proc. Natl. Acad. Sci. U.S.A.* 74, 3777–3781.

1190. SUBRAMANI, S. and SCHACHMAN, H. K. (1980) Mechanism of disproportionation of aspartate transcarbamoylase molecules lacking one regulatory subunit. *J. Biol. Chem.* 255, 8136–8143.

1191. SUDGEN, P. H., KERBEY, A. L., RANDLE, P. J., WALLER, C. A. and REID, K. B. M. (1979) Amino acid sequences around the sites of phosphorylation in the pig heart pyruvate dehydrogenase complex. *Biochem. J.* 181, 419–426.

1192. SUDGEN, P. H. and RANDLE, P. J. (1978) Regulation of pig heart pyruvate dehydrogenase by phosphorylation. Studies on the subunit and phosphorylation stoichiometries. *Biochem. J.* 173, 659–668.

1193. SÜDI, J. (1970) Temperature dependence of the mechanism of heat inactivation of pig lactate dehydrogenase isoenzyme H_2M_2. *FEBS Symposium* 18, 169–176.

1194. SÜDI, J. and KHAN, M. G. (1970) Transient dissociation of lactate dehydrogenase resulting in increased reactivity towards iodoacetamide. *FEBS Lett.* 6, 253–256.

1195. SÜMEGI, B., GYÓCSI, L. and ALKONYI, I. (1980) Interaction between the pyruvate dehydrogenase complex and citrate synthase. *Biochim. Biophys. Acta* 616, 158–166.

1196. SUTTLE, D. P. and STARK, G. R. (1979) Coordinate overproduction of orotate phosphoribosyltransferase and orotidine-5′-phosphate decarboxylase in hamster cells resistant to pyrazofurin and 6-azauridine. *J. Biol. Chem.* 254, 4602–4607.

1197. SUZUKI, K. and HARRIS, J. I. (1975) Hybridization of glyceraldehyde-3-phosphate dehydrogenase. *J. Biochem.* 77, 587–593.

1198. SWANLJUNG, P., FRIGERI, L., OHLSON, K. and ERNSTER, L. (1973) Studies on the activation of purified mitochondrial ATPase by phospholipids. *Biochim. Biophys. Acta* 305, 519–533.

1199. SWANN, A. C. and ALBERS, R. W. (1980) (Na^+, K^+)-ATPase of mammalian brain: Differential effects on cation affinities of phosphorylation by ATP and acetylphosphate. *Arch. Biochem. Biophys.* 203, 422–427.

1200. SWANSON, M. and PACKER, L. (1980) Effect of crosslinking cytochrome c oxidase. *Arch. Biochem. Biophys.* **204**, 30–40.

1201. SWISSA, M., WEINHOUSE, H. and BENZIMAN, M. (1976) Activities of citrate synthase and other enzymes of *Acetobacter xylinum in situ* and *in vitro. Biochem. J.* **153**, 499–501.

1202. SZABOLCSI, G., BISZKU, E. and SAJGÓ, M. (1960) Studies on D-glyceraldehyde-3-phosphate dehydrogenase. XVI. On the mechanism of sulfhydryl group blocking in PGAD. *Acta Physiol. Acad. Sci. Hung.* **17**, 183–193.

1203. SZABOLCSI, G. and CSEKE, E. (1981) On the molecular sieving property of the human erythrocyte membrane. Localization of some proteins within the cell. *Acta Biol. Med. Germ.* **40**, 471–477.

1204. SZENT-GYÖRGYI, A. (1963) Lost in the twentieth century. *Annu. Rev. Biochem.* **32**, 1–14.

1205. SZUNDI, I., SZELÉNYI, J. G., BREUER, J. H. and BÉRCZI, A. (1980) Interactions of haemoglobin with erythrocyte membrane phospholipids in monomolecular lipid layers. *Biochim. Biophys. Acta* **595**, 41–46.

1206. TAKEMOTO, L. J., HANSEN, J. S. and HOKIN, L. E. (1981) Phosphorylation of lens membrane: Identification of the catalytic subunit of Na^+, K^+-ATPase. *Biochem. Biophys. Res. Commun.* **100**, 58–64.

1207. TAMAKI, N., KIMURA, K. and HAMA, T. (1978) Studies on the oligomeric structure of yeast aldehyde dehydrogenase by cross-linking with bifunctional reagents. *J. Biochem.* **83**, 821–825.

1208. TAN, S.-T. and MARZLUF, G. A. (1979) Multiple intracellular peptidases in *Neurospora crassa. J. Bacteriol.* **137**, 1324–1332.

1209. TANNER, M. J. A. and GRAY, W. R. (1971) The isolation and functional identification of a protein from the human erythrocyte 'Ghost'. *Biochem. J.* **125**, 1109–1117.

1210. TAURO, P., HOLZNER, U., CASTORPH, H., HILL, F. and SCHWEIZER, E. (1974) Genetic analysis of non-complementing fatty acid synthetase mutants in *Saccharomyces cerevisiae. Mol. Gen. Genet.* **129**, 131–148.

1211. TAYLOR, M. B., BERGHAUSEN, H., HEYWORTH, P., MESSENGER, N., REES, L. J. and GUTTERIDGE, W. E. (1980) Subcellular localization of some glycolytic enzymes in parasitic flagellated protozoa. *Int. J. Biochem.* **11**, 117–120.

1212. TEAGUE, W. M., PETTIT, F. H., YEAMAN, S. J. and REED, L. J. (1979) Function of phosphorylation sites on pyruvate dehydrogenase. *Biochem. Biophys. Res. Commun.* **87**, 244–252.

1213. TELFORD, J. N., LAD, P. M. and HAMMES, G. G. (1975) Electron microscope study of native and crosslinked rabbit muscle phosphofructokinase. *Proc. Natl. Acad. Sci. U.S.A.* **72**, 3054–3056.

1214. TELLAM, R. and WINZOR, D. J. A. (1977) Self-association of α-chymotrypsin at low ionic strength in the vicinity of its pH optimum. *Biochem. J.* **161**, 687–694.

1215. TENENHOUSE, H. S., SCRIVER, C. R. and VIZEL, E. J. (1980) Alkaline phosphatase activity does not mediate phosphate transport in the renal-cortical brush-border membrane. *Biochem. J.* **190**, 473–476.

1216. THIRY, L. and HERVÉ, G. (1978) The stimulation of *Escherichia coli* aspartate transcarbamylase activity by adenosine triphosphate. Relation with the other regulatory conformational changes; A model. *J. Mol. Biol.* **125**, 515–534.

1217. THOMAS, C. B., ARNOLD, W. J. and KELLEY, W. N. (1973) Human adenine phosphoribosyltransferase. Purification, subunit structure, and substrate specificity. *J. Biol. Chem.* **248**, 2529–2535.

1218. THORNBURGH, B. N., WU, L. L. and GRIFFIN, C. C. (1978) Phosphofructokinase from *Escherichia coli:* Further evidence for identical subunits. *Can. J. Biochem.* **56**, 836–838.

1219. THORNER, J. W. and PAULUS, H. (1971) Composition and subunit structure of glycerol kinase from *Escherichia coli. J. Biol. Chem.* **246**, 3885–3894.

1220. THUSIUS, D. (1977) Does a functional relationship exist between the polymerization and catalytic activity of beef liver glutamate dehydrogenase? *J. Mol Biol.* **115**, 243–247.

1221. TIEMEIER, D. C. and MILMAN, G. (1972) Chinese hamster liver glutamine synthetase. Purification, physical and biochemical properties. *J. Biol. Chem.* **247**, 2272–2277.

1222. TIJANE, M. N., SEYDOUX, F. J., HILL, M., ROUCOUS, C. and LAURENT, M. (1979) Octameric structure of yeast phosphofructokinase as determined by crosslinking with disuccinimidyl β-hydromuconate. *FEBS Lett.* **105**, 249–253.

1223. TILL, U., KÖHLER, W., RUSCHKE, I., KÖHLER, A. and LÖSCHE, W. (1973) Compartmentation of orthophosphate and adenine nucleotides in human red cells. *Eur. J. Biochem.* **35**, 167–178.

1224. TILLMANN, W., CORDUA, A. and SCHRÖTER, W. (1975) Organization of enzymes of glycolysis and of glutathione metabolism in human red cell membranes. *Biochim. Biophys. Acta* **382**, 157–171.

1225. TITANI, K., KOIDE, A., HERMANN, J., ERICSSON, L. H., KUMAR, S., WADE, R. D., WALSH, K. A., NEURATH, H. and FISCHER, E. H. (1977) Complete amino acid sequence of rabbit muscle glycogen phosphorylase. *Proc. Natl. Acad. Sci. U.S.A.* **74**, 4762–4766.

1226. TOKUMITSU, Y. and UI, M. (1973) The phosphorylation of intramitochondrial AMP: A suggestion for the compartmentation of endogenous P_i pool. *Biochim. Biophys. Acta* **292**, 325–337.

1227. TOMKINS, G. M., YIELDING, K. L. and CURRAN, J. (1961) Steroid hormone activation of L-alanine oxidation catalyzed by a subunit of crystalline glutamic dehydrogenase. *Proc. Natl. Acad. Sci. U.S.A.* **47**, 270–278.

1228. TOSON, G., CONTESSA, A. R. and BRUNI, A. (1972) Solubilization of mitochondrial ATPase by phospholipids. *Biochem. Biophys. Res. Commun.* **48**, 341–347.

1229. TOUSTER, O. (1960) Essential pentosuria and the glucuronate-xylulose pathway. *Fed. Proc.* **19**, 977–983.

1230. TRAYER, I. P. (1981) Structure, function and evolution of the mammalian hexokinases. *Biochem. Soc. Trans.* **9**, 23–25.

1231. TRÄUBLE, H. (1976) Membrane electrostatics. In *Structure of Biological Membranes*. pp. 509–550. (ABRAHAMSSON, S. and PASCHER, I., eds), Plenum Press, New York.

1232. TRAUT, T. W. (1980) Significance of the enzyme complex that synthesizes UMP in Ehrlich ascites cells. *Arch. Biochem. Biophys.* **200**, 590–594.

1233. TRAUT, T. W. and JONES, M. E. (1977) Kinetic and conformational studies of the orotate phosphoribosyltransferase: Orotidine-5′-phosphate decarboxylase enzyme complex from mouse Ehrlich ascites cells. *J. Biol. Chem.* **252**, 8374–8381.

1234. TRAUT, T. W. and JONES, M. E. (1978) Mammalian synthesis of UMP from orotate: The regulation of and conformers of complex U. *Adv. Enzyme Regul.* **16**, 21–41.

1235. TRAUT, T. W. and JONES, M. E. (1979) Interconversion of different molecular weight forms of the orotate phosphoribosyltransferase orotidine-5′-phosphate decarboxylase enzyme complex from mouse Ehrlich ascites cells. *J. Biol. Chem.* **254**, 1143–1150.

1236. TRENTHAM, D. R., McMURRAY, C. H. and POGSON, C. I. (1969) The active chemical state of D-glyceraldehyde 3-phosphate in its reactions with D-glyceraldehyde 3-phosphate dehydrogenase, aldolase and triose phosphate isomerase. *Biochem. J.* **114**, 19–24.

1237. TRESGUERRES, M. E. F., INGLEDEW, W. M. and CÁNOVAS, J. L. (1972) Potential competition for 5-dehydroshikimate between the aromatic biosynthetic route and the catabolic hydroaromatic pathway. *Arch. Mikrobiol.* **82**, 111–119.

1238. TRUFFA-BACHI, P., VERON, M. and COHEN, G. N. (1974) Structure, function, and possible origin of a bifunctional allosteric enzyme *Escherichia coli* aspartokinase I–homoserine dehydrogenase I. *CRC Crit. Rev. Biochem.* **2**, 379–415.

1239. TSAI, I.-H., MURTHY, S. N. P. and STECK, T. L. (1982) Effect of red cell membrane binding on the catalytic activity of glyceraldehyde-3-phosphate dehydrogenase. *J. Biol. Chem.* **257**, 1438–1442.

1240. TSCHOPP, J. and KIRSCHNER, K. (1980) Kinetics of cooperative ligand binding to the Apo $β_2$ subunit of tryptophan synthase and its modulation by the α subunit. *Biochemistry* **19**, 4521–4527.

1241. TSUBOI, K. K., SCHWARTZ, S. M., BURRILL, P. H., KWONG, L. K. and SUNSHINE, P. (1979) Sugar hydrolases of the infant rat intestine and their arrangement on the brush border membrane. *Biochim. Biophys. Acta* **554**, 234–248.

1242. TSUJI, S., IMAHORI, K. and NONOMURA, Y. (1981) The quaternary structure of DNA-dependent RNA polymerase. *J. Biochem.* **89**, 1903–1912.

1243. TULINSKY, A., VANDLEN, R. L., KORIMOTO, C. N., MANI, N. V. and WRIGHT, L. H. (1973) Variability in the tertiary structure of α-chymotrypsin at 2.8 Å resolution. *Biochemistry* **12**, 4185–4192.

1244. TZAGOLOFF, A., MACINO, G. and SEBALD, W. (1979) Mitochondrial genes and translation products. *Annu. Rev. Biochem.* **48**, 419–441.

1245. TZAGOLOFF, A., MacLENNAN, D. H. and BYINGTON, K. H. (1968) Studies on the mitochondrial adenosine triphosphatase system. III. Isolation from the oligomycin-sensitive adenosine triphosphatase complex of the factors which bind F_1 and determine oligomycin sensitivity of bound F_1. *Biochemistry* **7**, 1596–1602.

1246. UENO, K., SEKIMUZU, K., OBINATA, M., MIZUNO, D. and NATORI, S. (1981) Stimulation of messenger ribonucleic acid synthesis in isolated nuclei by a protein that stimulates RNA polymerase-II. *Biochemistry* **20**, 634–640.

1247. ULLMANN, A. and PERRIN, D. (1970) Complementation in β-galactosidase. In *The Lactose Operon*. pp. 143–172. (BECKWITZ, J. R. and ZIPSER, D., eds)

1248. UMEZU, K., AMAYA, T., YOSHIMOTO, A. and TOMITA, K. (1971) Purification and properties of orotidine-5′-phosphate pyrophosphorylase and orotidine-5′-phosphate decarboxylase from bakers' yeast. *J. Biochem.* **70**, 249–262.

1249. U'PRICHARD, D. C., BYLUND, D. B. and SNYDER, S. H. (1978) (±)-[^{3}H] Epinephrine and (−)-[^{3}H] dihydroalprenolol binding to β_1- and β_2-noradrenergic receptors in brain, heart, and lung membranes. *J. Biol. Chem.* **253**, 5090–5102.

1250. URECH, K., DÜRR, M., BOLLER, TH., WIEMKEN, A. and SCHWENCKE, J. (1978) Localization of polyphosphate in vacuoles in *Saccharomyces cerevisiae*. *Arch. Microbiol.* **116**, 275–278.

1251. VÁDINEANU, A., BERDEN, J. A. and SLATER, E. C. (1976) Proteins required for the binding of mitochondrial ATPase to the mitochondrial inner membrane. *Biochim. Biophys. Acta* **449**, 468–479.

1252. VAISIUS, A. C. and HORGEN, P. A. (1980) The effects of several divalent cations on the activation of inhibition of RNA ploymerases II. *Arch. Biochem. Biophys.* **203**, 553–564.

1253. VALENTINE, R. C., SHAPIRO, B. M. and STADTMAN, E. R. (1968) Regulation of glutamine synthetase. XII. Electron microscopy of the enzyme from *Escherichia coli*. *Biochemistry* **7**, 2143–2152.

1254. VALENTINE, R. C., WRIGLEY, N. G., SCRUTTON, M. C., IRIAS, J. J. and UTTER, M. F. (1966) Pyruvate carboxylase. VIII. The subunit structure as examined by electron microscopy. *Biochemistry* **5**, 3111–3116.

1255. VALLEE, B. L. and ULMER, D. D. (1972) Biochemical effects of mercury, cadmium, and lead. *Annu. Rev. Biochem.* **41**, 91–128.

1256. VANCE, D. E. (1976) The rate-limiting reaction catalyzed by a multi-enzyme complex. The fatty acid synthetases. *J. Theor. Biol.* **59**, 409–413.

1257. VAN DAM, K. and VAN GELDER, B. F. (eds) (1977) *Structure and Function of Energy-transducing Membranes*. B.B.A. Library, Vol. 14. Elsevier, Amsterdam–Oxford–New York.

1258. VAN DEN BERG, L. and ROSE, D. (1959) Effect of freezing on the pH and composition of sodium and potassium phosphate solutions: the reciprocal system KH_2PO_4-Na_2HPO_4-H_2O. *Arch. Biochem. Biophys.* **81**, 319–329.

1259. VANDERKOOI, J. M., IEROKOMAS, A., NAKAMURA, H. and MARTONOSI, A. (1977) Fluorescence energy transfer between Ca^{++} transport ATPase molecules in artificial membranes. *Biochemistry* **16**, 1262–1267.

1260. VAS, M. and BATKE, J. (1981) Evidence for absence of an interaction between purified 3-phosphoglycerate kinase and glyceraldehyde-3-phosphate dehydrogenase. *Biochim. Biophys. Acta* **660**, 193–198.

1261. VAS, M., BERNI, R., MOZZARELLI, A., TEGONI, M. and ROSSI, G. L. (1979) Kinetic studies of crystalline enzymes by single crystal microspectrophotometry. Analysis of a single catalytic turnover in a D-glyceraldehyde-3-phosphate dehydrogenase crystal. *J. Biol. Chem.* **254**, 8480–8486.

1262. VAS, M. and BOROSS, L. (1972) Heat inactivation of D-glyceraldehyde-3-phosphate dehydrogenase apoenzyme. *Acta Biochim. Biophys. Acad. Sci. Hung.* **7**, 105–114.

1263. VAS, M., LAKATOS, S., HAJDÚ, J. and FRIEDRICH, P. (1981) Kinetic behaviour and oligomeric state of 3-phosphoglyceoryl-D-GAPD. *Biochimie* **63**, 89–96.

1264. VELICK, S. F., BAGGOTT, J. P. and STURTEVANT, J. M. (1971) Thermodynamics of nicotin amide–adenine dinucleotide addition to the glyceraldehyde 3-phosphate dehydrogenase of yeas and of rabbit skeletal muscle. An equilibrium and calometric analysis over a range of temperatures *Biochemistry* **10**, 779–786.

1265. VERON, M., SAARI, J. C., VILLAR-PALASI, C. and COHEN, G. N. (1973) The threonine-sensitive homoserine dehydrogenase and aspartokinase activities of *Escherichia coli* K12. Intra- and inter-subunit interactions between the catalytic regions of the bifunctional enzyme. *Eur. J. Biochem.* **38**, 325–335.

1266. VICKERS, L. P. (1981) Trimeric enzymes in carbamoylphosphate metabolism. *Trends Biochem. Sci.* **6**, xi–xii.

1267. VIGNY, M., BON, S., MASSOULIÉ, J. and LETERRIER, F. (1978) Active-site catalytic efficiency of acetylcholinesterase. Molecular forms in electrophorus, torpedo, rat and chicken. *Eur. J. Biochem.* **85**, 317–323.

1268. VIITANEN, P. and GEIGER, P. (1979) Heart mitochondrially bound hexokinase: Direct demonstration of compartmentation with the use of newly synthesized ATP. *Fed. Proc.* **38**, 561.

1269. VIMARD, C., ORSINI, G. and GOLDBERG, M. E. (1975) Renaturation of acid-denatured rabbit muscle aldolase. Existence and properties of a stable monomeric intermediate. *Eur. J. Biochem.* **51**, 521–527.

1270. VINCENT, J.-P. and LAZDUNSKI, M. (1972) Trypsin–pancreatic trypsin inhibitor association, Dynamics of the interaction and role of disulfide bridges. *Biochemistry* **11**, 2967–2977.

1271. VISSER, N., OPPERDOES, F. R. and BORST, P. (1981) Subcellular compartmentation of glycolytic intermediates in *Trypanosoma brucei*. *Eur. J. Biochem.* **118**, 521–526.

1272. VITTO, A. and GAERTNER, F. H. (1978) Proteolytic inactivation of a pentafunctional enzyme conjugate: Coordinate protection by the first substrate. *Biochem. Biophys. Res. Commun.* **82**, 977–981.

1273. VOGEL, H. J. and BONNER, D. M. (1954) On the glutamate-proline-ornithine interrelation in *Neurospora crassa*. *Proc. Natl. Acad. Sci. U.S.A.* **40**, 688–694.

1274. VOGEL, O., HOEHN, B. and HENNING, U. (1972) Molecular structure of the pyruvate dehydrogenase complex from *Escherichia coli* K-12. *Proc. Natl. Acad. Sci. U.S.A.* **69**, 1615–1619.

1275. VOGT, V. M., PEPINSKY, R. B. and RACKER, E. (1981) Src protein and the kinase cascade. *Cell* **25**, 827.

1276. VOLPE, J. J. and VAGELOS, P. R. (1976) Mechanisms and regulation of biosynthesis of saturated fatty acids. *Physiol. Rev.* **56**, 339–417.

1277. WACKER, H., LEHKY, P., VANDERHAEGHE, F. and STEIN, E. A. (1976) On the subunit structure of particulate aminopeptidase from pig kidney. *Biochim. Biophys. Acta* **429**, 546–554.

1278. WALSH, T. P., CLARKE, F. M. and MASTERS, C. J. (1977) Modification of the kinetic parameters of aldolase on binding to the actin-containing filaments of skeletal muscle. *Biochem. J.* **165**, 165–167.

1279. WALSH, T. P., WINZOR, D. J., CLARKE, F. M., MASTERS, C. J. and MORTON, D. J. (1980) Binding of aldolase to actin-containing filaments. *Biochem. J.* **186**, 89–98.

1280. WANG, C.-S. and ALAUPOVIC, P. (1980) Glyceraldehyde-3-phosphate dehydrogenase from human erythrocyte membranes. Kinetic mechanism and competitive substrate inhibition by glyceraldehyde 3-phosphate. *Arch. Biochem. Biophys.* **205**, 136–145.

1281. WANG, J. H., SHONKA, M. L. and GRAVES, D. J. (1965) Influence of carbohydrates on phosphorylase structure and activity. I. Activation by preincubation with glycogen. *Biochemistry* **4**, 2296–2301.

1282. WANGERMANN, G., YEDINTSOV, YU. M., IVANAITSKI, G. R., KUMINSKI, A. S., REICHELT, R. and TSYGANOV, M. A. (1976) Interpretation of electron microscopic images of crystalline aggregation of leucine aminopeptidase. *Stud. Biophys.* **60**, 241–242.

1283. WARD, K. B., HENDRICKSON, W. A. and KLIPPENSTEIN, G. L. (1975) Quaternary and tertiary structure of haemerythrin. *Nature* **257**, 818–821.

1284. WARREN, S. G., EDWARDS, B. P. P., EVANS, D. R., WILEY, D. C. and LIPSCOMB, W. N. (1973) Aspartate transcarbamoylase from *Escherichia coli:* Electron density at 5.5 Å resolution. *Proc. Natl. Acad. Sci. U.S.A.* **70**, 1117–1121.

1285. WARSHEL, A. (1978) Energetics of enzyme catalysis. *Proc. Natl. Acad. Sci. U.S.A.* **75**, 5250–5254.

1286. WARSHEL, A., LEVITT, M. and LIFSON, S. (1970) Consistent force field for calculation of vibrational spectra and conformations of some amides and lactam ring. *J. Mol. Spectrosc.* **33**, 84–99.

1287. WAWRZYNCZAK, E. J., PERHAM, R. N. and ROBERTS, G. C. K. (1981) Conformational mobility of polypeptide chains in the 2-oxo acid dehydrogenase complexes from ox heart revealed by proton NMR spectroscopy. *FEBS Lett.* **131**, 151–154.

1288. WEBB, J. L. (1963) *Enzyme and Metabolic Inhibitors.* Vol. 1, p. 373, Academic Press, New York.

1289. WEBER, A., GREGORI, C. and SCHAPIRA, F. (1976) Aldolase B in the liver of senescent rats *Biochim. Biophys. Acta* **444**, 810–815.

1290. WEBER, I. T., JOHNSON, L. N., WILSON, K. S., YEATES, D. G. R., WILD, D. L. and JENKINS, I. A. (1978) Crystallographic studies on the activity of glycogen phosphorylase b. *Nature* **274**, 433–437.

1291. WEBER, K. (1968) New structural model of *E. coli* aspartate transcarbamylase and the amino-acid sequence of the regulatory polypeptide chain. *Nature* **218**, 1116–1119.

1292. WEIGL, K. and SIES, H. (1977) Drug oxidations dependent on cytochrome P-450 in isolated hepatocytes. The role of the tricarboxylates and the aminotransferases in NADPH supply. *Eur. J. Biochem.* **77**, 401–408.

1293. WEIL, P. A., SEGALL, J., HARRIS, B., NG, S.-Y. and ROEDER, R. G. (1979) Faithful transcription of eukaryotic genes by RNA polymerase III in systems reconstituted with purified DNA templates. *J. Biol. Chem.* **254**, 6163–6173.

1294. WEINSTEIN, R. S., KHADADAD, J. K. and STECK, T. L. (1978) Ultrastructural characterization of proteins on natural surfaces of the red cell membrane. In *The Red Cell.* pp. 413–427. (BREWER G. J., ed.), Alan R. Liss, New York.

1295. WEISCHET, W. O. and KIRSCHNER, K. (1976) Steady-state kinetic studies of the synthesis of indoleglycerol phosphate catalyzed by the α subunit of tryptophan synthase from *Escherichia coli.* Comparison with the $\alpha_2\beta_2$-complex. *Eur. J. Biochem.* **65**, 375–385.

1296. WEISS, R. L. (1973) Intracellular localization of ornithine and arginine pools in *Neurospora. J. Biol. Chem.* **248**, 5409–5413.

1297. WEISS, R. L. (1976) Compartmentation and control of arginine metabolism in *Neurospora. J. Bacteriol.* **126**, 1173–1179.

1298. WEISS, R. L. and DAVIS, R. H. (1977) Control of arginine utilization in *Neurospora. J. Bacteriol.* **129**, 866–873.

1299. WEISSBACH, A. (1977) Eukaryotic DNA polymerases. *Annu. Rev. Biochem.* **46**, 25–47.

1300. WEITZMAN, P. D. J. (1973) Behaviour of enzymes at high concentration. Use of permeabilised cells in the study of enzyme activity and its regulation. *FEBS Lett.* **32**, 247–250.

1301. WEITZMAN, P. D. J. and HEWSON, J. K. (1973) *In situ* regulation of yeast citrate synthase. Absence of ATP inhibition observed *in vitro*. *FEBS Lett.* **36**, 227–231.

1302. WELCH, G. R. (1977) On the free energy "Cost of Transition" in intermediary metabolic processes and the evolution of cellular infrastructure. *J. Theor. Biol.* **68**, 267–291.

1303. WELCH, G. R. (1977) On the role of organized multienzyme systems in cellular metabolism: A general synthesis. *Prog. Biophys. Molec. Biol.* **32**, 103–191.

1304. WELCH, G. R. (1981) Thermodynamic-kinetic aspects of enzyme function in the "living state", (in press).

1305. WELCH, G. R. and GAERTNER, F. H. (1975) Influence of an aggregated multienzyme system on transient time: Kinetic evidence for compartmentation by an aromatic-amino-acid-synthesizing complex of *Neurospora crassa. Proc. Natl. Acad. Sci. U.S.A.* **72**, 4218–4222.

1306. WELCH, G. R. and GAERTNER, F. H. (1976) Coordinate activation of a multienzyme complex by the first substrate. Evidence for a novel regulatory mechanism in the polyaromatic pathway of *Neurospora crassa. Arch. Biochem. Biophys.* **172**, 476–489.

1307. WELCH, G. R. and GAERTNER, F. H. (1980) Multienzyme systems clarifying the nomenclature. *Trends Biochem. Sci.* **5**, No. 3, vii.

1308. WELCH, G. R. and GAERTNER, F. H. (1980) Enzyme organization in the polyaromatic-biosynthetic pathway: The arom conjugate and other multienzyme systems. *Curr. Top. Cell. Regul.* **16**, 113–162.

1309. WELCH, G. R. and KELETI, T. (1981) On the "cytosociology" of enzyme action *in vivo*: A novel thermodynamic correlate of biological evolution. *J. Theor. Biol.* **93**, 701–735.

1310. WENZEL, K.-W., KURGANOV, B. I., ZIMMERMANN, G., YAKOVLEV, V. A., SCHELLENBERGER, W. and HOFMANN, E. (1976) Cooperative association of erythrocyte phosphofructokinase. *Eur. J. Biochem.* **61**, 181–190.

1311. WENZEL, K.-W., MÜLLER, D., BÄR, J. and HOFMANN, E. (1978) Relationships between association state and enzymic activity of human erythrocyte phosphofructokinase. *Acta Biol. Med. Germ.* **37**, 519–526.

1312. WEYL, M. (1952) *Symmetry.* Princeton University Press, Princeton, New Jersey.

1313. WHANGER, P. D., PHILLIPS, A. T., RABINOWITZ, K. W., PIPERNO, J. R., SHADA, J. D. and WOOD, W. A. (1968) The mechanism of action of 5′-adenylic acid-activated threonine dehydrase. II. Protomer-oligomer interconversions and related properties. *J. Biol. Chem.* **243**, 167–173.

1314. WIELAND, O. H. (1975) On the mechanism of irreversible pyruvate dehydrogenase inactivation in liver mitochondrial extracts. *FEBS Lett.* **52**, 44–47.

1315. WIEMKEN, A. (1980) Compartmentation and control of amino acid utilization in yeast. In *Cell Compartmentation and Metabolic Channeling.* pp. 225–238. (NOVER, L., EYNEN, F. and MOTHES, K. eds), VEB Gustav Fischer Verlag, Jena, Elsevier/North-Holland Biomedical Press, Amsterdam–New York–Oxford.

1316. WIEMKEN, A., SCHELLENBERG, M. and URECH, K. (1979) Vacuoles: The sole compartments of digestive enzymes in yeast *(Saccharomyces cerevisiae). Arch. Microbiol.* **123**, 23–25.

1317. WIESINGER, H., BARTHOLMES, P. and HINZ, H.-J. (1979) Subunit interaction in tryptophan synthase of *Escherichia coli*: Calorimetric studies on association of α and β subunits. *Biochemistry* **18**, 1979–1984.

1318. WIKSTRÖM, M., KRAB, K. and SARASTE, M. (1981) Proton-translocating cytochrome complexes. *Annu. Rev. Biochem.* **50**, 623–655.

1319. WIKSTRÖM, M., KRAB, K. and SARASTE, M. (1982) *Cytochrome oxidase: A synthesis.* Academic Press, London–New York–Toronto–Sydney–San Francisco.

1320. WILLIAMS, C. H. and LAWSON, J. (1975) Monoamine oxidase. Further studies of inhibition by propargylamines. *Biochem. Pharmacol.* **24**, 1889–1891.

1321. WILLIAMS, L. G., BERNHARDT, S. and DAVIS, R. H. (1970) Copurification of pyrimidine-specific carbamyl phosphate synthetase and aspartate transcarbamylase. *Biochemistry* **9**, 4329–4335.

1322. WILSON, D. A. and CRAWFORD, I. P. (1965) Purification and properties of the B component of *Escherichia coli* tryptophan synthetase. *J. Biol. Chem.* **240**, 4801–4808.

1323. WILSON, J. E. (1968) Brain hexokinase. *J. Biol. Chem.* **243**, 3640–3647.

1324. WILSON, J. E. (1978) Ambiquitous enzymes: Variation in intracellular distribution as a regulatory mechanism. *Trends Biochem. Sci.* **3**, 124–125.

1325. WILSON, J. E. (1980) Brain hexokinase, the prototype ambiquitous enzyme. *Curr. Top. Cell. Regul.* **16**, 1–44.

1325a. WILSON, M. E., REID, S. and MASTERS, C. J. (1982) A comparative study of the binding of aldolase and glyceraldehyde-3-phosphate. *Arch. Biochem. Biophys.* **215**, 610–620.

1326. WINGARD, L. B., JR., KATCHALSKI-KATZIR, E. and GOLDSTEIN, L. (eds) (1976) *Applied Biochemistry and Bioengineering.* Vol. 1. *Immobilized Enzyme Principles,* Academic Press, New York–San Francisco–London.

1327. WLODAWER, A., SEGREST, J. P., CHUNG, B. H., CHIOVETTI, R., JR. and WEINSTEIN, J. N. (1979) High-density lipoprotein recombinants: Evidence for a bicycle tire micelle structure obtained by neutron scattering and electron microscopy. *FEBS Lett.* **104**, 231–235.

1328. WOHLRAB, H. (1979) Identification of the N-ethylmaleimide reactive protein of the mitochondrial phosphate transporter. *Biochemistry* **18**, 2098–2102.

1329. WOLD, F. (1981) *In vivo* chemical modification of proteins (post-translational modification). *Annu. Rev. Biochem.* **50**, 783–814.

1330. WOLOSIN, J. M., GINSBURG, H. and CABANTCHIK, Z. I. (1977) Functional characterization of anion transport system isolated from human erythrocyte membranes. *J. Biol. Chem.* **252**, 2419–2427.

1331. WOLPERT, J. S. and ERNST-FONBERG, M. L. (1975) A multienzyme complex for CO_2 fixation. *Biochemistry* **14**, 1095–1102.

1332. WOLPERT, J. S. and ERNST-FONBERG, M. L. (1975) Dissociation and characterization of enzymes from a multienzyme complex involved in CO_2 fixation. *Biochemistry* **14**, 1103–1107.

1333. WONG, J. T.-F. and ENDRÉNYI, L. (1971) Interpretation of nonhyperbolic behavior in enzymic systems. I. Differentiation of model mechanisms. *Can. J. Biochem.* **49**, 568–580.

1334. WOOD, W. I., PETERSON, D. O. and BLOCH, K. (1978) Subunit structure of *Mycobacterium smegmatis* fatty acid synthetase. Evidence for identical multifunctional polypeptide chains. *J. Biol. Chem.* **253**, 2650–2656.

1335. WOOSTER, M. S. and WRIGGLESWORTH, J. M. (1976) Modification of glyceraldehyde 3-phosphate dehydrogenase activity by adsorption on phospholipid vesicles. *Biochem. J.* **159**, 627–631.

1336. WRIGGLESWORTH, J. M., KEOKITICHAI, S., WOOSTER, M. S. and MILLAR, F. A. (1976) Modification of glyceraldehyde-3-phosphate dehydrogenase by adsorption to erythrocyte membranes and phospholipid vesicles. *Biochem. Soc. Trans.* **4**, 637–640.

1337. WRIGHT, B. E., TAI, A. and KILLICK, K. A. (1977) Fourth expansion and glucose perturbation of the Dictyostelium kinetic model. *Eur. J. Biochem.* **74**, 217–225.

1338. WYMAN, J. (1948) Heme proteins. *Adv. Protein Chem.* **4**, 407–531.

1339. YANG, Y. R. and SCHACHMAN, H. K. (1980) Communication between catalytic subunits in hybrid aspartate transcarbamoylase molecules: Effect of ligand binding to active chains on the conformation of unliganded, inactive chains. *Proc. Natl. Acad. Sci. U.S.A.* **77**, 5187–5191.

1340. YANOFSKY, C. and CRAWFORD, I. P. (1972) Tryptophan synthetase. In *The Enzymes.* (P. D. BOYER, ed.), Vol. 7, pp. 1–31. Academic Press, New York–London.

1341. YANOFSKY, C. and RACHMELER, M. (1958) The exclusion of free indole as an intermediate in the biosynthesis of tryptophan in *Neurospora crassa*. *Biochim. Biophys. Acta* **28**, 640–645.

1342. YEAMAN, S. J., HUTCHESON, E. T., ROCHE, T. E., PETTIT, F. H., BROWN, J. R., REED, L. J.,

WATSON, D. C. and DIXON, G. H. (1978) Sites of phosphorylation on pyruvate dehydrogenase from bovine kidney and heart. *Biochemistry* **17**, 2364–2370.

1343. YELTMAN, D. R. and HARRIS, B. G. (1977) Purification and characterization of aldolase from human erythrocytes. *Biochim. Biophys. Acta* **484**, 188–198.

1344. YELTMAN, D. R. and HARRIS, B. G. (1980) Localization and membrane association of aldolase in human erythrocytes. *Arch. Biochem. Biophys.* **199**, 186–196.

1345. YU, J. and STECK, T. L. (1975) Isolation and characterization of band 3, the predominant polypeptide of the human erythrocyte membrane. *J. Biol. Chem.* **250**, 9170–9175.

1346. YU, J. and STECK, T. L. (1975) Association of band 3, the predominant polypeptide of the human erythrocyte membrane. *J. Biol. Chem.* **250**, 9176–9184.

1347. ZEIRI, L. and REISLER, E. (1978) Uncoupling of the catalytic activity and the polymerization of beef liver glutamate dehydrogenase. *J. Mol. Biol.* **124**, 291–295.

1348. ZETINA, C. R. and GOLDBERG, M. E. (1980) A comparative study of the thermal inactivation of the isolated and associated domains within the β_2 subunit of *Escherichia coli* tryptophan synthetase. *J. Biol. Chem.* **255**, 4381–4385.

1349. ZIMMERMANN, G., SCHELLENBERGER, W., WENZEL, K.-W. and HOFMANN, E. (1978) Association behaviour of human erythrocyte phosphofructokinase. *Acta Biol. Med. Germ.* **37**, 527–535.

1350. ZS.-NAGY, I. (1978) A membrane hypothesis of aging. *J. Theor. Biol.* **75**, 189–195.

1351. ZS.-NAGY, I. (1979) The role of membrane structure and function in cellular aging: A review. *Mech. Age. Devl.* **9**, 237–246.

SUBJECT INDEX